SOUNDING SOLAR AND STELLAR INTERIORS

INTERNATIONAL ASTRONOMICAL UNION

UNION ASTRONOMIQUE INTERNATIONALE

SOUNDING SOLAR AND STELLAR INTERIORS

PROCEEDINGS OF THE 181ST SYMPOSIUM OF THE
INTERNATIONAL ASTRONOMICAL UNION,
HELD IN NICE, FRANCE,
SEPTEMBER 30–OCTOBER 3, 1996

EDITED BY

JANINE PROVOST

Observatoire de la Côte d'Azur, Nice, France

and

FRANÇOIS-XAVIER SCHMIDER

Université de Nice, France

KLUWER ACADEMIC PUBLISHERS

DORDRECHT / BOSTON / LONDON

A C.I.P. Catalogue record for this book is available from the Library of Congress.

ISBN 0-7923-4838-9 (HB)
ISBN 0-7923-4839-7 (PB)

Published on behalf of
the International Astronomical Union
by
Kluwer Academic Publishers, P.O. Box 17, 3300 AA Dordrecht, The Netherlands.

Sold and distributed in the U.S.A. and Canada
by Kluwer Academic Publishers,
101 Philip Drive, Norwell, MA 02061, U.S.A.

In all other countries, sold and distributed
by Kluwer Academic Publishers Group,
P.O. Box 322, 3300 AH Dordrecht, The Netherlands.

Printed on acid-free paper

TABLE OF CONTENTS

Introduction

Part I : Helioseismology :
From ground or space, a worldwide cooperation

Part II : Internal structure and rotation. Seismic inversions

Part III : Special session dedicated to Philippe Delache

Part IV : Inputs from helioseismology to solar physics

Part V : Asteroseismology : Theory and methods.

Part VI : Results on some selected types of stars.

Conclusion

List of Participants

Andersen Bo
Head, Space Science
Norwegian Space Centre
P.O. Box 85, Smestad
N-0309 Oslo
NORWAY
bo@admin.nsc.no

Andreev Alexander
Crimean Astrophysical Observatory
334413 Nauchny, Crimea
UKRAINE
andre@crao.crimea.ua

Andriyenko Olexa
Main Astronomical Observatory
National Academy of Sciences of Ukraine
Golosiiv
Kyiv-22 252650
UKRAINE
olexa@mao.gluk.apc.org

Appourchaux Thierry
ESTEC/ESA
P. O. Box 299
2200 AG Noordwijk
THE NETHERLANDS
thierry@walrus.so.estec.esa.nl

Audard Nathalie
Institut für Astronomie
Der Universität Wien
Türkenschanzstraße 17
A-1180 Vienne
AUTRICHE
nathalie@isaac.ast.univie.ac.at

Baglin Annie
Observatoire de Paris, Section Meudon
5 pl J Janssen
92190 Meudon
FRANCE
baglin@obspm.fr

Baldry Ivan
Astronomy Department
School of Physics
University of Sydney
NSW 2006
AUSTRALIA
baldry@physics.usyd.edu.au

Basu Sarbani
Teoretisk Astrofysik Center
Institute for Physics and Astronomy
University of Aarhus
DK-8000 Aarhus C
DENMARK
basu@obs.aau.dk

Baudin Frédéric
Instituto de Astrofisica de Canarias
38200 La Laguna
TENERIFE, SPAIN
fbaudin@iac.es

Belmonte Juan Antonio
Instituto de Astrofisica de Canarias
C/ Via Lactea S.N.
38200 La Laguna
TENERIFE, SPAIN
jba@iac.es

Bely-Dubau Françoise
Département Cassini
Observatoire de la Côte d'Azur
Grande Corniche, BP 229
06304 Nice Cedex 4
FRANCE
bely@obs-nice.fr

Berruyer Nicole
Département Cassini
Observatoire de la Côte d'Azur
Grande Corniche, BP 229
06304 Nice Cedex 4
FRANCE
nicole@obs-nice.fr

Bertello Luca
Head, Space Science
Norwegian Space Centre
P.O. Box 85, Smestad
N-0309 Oslo
NORWAY
bo@admin.nsc.no

Berthomieu Gabrielle
Département Cassini
Observatoire de la Côte d'Azur
Grande Corniche, BP 229
06304 Nice Cedex 4
FRANCE
bertho@obs-nice.fr

Bisnovatyi-Kogan G.
Space Research Institute
Profsoyuznaya 84/32
Moscow 117810
RUSSIA
gkogan@mx.iki.rssi.ru

Bogart Richard
Stanford University
CSSA-HEPL
Stanford CA 94305-4055
USA
rick@rick.stanford.edu

Bonnet R.
Agence Spatiale Européenne (ESA)
8-10 rue Mario Nikis
75738 Paris Cedex 15
FRANCE
bauer%esa.bitnet@vm.gmd.de

Boumier Patrick
Institut d'Astrophysique Spatiale
Universite Paris XI
Bat. 121
91405 Orsay Cedex
FRANCE
boumier@iaslab.ias.fr

Bradley Paul A.
XTA MS B220
Los Alamos National Laboratory
Los Alamos NM 87545
USA
pbradley@lanl.gov

Briot Danielle
Observatoire de Paris
61 Avenue de l'Observatoire
75014 Paris
FRANCE
Danielle.Briot@obspm.fr

Brown Timothy
High Altitude Observatory / NCAR
P.O. Box 3000
Boulder CO 80307
USA
brown@hao.ucar.edu

Brun Allan Sacha
DAPNIA, Service d'Astrophysique
Centre d'Etudes de Saclay
Orme des Merisiers
91191 Gif-sur-Yvette Cedex 01
FRANCE
Sacha.Brun@cea.fr

Bush Rock
Stanford University
CSSA - HEPL Annex B212
Stanford CA 94305-4085
USA
RBush@solar.stanford.edu

Cacciani Alessandro
Department of Physics
University of Rome "La Sapienza"
P.le A. Moro 2
00185 Rome
ITALY
cacciani@roma1.infn.it

Carlier François
Institut d'Astrophysique
Université de Liège
5, Avenue de Cointe
4000 Liège
BELGIQUE
carlier@astro.ulg.ac.be

Chan Kwing Lam
Mathematics Department
Hong Kong University of Sci.& Tech.
Clear Water Bay
Kowloon
HONG KONG
maklchan@usthk.ust.hk

Charra Jacques
Institut d'Astrophysique Spatiale
Universite Paris XI
Bat. 121
91405 Orsay cedex
FRANCE
charra@iaslab.ias.fr

Chiuderi Claudio
Department of Astronomy
University of Florence
Largo E. Fermi, 5
50125 Firenze
ITALY
chiuderi@arcetri.astro.it

Chou Dean-Yi
Physics Department
Tsong Hua University
Hsinchu 30043
TAIWAN ROC
chou@phys.nthu.edu.tw

Christensen-Dalsgaard Jörgen
Institute of Physics and Astronomy
University of Aarhus
DK-8000 Aarhus C
DENMARK
jcd@obs.aau.dk

Corbard Thierry
Département Cassini
Observatoire de la Côte d'Azur
Grande Corniche, BP 229
06304 Nice Cedex 4
FRANCE
corbard@obs-nice.fr

Crommelynck Dominique
Royal Meteorological Institute of Belgium
3 Avenue Circulaire
B-1180 Bruxelles
BELGIUM
Dominique.Crommelynck@oma.be

Cunha Margarida
Institute of Astronomy
Madingley Road
Cambridge CB3 OHA
ENGLAND
mcunha@ast.cam.ac.uk

Däppen Werner
Department of Physics & Astronomy
University of Southern California
Los Angeles CA 90089-1342
USA
dappen@usc.edu

DeForest Craig
Goddard Space Flight Center
Stanford University
Mailcode 682.3
Greenbelt MD 20771
U.S.A.
zowie@urania.nascom.nasa.gov

Delmas Christian
CERGA
Observatoire de la Côte d'Azur
32, avenue Copernic
06130 GRASSE
FRANCE
delmas@obs-azur.fr

Deubner Franz
Astronomisches Institut
Universität Würzburg
D-97074 Würzburg Am Hubland
GERMANY
deubner@astro.uni-wuerzburg.de

Di Mauro Maria
Isituto di Astronomia
Universita di Catania
Citta Universitaria - Viale A. Doria 6
I-95125 Catania
ITALY
dimauro@astrct.ct.astro.it

Dolez Noël
Laboratoire d'Astrophysique
Observatoire Midi-Pyrénées
14 Avenue Edouard Belin
31400 Toulouse
FRANCE
dolez@obs-mip.fr

Domingo Vicente
European Space Agency
NASA/Godard Space Flight Center
Mail Code 682
Greenbelt, Maryland 20771
USA
vdomingo@esa. nascom.nasa.gov

xii

Dziembowski Wojciech
Copernicus Astronomical Center
ul. Bartycka 18
03-610 Warszawa
POLAND
wd@camk.edu.pl

Foing Bernard H.
ESA Solar System Division
ESTEC/SO postbus 299
2200 AG Noordwijk
NETHERLANDS
bfoing@estec.esa.nl

Eff-Darwich Antonio M.
Solar and Stellar Physics Division
Harvard-Smithsonian Center for Astrophysics
Mail Stop 16, 60 Garden Street
Cambridge MA 02138
USA
adarwich@cfa.harvard.edu

Fossat Eric
Départment d'Astrophysique
Université de Nice
Parc Valrose
F-06108 Nice Cedex 2
FRANCE
fossat@ayalga.unice.fr

Elsworth Yvonne
HIROS
School of Physics & Space Research
University of Birmingham, Edgbaston
Birmingham B15 2TT
UNITED KINGDOM
ype@star.sr.bham.ac.uk

Frandsen Søren
Institute of Physics and Astronomy
University of Aarhus
Ny Munkegade, Bygn. 520
DK-8000 Aarhus C
DENMARK
srf@obs.aau.dk

Fierry-Fraillon David
Département d'Astrophysique
Université de Nice Sophia Antipolis
Parc Valrose
06108 Nice Cedex 2
FRANCE
fierry@irisalfa.unice.fr

Frisch Hélène
Département Cassini
Observatoire de la Côte d'Azur
Grande Corniche, BP 229
06304 Nice Cedex 4
FRANCE
frisch@obs-nice.fr

Finsterle Wolfgang
Physikalisch Meteorologisches Observatorium
World Radiation Center
Dorfstrasse 33
CH-7260 Davos Dorf
SUISSE
wolfgang@pmodwrc.ch

Fröhlich Claus
WRC/PMOD
Dorfstrasse 33
CH-7260 Davos Dorf
SUISSE
cfrohlich@obsun.pmodwrc.ch

Fleck Bernhard
Space Science Department of ESA
NASA/GSFC - Mailcode 682.3
Greenbelt, MD 20771
USA
bfleck@esa.nascom.nasa.gov

Gabriel Alan
Institut d'Astrophysique Spatiale
Universite Paris XI
Batiment 121
91405 Orsay Cedex
FRANCE
gabriel@ias.fr

Foglizzo Thierry
Service d'Astrophysique
CE - Saclay
91191 Gif-sur-Yvette Cedex
FRANCE
foglizzo@cea.fr

Garcia Rafael
Service d'Astrophysique
CE - Saclay
91191 Gif-sur-Yvette Cedex
FRANCE
rgarcia@ariane.saclay.cea.fr

Gautschy Alfred
Astronomisches Institut Basel
Venusstr. 7
CH-4102 Binningen
SUISSE
gautschy@astro.unibas.ch

Gavriousseva Elena
Institute for Nuclear Research
Academy of Sciences of Russia
Pr. 60th October Anniversary, 7a
Moscow 117312
RUSSIA
egavryuseva@solar.stanford.edu

Gelly Bernard
Départment d'Astrophysique
Universite de Nice Sophia Antipolis
Parc Valrose
F-06108 Nice Cedex 2
FRANCE
bernard@ayalga.unice.fr

Gibson Sarah
NASA Goddard Space Flight Center
Code 682, NASA GSFC
Greenbelt, MD 20771
USA
gibson@hao.ucar.edu

Gizon Laurent
School of Mathematics and Statistics
University of Sydney
Sidney NSW 2006
AUSTRALIA
gizon@irisalfa.unice.fr

Gonczi Georges
Département Cassini
Observatoire de la Côte d'Azur
Grande Corniche, BP 229
06304 Nice Cedex 4
FRANCE
gonczig@obs-nice.fr

Gonzalez Jean-François
Centre de Recherche Astronomique de Lyon
Ecole Normale Supérieure de Lyon
46 Allée d'Italie
69364 Lyon Cedex 07
FRANCE
jfgonzal@cral.ens-lyon.fr

Gonzalez Hernandez I.
Instituto de Astrofisica de Canarias
Via Lactea s/n
38200 La Laguna
TENERIFE, SPAIN
jpr@iac.es

Gough Douglas
Institute of Astronomy
University of Cambridge
Madingley Road
Cambridge CB3 OHA
U.K.
douglas@ast.cam.ac.uk

Goupil Marie Jo
DASGAL
Observatoire de Paris, Section Meudon
5 pl J Janssen
92190 Meudon
FRANCE
Goupil@mesiob.obspm.fr

Grabowski Udo
Astronomisches Institut
Basel Universität
Venusstrasse 7
CH-4102 Binningen
SUISSE
gra@astro.unibas.ch

Grec Gérard
Department G.D. Cassini
Observatoire de la Côte d'Azur
Grande Corniche - BP 229
06304 Nice Cedex 4
FRANCE
grec@obs-nice.fr

Guzik Joyce Ann
XTA MS B220
Los Alamos National Laboratory
Los Alamos
NM 87545-2345
USA
joy@lanl.gov

Haber Deborah
JILA
University of Colorado
Campus Box 440
Boulder CO 80309
USA
dhaber@solarz.colorado.edu

Harvey John
National Solar Observatory
P. O. Box 26732
Tucson AZ 85726
USA
jharvey@noao.edu

Heiter Ulrike
Institut für Astronomie
Der Universität Wien
Türkenschanzstraße 17
A-1180 Vienne
AUTRICHE
heiter@astro.ast.univie.ac.at

Henney Carl
Department of Physics and Astronomy
UCLA
8371 Math Sciences Building
Los Angeles CA 90095-1562
USA
henney@helios.astro.ucla.edu

Hernandez Mario M.
Instituto de Astrofisica de Canarias
Via Lactea s/n
38200 La Laguna
TENERIFE, SPAIN
mmanuel@ll.iac.es

Hill Frank
NSO/NOAO/GONG
P. O. Box 26732
Tucson AZ 85726-6732
USA
fhill@noao.edu

Hiremath Kampalayya
c/o Prof. H. Shibahashi
Dpt. Astronomy - University of Tokyo
Bunkyo-ku
Tokyo 113
JAPAN
hiremath@dept.astron.s.u-tokyo.ac.jp

Hoeksema Todd
CSSA-ERL 328
Stanford University
HEPL 213B
Stanford CA 94305-4085
USA
todd@solar.stanford.edu

Houdek Günter
Institut für Astronomie
Universität Wien
Turkenschanzstr. 17
A-1180 Wien
AUSTRIA
hg@shiva.edvz.tuwien.ac.at

Howe Rachel
Astronomy Unit, School of Math. Sci.
Queen Mary & Westfield College
University of London, Mile End Road
London E14NS
U.K.
r.howe@qmw.ac.uk

Isaak George R.
HIROS, School of Physics & Space Research
University of Birmingham
Edgbaston
Birmingham B15 2TT
U.K.
gri@star.sr.bham.ac.uk

Janot-Pacheco Eduardo
Instituto Astronômico e Geofisica da USP
Caixa Postal 9638
01065-970 Sao Paulo
BRASIL
janot@sismo.iagusp.usp.br

Jerzykiewicz Mikolaj
Wroclaw University Observatory
ul. Kopernika 11
51-622 Wroclaw
POLAND
mjerz@astro.uni.wroc.pl

Jimenez Antonio
Instituto de Astrofisica de Canarias
c/La Hornera s/n
E-38200 La Laguna
TENERIFE, SPAIN
ajm@ll.iac.es

Jurcsik Johanna
Konkoly Observatory
Hungarian Academy of Sciences
P.O.Box 67
Budapest XII H-1525
HUNGARY
jurcsik@buda.konkoly.hu

Kennedy James
GONG Project
National Solar Observatory
P. O. Box 26732
Tucson AZ 85726-6732
USA
kennedy@noao.edu

Korzennik Sylvain G.
Solar &Stellar Physics Division
Harvard Smithsonian Ctr. for Astrophysics
Mail Stop 16, 60 Garden Street
Cambridge MA 02138
USA
skorzennik@cfa.harvard.edu

Kosovichev Alexander
HEPL A204
Stanford University
MC 4085
Stanford CA 94305
USA
sasha@quake.stanford.edu

Kotov Valery A.
Crimean Astrophysical Observatory
334413 Nauchny
CRIMEA, UKRAINE
vkotov@crao.crimea.ua

Kovàcs Géza
Konkoly Observatory
Hungarian Academy of Sciences
P.O.Box 67
Budapest XII H-1525
HUNGARY
kovacs@buda.konkoly.hu

Kuhn Jeff
National Solar Observatory
Sacramento Peak - PO Box 62
Sunspot, NM 88349
USA
jkuhn@sunspot.noao.edu

Küker Manfred
Astrophysikalisches Institut Postdam
An der Sternwarte 16
D-14482 Postdam
GERMANY
mkueker@aip.de

Kumar Pawan
Dept. of Physics, 6-211
MIT
77 Mass. Avenue
Cambridge MA 02139
USA
pk@brmha.mit.edu

Kurtz Don
Astronomy Department
Inst. of Theoretical Physics & Astrophysics
University of Cape Town
Rondebosch 7700
SOUTH AFRICA
dkurtz@physci.uct.ac.za

Laclare Francis
CERGA
Observatoire de la Côte d'Azur
2130 Route de l'Observatoire
06460 Saint Vallier de Thiey
FRANCE
laclare@obs-azur.fr

Lazrek Mohamed
Département G.D.Cassini
Observatoire de la Côte d'Azur
Grande Corniche, BP 229
06304 Nice Cedex 4
FRANCE
lazrek@obs-nice.fr

Lazrek Mohamed
CNCPRST
B.P. 8027
10102, Rabat
MAROC
(Adresse permanente)

Lebreton Yveline
Observatoire de Paris, Section Meudon
5 pl J Janssen
92195 Meudon Principal Cedex
FRANCE
Yveline.Lebreton@obspm.fr

Leibacher John
NSO/GONG
P. O. Box 26732
Tucson AZ 85726
USA
jleibacher@noao.edu

Leifsen Torben
Institute of Theoretical Astrophysics
University of Oslo
P. O. Box 1029
Blindern, N-0315 Oslo
NORWAY
tleifsen@astro.uio.no

Leister Nelson Vani
Instituto Astronomico e Geofisico - USP
Av. Miguel Stefano 4200
04301-904 Sao Paulo
BRASIL
leister@astro1.iagusp.usp.br

Liu Yan Ying
Beijing Astronomical Observatory
Chinese Academy of Sciences
Zhong-Guan-Cun
100080 Beijing
P. R. CHINA
liu@mesioa.obspm.fr

Löffler Wolfgang
Astronomisches Institut
Basel Universität
Venusstrasse 7
CH-4102 Binningen
SUISSE
loeffler@sirrah.astro.unibas.ch

Lopes Ilidio Pereira
The Observatories
Institute of Astronomy
Madingley Road
Cambridge CB3 OHA
ENGLAND
lopes@ast.cam.ac.uk

Loudagh Said
Department of Physics
National Tsing Hua University
Kuang Fu Road
Hsinchu, Taiwan 300
REPUBLIC OF CHINA
loudagh@phys.nthu.edu.tw

Marchenkov Konstantin
Astronomy Unit
Queen Mary and Westfield College
Mile End Road
London E1 4NS
U.K.
K.I.Marchenkov@qmw.ac.uk

Martin Vera A.F.
CERGA
O.C.A.
Avenue Copernic
06130 Grasse
FRANCE
vera@ocar01.obs-azur.fr

Martinez Peter
Department of Astronomy
University of Cape Town
Rondebosch 7700
SOUTH AFRICA
peter@uctvax.uct.ac.za

Mathias Philippe
Département Fresnel
Observatoire de la Côte d'Azur
Grande Corniche, BP 229
06304 Nice Cedex 4
FRANCE
mathias@obs-nice.fr

Matias José
DASGAL
Observatoire de Paris
5 pl J Janssen
92190 Meudon
FRANCE
Jose.Matias@obspm.fr

Matthews Jaymie
Department of Physics & Astronomy
University of British Columbia
Vancouver
British Columbia V6T 1Z4
CANADA
matthews@astro.ubc.ca

Medupe Rodney
Institut for Fysik og Astronomi
Aarhus Universitet
Building 520 - Ny Munkegade
DK 8000 Aarhus C
DENMARK
thebe@obs.aau.dk

Michel Eric
DASGAL
Observatoire de Paris
Place J. Janssen
F-92195 Meudon
FRANCE
Eric.Michel@mesiob.obspm.fr

Miesch Mark
Joint Institute for Laboratory Astrophysics
University of Colorado
Campus Box 440
Boulder CO 80309
USA
miesch@solarz.colorado.edu

Mironova Irina
Sternberg Astronomical Institute
Moscow State University
Universitetskii prospect., 13
Moscow 119899
RUSSIA
mir@sai.msu.su

Missana Marco
Osservatorio Astronomico di Brera
Via Brera 28
20121 Milano
ITALY
missana@brera.mi.astro.it

Morel Pierre
Département Cassini
Observatoire de la Côte d'Azur
Grande Corniche, BP 229
06304 Nice Cedex 4
FRANCE
morel@obs-nice.fr

Moskalik Pawel
Copernicus Astronomical Center
ul. Bartycka 18
00-716 Warsaw
POLAND
pam@alfa.camk.edu.pl

Mosser Benoit
Institut d'Astrophysique de Paris
98 bld Arago
75014 Paris
FRANCE
mosser@iap.fr

Nesis Anastasios
Kiepenheuer-Institut für Sonnenphysik
Schöneckstrasse 6
D-79104 Freiburg
GERMANY
nesis@kis.uni-freiburg.de

Neuforge-Verheecke C.
Institut d'Astrophysique
5, Avenue de Cointe
4000 Liège
BELGIQUE
neuforge@kometh.astro.ulg.ac.be

Nigam Rakesh
Stanford University
HEPL Annex B 201
Stanford CA 94305-4085
U.S.A.
rakesh@quake.Stanford.edu

Oliviero Maurizio
Dipartimento di Scienze Fisiche
Universita de Napoli "Federico II"
Mostra d'Oltremare pad. 19
I-80125 Napoli
ITALY
oliviero@na.infn.it

Osaki Yoji
Department of Astronomy
School of Science, University of Tokyo
Bunkyo-ku
Tokyo 113
JAPAN
osaki@dept.astron.s.u-tokyo.ac.jp

Pallé Pere
Instituto de Astrofisica de Canarias
Via Lactea
38205 La Laguna
TENERIFE, SPAIN
plp@ll.iac.es

Pantel Alain
Départment d'Astrophysique
Université de Nice
Parc Valrose
F-06108 Nice Cedex 2
FRANCE
pantel@irisalfa.unice.fr

Pap Judit M.
Astronomy and Astrophysics Division
University of California
405 Hilgard Avenue
Los Angeles CA 90095
USA
pap@bonnie.astro.ucla.edu

Paparo Margit
Konkoly Observatory
Hungarian Academy of Sciences
PO Box 67
Budapest XII H-1525
HUNGARY
paparo@ogyalla.konkoly.hu

Paterno Lucio
Istituto di Astronomia
Universita di Catania
Citta Universitaria, Viale Andrea Doria 6
I-95125 Catania
ITALY
lpaterno@astrct.ct.astro.it

Patron Jesùs
Instituto de Astrofisica de Canarias
C/ Via Lactea s/n
38200 La Laguna
TENERIFE, SPAIN
jpr@ll.iac.es

Paunzen Ernst
Institut für Astronomie
University Vienna
Türkenschanzstrasse 17
1180 Vienna
AUSTRIA
paunzen@astro.ast.univie.ac.at

Pecker J.C.
Collège de France, Annexe
3 Rue d'Ulm
75231 Paris Cedex 5
FRANCE

Pigulski Andrzej
Wroclaw University Observatory
ul. Kopernika 11
51-622 Wroclaw
POLAND
pigulski@astro.uni.wroc.pl

Pijpers Franck P.
Theoretical Astrophysics Center
Institute for Physics and Astronomy
Aarhus University, Bldg. 520 Ny Munkegade
8000 Aarhus C
DENMARK
fpp@obs.aau.dk

Ponyavin Dmitri I.
Institute of Physics
University of St. Petersburg
St. Petersburg 198904
RUSSIA
ponyavin@snoopy.niif.spb.su

Poppe Paulo Cesar Da Rocha
CERGA
O.C.A.
Avenue Copernic
06130 Grasse
FRANCE
poppe@ocar01.obs-azur.fr

Poretti Ennio
Osservatorio Astronomico di Brera
Via E. Bianchi 46
I-22055 Merate (CO)
ITALY
poretti@merate.mi.astro.it

Probhas Raychaudhuri
Dept. of Applied Mathematics
Calcutta University
92, APC Road
Calcutta 700009
INDIA
probhas@cubmb.ernet.in

Provost Janine
Département Cassini
Observatoire de la Côte d'Azur
Grande Corniche, BP 229
06304 Nice Cedex 4
FRANCE
provost@obs-nice.fr

Rabello-Soares Maria Cristina de Assis
Instituto de Astrofisica de Canarias
Calle Via Lactea, s/n
38200 La Laguna
TENERIFE, SPAIN
crs@iac.es

Regulo Clara
Instituto de Astrofisica de Canarias
Via Lactea s/n
38200 La Laguna
TENERIFE, SPAIN
crr@iac.es

Renaud Catherine
Department G.D. Cassini
Observatoire de la Côte d'Azur
Grande Corniche - BP 229
06304 Nice Cedex 4
FRANCE
renaud@obs-nice.fr

Richard Olivier
Observatoire Midi-Pyrénées
14 Avenue Edouard Belin
31400 Toulouse
FRANCE
richard@obs-mip.fr

Rieutord Michel
Laboratoire d'Astrophysique
Observatoire Midi-Pyrénées
14 Avenue Edouard Belin
31400 Toulouse
FRANCE
rieutord@astro.obs-mip.fr

Robillot Jean-Maurice
Observatoire de Bordeaux
2, rue de l'Observatoire - BP 89
33270 Floirac
FRANCE
robillot@observ.u-bordeaux.fr

Rode Monika
Institute for Astronomy
Tuerkenschanzstr. 17
A-1180 Vienna
AUSTRIA
rode@astro.ast.univie.ac.at

Roques Sylvie
Laboratoire d'Astrophysique
Observatoire Midi-Pyrénées
14 Avenue Edouard Belin
31400 Toulouse
FRANCE
roques@obs-mip.fr

Rosenthal Colin
Teoretisk Astrofysik Center
Aarhus Universitet
DK-8000 Aarhus C
DENMARK
rosentha@obs.aau.dk

Rouse Carl A.
Rouse Research Incorporated
627 15th Street
Del Mar, CA 92014-2524
USA
u1164@sdsc.edu

Roxburgh Ian Walter
Astronomy Unit
Queen Mary and Wesfield College
Mile End Rd
London E1 4NS
UK
I.W.Roxburgh@qmw.ac.uk

Schatzman Evry
DASGAL, Bâtiment Copernic B
Observatoire de Paris-Meudon
5 pl J Janssen
92195 Meudon Principal Cedex
FRANCE
schatzman@mesiob.obspm.fr

Schenker Klaus
Astronomisches Institut
Basel Universität
Venusstrasse 7
CH-4102 Binningen
SUISSE
schenker@astro.unibas.ch

Scherrer Philip
Stanford University
HEPL Annex B211
Stanford CA 94305-4085
USA
pscherrer@solar.stanford.edu

Schmider François-Xavier
Départment d'Astrophysique
Université de Nice
Parc Valrose
F-06108 Nice Cedex 2
FRANCE
fxs@ayalga.unice.fr

Schou Jesper
Stanford University
HEPL Annex A201
Stanford CA 94305-4085
USA
jschou@solar.stanford.edu

Sekii Takashi
Institute of Astronomy
University of Cambridge
Madingley Road
Cambridge CB3 OHA
U.K.
sekii@ast.cam.ac.uk

Severino Giuseppe
Osservatorio Astronomico di Capodimonte
Via Moiariello, 16
I-80131 Napoli
ITALY
severino@astrna.na.astro.it

Shibahashi Hiromoto
Department of Astronomy
School of Science, University of Tokyo
Bunkyo-ku
Tokyo 113
JAPAN
shibahashi@astron.s.u-tokyo.ac.jp

Szabados Laszlo
Konkoly Observatory
 Hungarian Academy of Sciences
P.O. Box 67
H-1525 Budapest XII
HUNGARY
szabados@ogyalla.konkoly.hu

Talon Suzanne
DASGAL
Observatoire de Paris-Meudon
5 Place Jules Janssen
92195 Meudon Cedex
FRANCE
talon@mesioc.obspm.fr

Tamazian Vakhtang
Astronomical Observatory Ramon Maria Aller
University of Santiago de Compostela
PO Box 197
15706 Santiago de Compostela
SPAIN
oatamaz@usc.es

Thompson Michael J.
Astronomy Unit, School of Math. Sci.
Queen Mary & Westfield College
University of London, Mile End Road
London E14NS
U.K.
mthompson@solar.stanford.edu

Tomczyk Steven
High Altitude Observatory
HAO/NCAR
1850 Table Mesa Dr.
Boulder CO 80307
USA
tomczyk@hao.ucar.edu

Toner Clifford
National Solar Observatory
P. O. Box 26732
Tucson AZ 85726-6732
USA
toner@noao.edu

Toomre Juri
JILA/University of Colorado
Boulder CO 80309-0440
USA
jtoomre@jila.colorado.edu

Toutain Thierry
Département Cassini
Observatoire de la Côte d'Azur
Grande Corniche, BP 229
06304 Nice Cedex 4
FRANCE
toutain@obs-nice.fr

Tran Minh Françoise
DASGAL
Observatoire de Paris-Meudon
92195 Meudon Cedex
FRANCE
tranmin@obspm.fr

Turck-Chieze Sylvaine
Service d'Astrophysique
DAPNIA/DSM
CE Saclay
91190 Gif Sur Yvette Cedex
FRANCE
turck@ariane.saclay.cea.fr

Van Hoolst Tim
Instituut voor Sterrenkunde
Celestijnenlaan 200B
B-3001 Heverlee
BELGIUM
tim@ster.kuleuven.ac.be

Vandakurov Yuri V.
Dpt. of Plasma and Nucl. Phys. & Astrophys.
A.F. Ioffe Physical Technical Institute
Polytechnitcheskaja 26
194021 St. Petersburg
RUSSIA
van@mhd.ioffe.rssi.ru

Vauclair Gérard
Observatoire de Midi-Pyrénées
Observatoire de Toulouse
14 Avenue Edouard Belin
31400 Toulouse
FRANCE
gerardv@srvdec.obs-mip.fr

Vauclair Sylvie
Observatoire de Midi-Pyrénées
Observatoire de Toulouse
14 Avenue Edouard Belin
31400 Toulouse
FRANCE
svcr@obs-mip.fr

Vigouroux Anne
Département Cassini
Observatoire de la Côte d'Azur
Grande Corniche, BP 229
06304 Nice Cedex 4
FRANCE
gonczig@obs-nice.fr

Viskum Michael
Institute of Physics and Astronomy
Ny-Munkegade
DK-8000 Arhus C
DENMARK
mv@obs.aau.dk

Von Rekowski Brigitta
Astrophysikalisches Institut Postdam
An der Sternwarte 16
D-14482 Postdam
GERMANY
bvrekowski@aip.de

Vorontsov Sergei V.
Astronomy Unit
Queen Mary and Westfield College
Mile End Road
London E1 4NS
UK
S.V.Vorontsov@qmw.ac.uk

Wehrli Christoph
World Radiation Center
Physikalisch-Meteorologisches
Obs., Dorfstrasse 33
CH-7260 Davos Dorf
SWITZERLAND
chwehrli@pmodwrc.ch

Willems Bart
Institut Voor Sterrenkunde
Celestijnenlaan 200 B
B-3001 Heverlee
BELGIUM
conny@ster.kuleuven.ac.be

Worrall G.
Birdswood, Eardisley
Hereford HR3 6NJ
ENGLAND

Yerle Raymond
Observatoire de Midi-Pyrénées
Observatoire de Toulouse
14 Avenue Edouard Belin
31400 Toulouse
FRANCE

Zahn Jean-Paul
Observatoire de Paris, Section Meudon
5 pl J Janssen
92190 Meudon
FRANCE
zahn@obspm.fr

Zaqarashvili Teimuraz
Dept. Theoretical Astrophysics
Abastumani Astrophysical Observatory
A. Kazbegi av. 2a
380060 Tbilisi
REPUBLIC OF GEORGIA
tm@dtapha.kheta.ge

Zharkova Valentina
Physics & Astronmy Department
University of Glasgow
University Avenue
G12 8QQ Glasgow
UK (SCOTLAND)
valja@astro.gla.ac.uk

Zhugzhda Yuzef
Kiepenheuer Institut für SonnenPhysik
Schöncckstrassc 6
D-79104 Freiburg
GERMANY
yuzef@kis.uni-freiburg.de

PREFACE

Ce volume contient les revues invitées et les présentations orales du Symposium UAI 181 *Sounding Solar and Stellar Interiors*, tenu à Nice du 30 septembre au 3 octobre 1996. Les posters présentés à cette conférence sont publiés dans un volume séparé.[1]

Depuis le lancement avec succès du satellite SoHO le 2 Décembre 1995, et après des années d'efforts importants dans le domaine de l'hélio- et l'astérosismologie, il a semblé approprié de tenir un symposium dédié aux conséquences de la sismologie pour notre connaissance de la structure interne du Soleil et des étoiles. Le but était de présenter à la communauté entière les nouveaux résultats de l'héliosismologie obtenus aussi bien par les expériences spatiales que par les réseaux existants au sol, ainsi que les avancées théoriques dans les domaines de la structure et de la physique de l'intérieur solaire.

Ce but a été largement atteint: alors que SoHO ne fournissait des données que depuis environ 6 mois, la qualité des données héliosismologiques a été largement améliorée, de nouvelles techniques ont été élaborées comme la "téléchronosismologie" (nom proposé par D. Gough dans sa conclusion pour l'analyse temps-distance), et de nouveaux mécanismes physiques ont été proposés pour tenter de réduire les différences persistantes entre le Soleil observé et les modèles standard.

En même temps, plusieurs projets et résultats en astérosismologie attiraient l'attention de la communauté. Bien qu'encore dans l'enfance, comparée à l'héliosismologie, cette nouvelle branche connaît un essor manifeste. Le symposium a présenté aussi bien les interprétations théoriques d'observations déja existantes que les implications des résultats attendus des projets futurs pour notre connaissance de l'évolution stellaire. A cette occasion, un espoir est né de voir cette jeune discipline prendre son essor dans l'espace avec la mission COROT, qui sera le premier satellite dédié à la sismologie des étoiles et à la détection des planètes extra-solaires.

Ce colloque était dédié à la mémoire de Philippe Delache, décédé en octobre 1994, qui contribua activement à l'hélio- et l'astérosismologie. Pendant ce meeting une session a été consacrée aux progrès récents dans les domaines qui ont beaucoup intéressé Philippe durant sa vie. Les contributions à ce symposium ont montré que ces sujets restent toujours d'actualité,

---

[1]Ce volume est publié par l'Observatoire de la Côte d'Azur et l'Université de Nice et peut être commandé par e-mail à poster-IAU181@irisalfa.unice.fr

xxiv

en particulier la recherche des modes de gravité, les variations du Soleil à longues périodes et la sismologie de Jupiter.

Le nombre des participants, en particulier des jeunes chercheurs, et la qualité des exposés oraux et des posters ont contribué à faire de ce Symposium un vif succès scientifique. Le bouillonement d'idées théoriques et de résultats d'observations qui ont été présentés montre l'importance et la maturité de la sismologie solaire et stellaire pour l'astrophysique.

Le Comité d'Organisation est heureux de remercier pour leur soutien moral et financier l'Union Astronomique Internationale, le réseau européen ANTENA de la CEE, le Centre National de Recherche Scientifique, le Centre National d'Etudes Spatiales, l'Agence Spatiale Européenne, le Ministère de l'Education Nationale, le Ministère des Affaires Etrangères, la Région Alpes Provence Côte d'Azur, le Conseil Général des Alpes Maritimes, la Mairie de Nice et l'Observatoire de la Côte d'Azur.

Le Comité d'Organisation Scientifique présidé par R. Bonnet (ASE, France), était composé de C. Chiuderi (Florence, Italie), G. Debouzy (CNES, France), S. Frandsen (Aarhus, Danemark), C. Fröhlich (Davos, Suisse), A. Gabriel (Paris, France), D. Gough (Cambridge, UK), J. Leibacher (Tucson, USA), J.C. Pecker (Paris, France), T. Roca Cortés (La Laguna, Espagne), E. Schatzman (Meudon, France), P. Scherrer (Stanford, USA), S. Vorontsov (Londres, UK). Nous leur sommes très reconnaissants pour leurs suggestions et leur participation active.

Le Comité d'Organisation Local était composé de : A. Baglin (Meudon), F. Bely Dubau (Nice), N. Berruyer (Nice), V. Chéron (Nice), R. Feldman (Nice), E. Fossat (Nice), F. Laclare (Nice), D. Mekarnia (Nice), J. Pacheco (Nice), J. Provost (Nice), F.X. Schmider (Nice), G. Vauclair (Toulouse), J.P. Zahn (Meudon). La Présidente en était Gabrielle Berthomieu dont nous tenons à souligner le rôle majeur dans la réussite de ce Symposium.

Le Comité tient à remercier tout spécialement Christiane Caseneuve, Valérie Chéron, Renata Feldman, Bernard Gelly, Marie-Claude Pophillat and Nicole Selig pour leur soutien efficace dans les différents aspects de la conférence et dans l'édition des actes de ce colloque.

Les éditeurs
Janine Provost et François-Xavier Schmider

PREFACE

This volume contains the invited review papers and oral presentations of the IAU 181 Symposium *Sounding Solar and Stellar Interiors*, held at Nice from September 30 to October 3, 1996. The posters presented at this Symposium are published in a separate issue.[1]

Since the successful launching of the Solar Heliospheric Observatory satellite SoHO on December 2, 1995, and after many years of widespread effort in the fields of helio- and astero-seismology, it seemed appropriate to hold a symposium devoted to the implications of seismology for the internal structure of the Sun and stars. The aim was to present to the whole community the new results in helioseismology obtained by space experiments as well as by already existing networks, and the theoretical advances in the fields of internal structure and physics of the Sun.

This goal has been fully achieved: after only 6 months of SoHO operation, the quality of available data has been significantly improved, new techniques like time-distance analysis (or telechronoseismology, as suggested by D. Gough) have been proposed, and physical mechanisms envisioned to explain the remaining discrepancies between the observed Sun and the 'standard' models.

At the same time, new results concerning seismological studies of different types of stars have been available, and several projects, at different levels of development, required the attention of the community. Although still in its infancy, as compared to helioseismology, asteroseismology demonstrates a reinforced activity. The symposium dealt with the theoretical interpretation of the already existing observations and the implication of the expected results of the future projects in our knowledge of stellar evolution. During the meeting was presented the space project COROT, which will represent a unique opportunity to sound the stellar interiors by providing almost continuous data of high quality.

This meeting was devoted to the memory of Philippe Delache, who passed away in October 1994 and who contributed actively to the development of both helio and asteroseismology. During the meeting, a session was dedicated to recent advances in those topics which especially interested Philippe during his lifetime. The contributions showed that most of these topics remained of major importance at present, like for instance the search

---

[1] The poster volume is published by the *Observatoire de la Côte d'Azur* and the *Université de Nice* and can be purchased by sending an e-mail to poster-IAU181@irisalfa.unice.fr

for gravity modes, the long-term variations of the Sun, and the jovian seismology.

The large assistance, including a lot of young researchers, as well as the high quality of the oral presentations and posters, contributed to make this Symposium a real scientific success. The new theoretical ideas and the observational results presented have shown the importance and the maturity of the rather young field of helio- and asteroseismology for astrophysics in particular and physics in general.

The Organising Committee is glad to acknowledge for their moral and financial support the International Astronomical Union, the EEC network ANTENA, the *Centre National de Recherche Scientifique*, the *Centre National d'Etudes Spatiales*, the European Space Agency, the *Ministère de l'Education Nationale*, the *Ministère des Affaires Etrangères*, the *Région Alpes Provence Côte d'Azur*, the *Conseil Général des Alpes Maritimes*, the *Mairie de Nice*, the *Observatoire de la Côte d'Azur*.

The Scientific Organising Committee, chaired by R. Bonnet (ESA, France) was composed of C. Chiuderi (Florence, Italy), G. Debouzy (CNES, France), S. Frandsen (Aarhus, Denmark), C. Fröhlich (Davos, Switzerland), A. Gabriel (Paris, France), D. Gough (Cambridge, UK), J. Leibacher (Tucson, USA), J.C. Pecker (Paris, France), T. Roca Cortés (La Laguna, Spain), E. Schatzman (Meudon, France), P. Scherrer (Stanford, USA), S. Vorontsov (London, UK). We are very grateful to them for suggestions and active participation.

The Local Organising Committee was composed of A. Baglin (Meudon), F. Bely Dubau (Nice), N. Berruyer (Nice), V. Cheron (Nice), R. Feldman (Nice), E. Fossat (Nice), F. Laclare (Nice), D. Mekarnia (Nice), J. Pacheco (Nice), J. Provost (Nice), F.X. Schmider (Nice), G. Vauclair (Toulouse), J.P. Zahn (Meudon). It was chaired by Gabrielle Berthomieu, whose major responsibility insured the success of the Symposium.

The committee wishes to thank especially Christiane Caseneuve, Valérie Chéron, Renata Feldman, Bernard Gelly, Marie-Claude Pophillat and Nicole Selig for their valuable support in the different parts of the Conference and the edition of these proceedings.

The editors
Janine Provost and François-Xavier Schmider

PHILIPPE DELACHE
1937-1994

Philippe Delache, scientifique, administrateur, magicien, conteur, homme de grande culture, pêcheur de truites, qu'est-ce qui se dégage de sa personnalité, bientôt deux ans après qu'il nous ait quittés? Il reste bien difficile de répondre, tant ces traits de caractère semblent indissociables. Je crois que la plus grande partie de la vie de Philippe a été dominée par son goût pour le Merveilleux et pour le Beau, et par sa capacité personnelle d'émerveillement. Beaucoup de scientifiques restent de grands enfants, mais ceci était encore bien plus vrai pour Philippe.

Toute sa carrière scientifique, depuis ses débuts en recherche spatiale jusqu'à ses multiples participations aux aventures hélio et astérosismologiques, sans oublier Jupiter, repose sur ce goût du merveilleux. Sa réputation, largement justifiée, d'homme de grande culture et son amour de l'art également. Ses talents de pêcheur de truite et de cuisinier aussi, sans aucun doute. Qui, enfin, n'est pas resté muet d'admiration devant ses tours de cartes et autres prestidigitations en tous genres?

Né le 8 Octobre 1937 à Semur en Auxois, en Bourgogne, il venait, en Octobre 1994, de fêter ses 57 ans.

Quand il sort agrégé de l'Ecole Normale Supérieure, en 1960, il est aux côtés de Pierre Léna, Roger Bonnet, François Roddier, Françoise Praderie et Marie-Lyse Lory, sans oublier Solange, qui n'est pas encore son épouse. Une belle brochette, pourrait-on dire! Il entre au CNRS dans le service d'aéronomie de Jacques Blamont, où il travaille à la préparation d'expériences spatiales pour l'étude du rayonnement Lyman Alpha de l'atmosphère solaire. Mais même à cette période d'enthousiasme pour la recherche spatiale encore naissante, les jeunes chercheurs bouillonnant d'impatience y rencontraient déjà souvent la frustration des longs délais avant résultats, ou de pas de résultats du tout quand la fusée ne partait pas dans la bonne direction.... C'est finalement comme maître-assistant au Collège de France sous la direction de Jean-Claude Pecker qu'il soutient en 1967 une thèse théorique sur la diffusion des éléments dans la couche de transition chromosphère-couronne de l'atmosphère solaire. De cette époque est demeurée entre Philippe et Jean-Claude une solide et durable amitié pleine de complicité culturelle et intellectuelle.

S'il a quitté la recherche spatiale pour faire aboutir sa thèse, Philippe n'en a pas moins gardé pendant toute sa carrière une tendresse particulière pour son côté merveilleux. Ceux qui l'ont entendu raconter un lancement raté de fusée Véronique à Colomb-Béchar en sont facilement convaincus. Pour les autres, on pourrait faire une liste des responsabilités que Philippe a assumées dans le spatial: Membre du Comité des programmes scientifiques du CNES, Président du groupe Astronomie, membre de plusieurs groupes de travail d'astronomie à l'ESA, rapporteur du projet Hipparcos (il a été invité à assister au lancement à Kourou), paternité partagée avec Roger Bonnet de l'héliosismologie à bord de SoHO. Il a également assumé bien d'autres responsabilités, pas forcément liées directement à la recherche spatiale: membre élu de la Section 18, puis du CNAP, Président de l'ADION, fondateur et rédacteur en chef du JAF, membre du CNFA, et j'en passe, la liste pourrait paraître fastidieuse.

En 1967, suivant avec quelques autres la venue à Nice de Jean-Claude Pecker, il commence par un poste de Professeur à l'Université, puis d'Astronome à l'Observatoire. C'est donc en 1967, alors que j'étais étudiant de DEA, que j'ai connu Philippe. Un de mes premiers souvenirs de l'époque se rapporte aux haut-parleurs Ellipson. Philippe nous parlait un jour d'un certain Léon, ingénieur acousticien de génie qui fabriquait ces fameuses boules en plâtre qui équipaient déjà sa chaîne Hi-Fi et celles de Roger Bonnet, François Roddier et quelques autres. Très peu de temps plus tard, Gérard Grec et moi-même avions les mêmes, et ces merveilleuses enceintes acoustiques sont toujours chez nous aujourd'hui, près de 30 ans plus tard.

Directeur de l'Observatoire de Nice de 1969 à 1972, puis encore un an en 1975, il fut ensuite un des principaux artisans du regroupement Nice-Cerga devenu OCA. En 1988, il a vainement essayé de me convaincre d'être candidat à la direction de l'OCA. En fait, je crois que Philippe avait envie d'y replonger lui-même. Cet exercice de la Direction lui plaisait vraiment, et il avait une revanche à prendre sur ses expériences précédentes. Mais il avait déjà, à cette date, subi ses deux interventions cardiaques, et il savait pertinemment, pour l'avoir déjà pratiqué, combien le métier de Directeur peut user quelqu'un en profondeur.

En parallèle avec toutes ces activités de responsabilité, Philippe a toujours continué à s'émerveiller pour la recherche. Quelqu'un m'a dit un jour: "Philippe n'est jamais aussi bon scientifique que quand il est vraiment saturé de charges administratives". C'était un peu vrai.

Pendant les années 70, c'est le transfert du rayonnement qu'il a d'abord pratiqué lui-même, puis vers lequel il a orienté l'activité du petit groupe d'Hélène Frisch à Nice: transfert de rayonnement dans les atmosphères stellaires et solaire, en milieu inhomogène et dépendant du temps. En 1980, le succès de notre expédition au Pôle Sud avec Gérard Grec l'a fasciné. Encore une fois pour le côté merveilleux de cette nouvelle science si prometteuse. Et c'est ce côté merveilleux qu'il va lui-même contribuer à promouvoir et développer. C'est à ce moment-là qu'il retrouve Roger Bonnet et qu'ensemble ils proposent à l'ESA de mettre un satellite au point de Lagrange du système Soleil - Terre pour refaire là-haut la "manip" Grec-Fossat, hors atmosphère et pendant bien plus longtemps. Le projet s'appelait alors DISCO. Allé assez loin dans le processus de sélection, il n'est cependant pas arrivé jusqu'au bout mais il a ensuite été repris dans un ensemble plus vaste, nommé SoHO, dont vous allez probablement entendre un peu parler pendant ce symposium. La contribution de Philippe à toutes les phases de cette aventure a été tout à fait essentielle non seulement par ses idées scientifiques, mais également par ses talents diplomatiques quand certaines tensions pointaient ici ou là.

En parallèle se sont développés les programmes d'héliosismologie au sol, dans lesquels Philippe s'est très largement engagé. Quand on essaie de faire un peu le tour des programmes en cours d'exploitation actuellement (GONG, IRIS, BISON, Antarctique, etc...) on s'aperçoit que Philippe a joué un rôle, et dans presque tous les cas un rôle très important, dans chacun d'eux. C'est pendant ces années "héliosismiques" de son activité scientifique que j'ai eu le plus souvent l'occasion de travailler avec lui, en grande partie par l'intermédiaire de la thèse d'Etat marocaine de Mohamed Lazrek, que nous avons co-encadrée. Elle nous a donné l'occasion d'écrire quelques articles ensemble, et surtout de passer des samedis matins à parler de science. Dans ces occasions, le sourire si malicieux de Philippe, que j'ai déjà évoqué, s'illuminait avec un plaisir purement enfantin quand il nous montrait, par un tour de passe-passe informatique, l'aboutissement de sa dernière idée de la nuit précédente.

Dans cette permanente démarche qu'était la recherche du merveilleux, c'est toujours vers le plus difficile que son activité de recherche s'est prioritairement orientée. Par exemple, si l'héliosismologie a connu un grand succès en mesurant un grand nombre de modes de pression, le Soleil devrait en principe aussi osciller dans une autre gamme de fréquences, correspondant aux modes de gravité, les fameux modes g. Ils sont, ce n'est un secret pour personne ici, considérablement plus difficiles à détecter par l'observation et à étudier par l'analyse. Avec Phil Scherrer d'abord, puis avec Claus Fröhlich, Philippe s'est fait dans les années 80 une réputation d'ambassadeur des modes g. Aujourd'hui, leur quête reste un des objectifs majeurs des 4 instruments d'héliosismologie embarqués à bord de SoHO et l'un des attraits de notre symposium ....

Et puisque le Soleil oscille, cela pourrait bien se voir sur son diamètre. Depuis une vingtaine d'années, ce diamètre est mesuré par Francis Laclare à Calern. En comparant ses variations à celles d'autres paramètres (activité, flux de neutrinos), Philippe fouille ces données en mettant en oeuvre, bien entendu, des méthodes d'analyse plus originales que la sempiternelle transformation de Fourier. L'analyse par ondelettes semble prometteuse en ce domaine, et il propose d'élargir le champ de cette étude comme sujet de thèse pour Anne Vigouroux, qui a repris ce flambeau aujourd'hui.

Je dois bien entendu également mentionner les oscillations de Jupiter, découvertes par F.X. Schmider qui avait été stagiaire de DEA de Philippe quelques années plus tôt. Elles ont, tout comme celles du Soleil, excité la curiosité et la sagacité de Philippe, et il s'est très rapidement plongé dans ce nouveau bain sismologique, aussi bien du côté du théoricien (sujet de stage de Benoît Mosser) que de celui des observateurs, en participant avec Djamel Mekarnia et Jean Gay à une campagne d'observation au CFHT.

Les prolongements stellaires et planétaires de cette extraordinaire moisson que nous a apporté l'héliosismologie ont fait l'objet d'une opération de la Communauté européenne, baptisée ANTENA (A New Technology European Network for Asteroseismology). Encore une fois, Philippe a été, avec Teo Roca Cortés, l'un des maîtres d'oeuvre du montage de ce projet, qui prévoyait, avec financement, l'organisation d'une importante réunion internationale à Nice. La dimension, la structure, et autres détails de cette réunion restaient totalement à définir, et Philippe s'était engagé à prendre cette affaire en mains. Modestement mais de leur mieux, les membres des deux comités, scientifique et local, se sont efforcés d'organiser ce symposium. Nous voici donc tous réunis à Nice, en pensant au plaisir qu'aurait notre ami Philippe de voir à quel point la dimension des résultats scientifiques disponibles aujourd'hui arrive à dépasser toutes les espérances, peut-être même celles qui étaient les siennes.

Er*c Fossat   30 septembre 1996

INTRODUCTION

SOUNDING SOLAR AND STELLAR INTERIORS: GENERAL INTRODUCTION

JOHN LEIBACHER
National Solar Observatory
950 North Cherry Avenue, Tucson, Arizona 85719, USA

1. An introduction to this Introduction - Helioseismology is great

in the many senses of the word. *It is a tremendous amount of fun* with a really large number of interesting problems, vigorous dialectic between observations and theory, rapid progress, fantastic observational opportunities, and extraordinarily stimulating colleagues. *It is grand*, in its phenomenally rapid development, the challenges that it presents, and the importance of the problems that it addresses, as well as its promise for the future. *It is doing very, very well* as seen in the vitality of the community of researchers pursuing it, in the major investments made in acquiring beautiful new data to fuel its continuing progress, and in the many new problems that open before us as the current ones achieve a measure of "understanding".

This Symposium marks a significant milestone in the development of helioseismology – the beginning of observations from *GONG* and *SoHO* – and it may be difficult to recall that, in many senses, the discipline began but a scant twenty years ago, here in Nice at a conference entitled "Physique des Mouvements dans les Atmosphères Stellaires" at which Franz-Ludwig Deubner first presented his remarkable observational demonstration that the "five-minute oscillations" were indeed normal modes of oscillation of the solar interior [Figure 1]. Contrast that landmark observation with the state of the art today [Figure 2] to see how dramatic has been our progress.

This meeting also marks the passing of time and the loss of two of the truly great personalities of the field of the "physique des mouvements dans les atmosphères stellaires" who participated very actively in that meeting: Philippe Delache and Dick Thomas. They were truly great figures in their contribution of new ways of thinking about the Universe, in their efforts – and their successes – in encouraging those around them to pursue these

1

J. Provost and F.-X. Schmider (eds.), Sounding Solar and Stellar Interiors, 1-12.
© *1997 IAU. Printed in the Netherlands.*

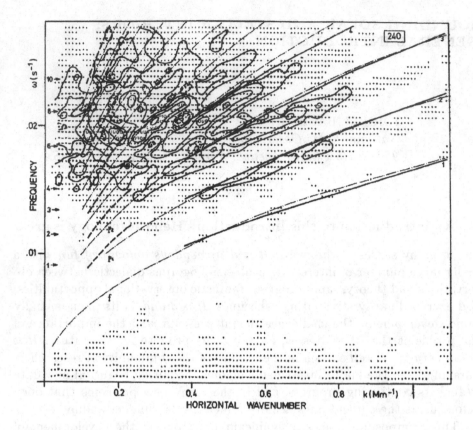

Figure 1. Franz-Ludwig Deubner's landmark $k - \omega$ diagram (Deubner, 1976) first
presented at Nice.

new avenues of exploration, and in their personal generosity. They were my
mentors and friends, and they represent why helioseismology is great.

The articles that follow provide the new knowledge that is in the process
of emerging, and which is the meat of these proceedings – why we are here.
Let me just preface these contributions with my own perspective on where
we have come from, what we are doing, and what the future may hold.
And, of course, Douglas will have the last word!

2. Where have we come from?   Helioseismology is an accident

but an inspired one, a beautiful example of scientific discovery, serendipity
at its best. As I view it, in the late 1950's a great deal of interest was being
devoted in all of the sciences to characterizing turbulence as it manifested
itself in the phenomena which they were dedicated to describing, and there
was a general awakening to the importance of non-thermal processes in all

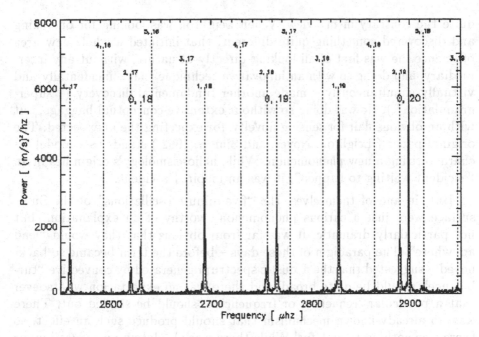

Figure 2. A snippet from a modern $\ell - \nu$ diagram, with the resonant modes identified by their spherical harmonic degree(ℓ) and their radial order (n).

of astrophysics. While many investigations had been devoted to the spatial distribution of "turbulent" motions on the Sun, which as a result of the assumption of ergodicity should have completely characterized the motions, Bob Leighton, a physicist who "didn't know any better" came up with a new scheme for looking at the "turbulent" velocity field of the solar surface that captured both its spatial *and* its temporal variations, and immediately noticed that the "turbulent" motions were, in fact, quite periodic, with a period close to five minutes.

As we now know, the "five-minute oscillations" of the solar surface are an extremely robust phenomenon. They are readily visible in virtually every measured parameter of the Sun's surface, and very simple instrumentation is capable of exhibiting it. It seems *so* obvious, even trivial, now that it is difficult to imagine how many very skilled observers of nature passed so close to this fundamental discovery without stumbling on to it. It is a bit reminiscent of the (apocryphal?) story of Columbus' challenging the detractors of his discovery of the new world to balance an egg on its end. It *is* trivial once the insight has been achieved, but how to stumble upon it??? Without – hopefully – belaboring the point, or romanticizing the mythology of the foundations of our discipline too much (nor trivializing it), let me emphasize that this major discovery was not "dumb luck"; it was quite the contrary in my view. When Leighton was looking for one thing and discovered something quite different, that initiated a whole new area of science, he was first of all looking directly at nature, without any intermediary, and doing so with an innovative technique that – incidentally and virtually simultaneously – made another fundamental discovery ("supergranulation"). He was doing so without extensive conceptual baggage, but with an obvious flair for sensing novelty, for expecting the unexpected. The original paper (Leighton, Noyes, and Simon, 1962) stands as a model for characterizing a new phenomenon. While helioseismology's origin lies in an "accident waiting to happen", it was an inspired accident.

But, in and of themselves, the "five-minute oscillations" of the Sun's surface were just a curious phenomenon, worthy of an explanation, but not particularly dramatic. It was far from obvious that they would "lead anywhere". The paradigm of those days – before the term became so hackneyed – suggested that the acoustic spectrum generated by convective "turbulence" should be quite broad, and there was no expectation whatsoever that a particular frequency, or frequencies, should be singled out. There was no already-known mechanism that should produce such an effect, so there was nothing to look for! While there was absolutely no suggestion, or even a hint, that the phenomenon of "five-minute oscillations" might exist prior to their discovery, a plethora of distinct theoretical models sprang into existence quite rapidly, all capable of "explaining" the unanticipated oscillations in very simple, straightforward fashions. [I will stop putting "explain" in quotation marks; my point being simply that we really do abuse the term in conversational science as we *describe* phenomena, hopefully with ever increasing validity, but it is rare that we can categorically *explain* them.] It is truly remarkable that all of these good ideas about how a stellar atmosphere might behave had not been pursued prior to the observational discovery; that they had not been advanced on the basis of pure thought. It rather gives the lie to the assertion attributed to Eddington that an astronomer on a cloud-covered planet should be capable of deriving a complete description of the Universe. Inductive logic would appear to be no match for a novel technique and an open mind. The converse of our lack of prior consideration of all of these mechanisms, in the absence of a good – that is observation-driven – reason to do so, is our possibly hasty discarding of all of the mechanisms that appeared as providing plausible descriptions of the "five-minute oscillations" but that were not supported by the observed modal structure of the spatial-temporal power spectrum. If they were reasonable things to be happening, they probably are! Some place in the Universe, maybe even in the Sun, they should be operating and we should be looking for their signatures, but I digress.

There were many, many more "constructive accidents", or incidents of "falling forward" in the early history of helioseismology. Bob Stein and I set out to demonstrate that the five-minute oscillations could be trapped in the photosphere-chromosphere temperature minimum, but the calculations were plagued by oscillations trapped below the visible surface, which we tried vainly to get rid of – since we just "knew" that they were artifacts – for the longest time before realizing that they were the answer, knocking us between the eyes. I recall Roger Ulrich (1970) acknowledging that he too was surprised initially when his calculations unexpectedly grew in time, before convincing himself that this was in fact a reasonable result of the κ-mechanism and a plausible description of the observed phenomenon. It is also worth recalling that there were very good reasons for believing that sounds waves trapped below the surface could not possibly maintain phase coherence and thus form resonant, normal modes.

Deubner's beautiful demonstration [Figure 1] that the oscillations could indeed be described as normal modes of the solar interior was possibly the first "correct" move in answering "where have we come from?", but I hazard to guess that if we pushed him a bit, there were a few "accidents" along the way to his major contribution to the development of our discipline which – in some sense – closed out a nearly textbook case of the idealized "scientific method": 1) an unanticipated phenomenon was discovered, quite "by chance", 2) a variety of models were advanced to describe the observation, with testable predictions, 3) the predictions of one of the models were, rather surprisingly, confirmed by subsequent observations.

Interesting though this phenomenon and its description might be, our discipline would not exist had the story concluded there. But Deubner's seminal work already showed very clearly that while the oscillatory power did exhibit resonant ridges of power in the "diagnostic diagram" or "$k - \omega$ diagram" as the two-dimensional power spectra were called in those antediluvian days, there were very significant differences between the observed frequencies and those calculated by Charlie Wolff (1972), and Hiroyasu Ando and Yoji Osaki (1975) utilizing the then-accepted models of the solar interior. It was this fourth step – the appreciation that not only was the description of the observed five-minute oscillations as sound waves trapped below the solar surface a good model of the phenomenon, but that the remaining differences between the predictions and observations could be used as an effective tool to test, and thus modify, our relatively poorly constrained models of solar, and stellar, structure that helioseismology really came into existence. It is frightfully dangerous, and terribly simplistic – if not simple-minded! – to try to distill the work of dozens of researchers to a few steps like this. This is a cartoon sketched to provide some context for the Symposium, not a history! The cartoon would, however, be dramatically incomplete without a caption, and that would be the term

"helioseismology" itself, which Douglas Gough bludgeoned us into accept-
ing, and which has served so nicely to describe our discipline's content
as both a tool and a science. At the same time that we are studying the
physics of the oscillations as phenomena in their own right, we are using
our working understanding of them as a tool to explore solar structure.

3. Where are we now?   Helioseismology is Nice

of course! It is particularly appropriate to be holding this Symposium in
Nice, and most fitting – though tragic – to be honoring Philippe, who himself had started the planning for the meeting. If the 1975 conference on the
"Physique des Mouvements dans les Atmosphères Stellaires" was a landmark for helioseismology, the real potential was demonstrated the next year
– again here in Nice – at IAU Colloquium Number 36 entitled "The Energy
Balance and Hydrodynamics of the Solar Chromosphere and Corona" organized by Philippe and Roger Bonnet, where Douglas Gough (1977) pointed
out that the discrepancy between the observed and predicted frequencies
could be accounted for by a substantial ($\approx 50\%$) increase in the depth
of the convection zone; the first major contribution of helioseismology. The
pace of new results has not slackened since.

Helioseismology is very nice indeed when one reflects on what it has
achieved in less than a twenty two year Hale cycle what assets have been
brought to bear to advance the science further, and the community of interest that it has engendered.

After discovering the discrepancy in the depth in the convection zone,
helioseismology went on – in very short order – to demonstrate that the
deep interior could not be rotating sufficiently rapidly to give rise to a
gravitational quadrupole moment sufficiently large to invalidate the support given to general relativity by the precession of Mercury's orbit (as
had been ardently proposed), that the interior could not possibly be as
cool as some *ad hoc* theories – proposed to address the observed deficit of
neutrinos – would have required, and that the descriptions of the opacity
and the equation of state of matter at the temperatures and densities representative of stellar interiors were substantially incomplete; *big* issues for all
of astrophysics, catapulting helioseismology to a position of great visibility
in astronomy, physics, and before the general public.

These "big physics" issues continue to occupy a very prominent position in helioseismology, of course, as we obtain better data, develop better
analysis techniques, and advance our understanding of the problems themselves. However, a nice indication of the overall vitality of the discipline has
been the continuing addition of new areas of investigation, *e.g.* sunspot seismology and more generally "local helioseismology", time-distance method-

ologies, temporal variations of the amplitudes, lifetimes, and frequencies of the normal modes on scales from hours to the eleven-year activity cycle, and the physics of the modes themselves as seen, for example, in the asymmetries of the normal mode resonances and in the differences between the atmospheric responses at different heights, or in different diagnostics. In large measure, helioseismology has been defined by its promise to address some of the most major issues in contemporary astrophysics, to which it has beautifully responded, but it continues to grow in importance because of its ability to open new dimensions to the existing questions, and – *most importantly* – by continuing to pose exciting, new questions of broad interest.

At the same time that we congratulate ourselves on the very rapid strides forward that we have made and the ease with which we have succeeded, it may be prudent to reflect on the rigor that we may have put aside "for just the moment" as we rushed forward. Our rough analyses, quite often using techniques developed for other problems and cobbled together to apply to our new data and scientific questions, have been remarkably useful and successful. However, just to give an example or two, the decomposition of our images into spherical harmonics assuming that the oscillatory velocities are purely radial or that the state variables do not display any center-to-limb variations is certainly not correct, and may well give rise to systematic errors of potentially significant magnitude. Ignoring the frequency asymmetries of the resonance peaks in our estimation of eigenfrequencies is almost certainly guaranteed to change our structural inferences systematically, and the apparent difference in the asymmetries seen in velocity and intensity spectra may by signaling even more serious difficulties. Our use of simple fourier transforms and rudimentary time series analyses has our colleagues in statistics grinding their teeth, and the list of "thin ice" goes on and on. In addition to the simple work-arounds of these first-generation problems – that, I should hasten to point out, have served us remarkably well – there are many areas in our analyses where we have simply not had the time to develop more sophisticated approaches to reap the maximum benefit from the data. I am thinking of mode leakage in the spherical harmonic decomposition, or optimal use of all of the individual frequencies in our inversions. Once again, I am not suggesting that we have been sloppy, but rather that the "quick and dirty" techniques worked so well, and gave such interesting results, that we were justified in rushing forward, but now we should reflect upon the assumptions made and short cuts taken, and critically evaluate our methodologies.

The extraordinary scientific potential, and support for fulfilling it, has engendered a correspondingly impressive array of observational assets summarized in Table 1.

TABLE 1. Some Current
Helioseismology Projects

BiSON
GONG
IRIS
TON
SoHO - GOLF
SoHO - SOI/MDI
SoHO - VIRGO

and single site programs
Crimea
El Teide
Haleakala
Kitt Peak
Mauna Loa
Mount Wilson
Napoli
Roma

Just for the record – it is unlikely to be lost on the participants at this Symposium – this armada of telescopes is required by the challenges of the current observational requirements and particularly by the subtlety of the measurements and the identification and removal of observational, or diagnostic-related, artifacts. While the basic oscillation signal is extremely robust, the current state of the art depends on the combining and differencing of many, many individual measurements; identifying and overcoming the tiny, tiny systematic errors is of paramount importance. We desperately need the independent, and complementary, observations to believe any of this!

The final aspect of where we are – our current assets – is the vigorous community as witnessed by the number and demographics of the participants at this meeting. For a discipline that literally did not exist twenty years ago, it is I believe unique in the annals of solar physics, and I hope the participants can sense the momentum of the discipline of which they are a part, and exploit the opportunity that this confluence of exciting and important scientific questions, extraordinary observational capabilities, and brilliant talent represents.

The community has developed in a rather harmonious way, I believe. Data exchange is very free and open to all, and this is a tremendous strength of course. While there remain many problems where a single, isolated investigator can work effectively and contribute significantly, I believe that the nature of many of the problems is such, and we have grown together in such a way, that the most effective way to attack many of the most "important" problems is through substantial collaborations, and this has been very nicely demonstrated in the several "hare and hounds" activities that have tested various inversions techniques, as well as the team publications of first results – where sundry analysis techniques were very nicely compared and a syntheses emerged that was – in may senses – a significantly more useful advancement of the subject than the sum of the individual works.

The large quantity of excellent and widely available data, the complexity and subtlety of the analysis, and the wide range of physical problems being addressed rather inevitably leads to some interesting changes in the way that the community works. Glancing at my bookshelf, I count seventeen major, international conferences devoted to helioseismology in the last twenty years, and I may well have overlooked one or two. Helioseismology has become something of a separate discipline within solar physics and astrophysics, and while the size and productivity of our activity is wonderful, we really must guard against becoming isolated from our colleagues in closely related fields. We are scientists, problem setters, not just problem solvers – although that is a lot of fun too!

4. In the guise of a conclusion – Helioseismology is just beginning

to become an established science, entering its second generation. Let me use the conclusion of this Introduction to put forward a few questions that hopefully will all be well answered within far less than the twenty years that separate us from the beginnings of helioseismology. And, even then, I think that we are justified in the well-founded assumption that they will be replaced with more, exciting, new questions derived from their answers.

- What improvements in our techniques of analysis of the data should be pursued?
- How can we render more certain our inferences from the data?
- What remains unknown in our description and modelling of the oscillations?
- What remains to be learned about the microphysics of stellar structure?
- What major uncertainties about the macrophysics of stellar structure are becoming amenable to attack?
- How well can we describe the phenomena of the Sun?

- Can g-modes by unambiguously identified in the Sun?
- How can we best identify p-mode analogues unambiguously in the Sun?

Our techniques for inferring parameters – transforming measurements of light emitted by the surface of the Sun into numbers that can be directly compared to theoretical predictions, or in some cases directly input to adjust the theory – have been spectacularly successful, while at the same time they have been rather crude. I am thinking here of better ways of fitting the observations to our model (that they arise from a superposition of independently excited normal modes of oscillation of a strongly inhomogeneous, differentially rotating fluid), and *vice versa*.

Whatever the techniques for relating observations to models, there will remain uncertainties ("noise") and outright lack of knowledge (temporal and spatial incompleteness) in our observations and a major challenge for the second generation of helioseismology will be to render these inferences less biased by our techniques and optimally representing constraints on the observational model.

It is only in the last couple of years that a good model for the excitation of the modes has emerged, and it is quite remarkable in hindsight to see what great strides were made using the frequencies of the modes with extremely little idea of how they were actually driven – thank goodness for linear phenomena! But, our knowledge of the physics of their excitation – and damping – as well as their behaviour through the visible atmosphere still leaves a great deal to be desired, and I anticipate considerable progress in this area in the immediate future; *e.g.* how does the energy in a mode vary with time, to what extent do excitation "events" couple various modes together or imprint similar signatures on different modes.

With the advent of helioseismic probing of the internal conditions of the Sun, deficiencies in our description of matter in the interesting temperature and density regimes became quickly apparent. As the precision of the measurements and the robustness of the inversions improves, how much further can we tighten these constraints and how much further can we extend the ranges of temperatures and densities over which they obtain? People really want to know!

A number of very significant astrophysical processes (*e.g.* rotational shear induced mixing, penetrative convection, diffusion, circulation currents) have become fairly directly accessible to our sounding using helioseismology. I think that it is fair to assume that our first attempts in these areas will not be uniformly successful and that we have a lot of changes in store. The elucidation of these processes may be one of helioseismology's most important demonstrations of the Sun's role as an astrophysical laboratory.

Even were the macrophysics of stellar internal structure to be fairly well understood, it is a considerable leap forward to be able to characterize the major phenomena of the solar interior such as differential rotation, the generation of magnetic fields and the twenty two year cycle of magnetic activity. For generations to come, the Sun will continue to provide the one place in the Universe where we can study these intriguing phenomena.

While helioseismology was at its outset driven by observational discoveries for which we were entirely unprepared, we have found ourselves for some time now with two major theoretically predicted phenomenon that have not been established observationally: internal gravity modes, and p-mode analogues in other, solar-type, stars. The search for these "holy grails" has provided a sort of frontier spirit to helioseismology. While all of the advances back in the better understood territory of p-mode helioseismology have been moving forward, the lure of these uncharted wildernesses continues to be enormously seductive. There is very little question that both must surely exist – the physics is quite straight-forward – and there have been numerous suggestions of their identification over the years, but their wileyness has become really quite frustrating on one hand, and enervating on the other. The potential new knowledge represented by each of them is staggering, as the number of contributions devoted to them at this Symposium bears witness. Gravity mode frequencies promise dramatically improved diagnostics of the deep solar interior, as well as all of the new insights that we are confidant that the existence of a new wave mode will offer.

In many way, we are better prepared – in theory at least – to deal with the implications of p-mode measurements on other stars, given their close analogy to the physics of the modes within the Sun; however the potential for surprises may be even greater. In some sense, we are calibrating the values of the parameters and processes that control stellar structure for one set of conditions with our helioseismic sounding. Asteroseismology offers the possibility of measuring the gradients of the parameters and processes!

It is simple enough to look back and see how helioseismology has developed from nothing over the past two decades, and we can all marvel at the sorts of questions that we can pose today and the precision with which we can offer answers. Yet the promise of the physics that we have at hand and the data that we are now in the process of acquiring is astonishing by comparison, while the potential of g-modes and of stellar p-modes is staggering. And, it may not take another twenty years to fulfill their promise, and open the next chapter!

References

Ando, H., and Osaki, Y., 1975, *Pub. Astron. Soc. Japan* **27**, 581-603

Deubner, F.-L. 1976, in *Physique des Mouvements dans les Atmosphères Stellaires*, Paris, Éditions du CNRS, eds. R. Cayrel and M. Steinberg, 259-261

Gough, D.O. 1977, in *The Energy Balance and Hydrodynamics of the Solar Chromosphere and Corona*, G. de Busaac, Clermont-Ferrand, eds. R.-M. Bonnet and Ph. Delache, 3-36

Leighton, R.B., Noyes, R.W., and Simon, G.W., 1962, *Astrophysical J.* **135**, 474-499

Ulrich, R.K., 1970, *Astrophys. J.* **162**, 993-1002

Wolff, C.L., 1972, *Astrophys. J.* **177**, L87-L92

HELIOSEISMOLOGY : FROM GROUND OR SPACE
A WORLDWIDE COOPERATION

THE STATE OF THE ART IN HELIOSEISMIC GROUND-BASED EXPERIMENTS

PERE L. PALLÉ

Instituto de Astrofísica de Canarias, 38205 La Laguna
Tenerife, Spain

Abstract. The new results obtained from the observation of solar oscillations over the past decade, have a direct impact on our knowledge of the Sun's interior. As a consequence, a great interest in helioseismology has arisen and is reflected in the development of new observational projects as well as new analyse and inversion techniques. In this review we will describe the present ground-based observational programmes, which, unlike the space ones, are mostly designed to produce high quality data over very long time spans (up to solar cycle time scales). The characteristics of the various observational programmes, single-site and network, will be described together with their performances, the main results obtained up to now, and some other logistical aspects.

1. Introduction

At present, helioseismology observations are carried out from both space and ground-based observatories, and by both means it is possible to overcome the main difficulty for progress in helioseismology: the non-continuity of observations. By establishing ground-based networks at appropriate locations on the Earth (like *BiSON, GONG, IRIS,* etc.) and spacecraft in adequate orbits (like *SOHO*), it is possible to continuously observe the Sun for long periods of time (months, years). It is not now the right time for establishing a competition between these two types of observing modes; each one has clear advantages and disadvantages. On the contrary, it is hoped that real progress in helioseismology will take place only from the extensive and precise analysis, as well as comparison, of both types of data. Since this particular review is devoted to ground-based observational projects, let us point out two distinct aspects of this kind of observation: the *control* and the

15

J. Provost and F.-X. Schmider (eds.), Sounding Solar and Stellar Interiors, 15-29.

lifetime. The possibility of repairing, modifying or even re-designing ground-based experiments leads to the possibility of having them operational over very long periods of time (many solar cycles). This is, at present, not possible with space experiments. In addition, the former offer the possibility of performing simultaneous observations which could allow for detection of annoying systematic errors. In summary, as pointed out recently (Harvey 1995), the combination of already existing space and ground data will accelerate the transformation of helioseismology from an *Art* into a *Science.*

In the following sections we will describe two groups of ground-based experiments: those having more than one *identical* operational instrument at different places (networks) and those having a single one (single site). For each of these groups we will compare the different characteristics and performances.

2. Helioseismology networks

Performing solar seismology observations from two or more sites simultaneously with similar instruments, fulfills various objectives: it not only provides with an uninterrupted series of observations but also serves to detect systematic errors and, most importantly, to prove the solar origin of any detected signal. Of course, since this last objective is very well accomplished today in the case of the solar origin of the 5 minute oscillation, the former ones become more important. Historically, it was in 1978 when the Birmingham-Tenerife group performed simultaneous solar oscillation observations from two different observatories (Claverie *et al.* 1979). Although the two sites were 2500 km distant, they were separated by only 1 hour in longitude. However, the detected signal (see Figure 1) definitively proved the solar origin of the 5 minute signal and also served to anticipate the network strategy.

After that big success, the Nice University group performed observations at the South Pole in the winter of 1979–80 to make use of the Sun's continuous presence during southern summer, in order to obtain continuous observations. As a result, they obtained 5 full days of continuous data (Grec, Fossat & Pomerantz 1980) used to obtain the best power spectrum of solar p-modes at that time. Moreover the strategy of the Birmingham-Tenerife group proceeded according to the network concept. As a result, in the summer of 1981 they ran two identical instruments at Haleakala (Hawaii) and Tenerife obtaining a superb $\sim 60\%$ duty cycle over three months. The clean power spectrum of solar p-modes obtained from these observations, was used as a reference until very recently. The most important of these historical facts is that in 1981 the "network" became well proven and feasible, allowing other groups all around the world to develop the different projects they had in

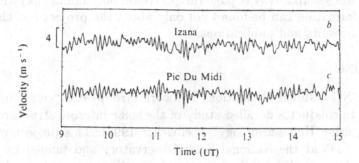

Figure 1. Comparison of the solar velocity residuals, showing the signature of the 5 minute oscillations, obtained simultaneously at the Observatorio del Teide (Izaña, Tenerife) and Observatoire de Pic du Midi (France) on 6 August 1978 (Claverie *et al.* 1979).

mind. At present, the operational helioseismology projects, in alphabetical order, are the following.

2.1. BISON

The BiSON (**Bi**rmingham **S**olar **O**scillations **N**etwork), was the first to become fully operational and has been conceived as "a global network of resonant-scattering spectrometers observing the low-ℓ solar p-modes" (Chaplin *et al.* 1996a). The zero resolution (Sun as a star) basic instrument consists of a spectrometer using the resonant scattering technique with potassium cells (KI 769.9 nm) to obtain very precise and stable measurements of the radial velocity of the Sun, from which the signals from solar oscillations of degree $\ell \leq 4$ are obtained. The project was fully funded by the PPARC (Particle Physics and Astronomy Research Council) to construct, deploy and operate a six-site network and became fully operational in 1993 once the last node was installed (Narrabri, Australia) late in 1992 (see Table 1 and Figure 2 for the node location). The whole network deployment took more than 10 years, after the first two sites were established in 1981. This is why the network consists of 3 different generations of instruments, the first one being the "Mark-I" at the Observatorio del Teide built in 1974. In spite of this apparent weakness, the quality of the actual data is the best among the low-degree helioseismic data sets.

From the scientific point of view, the contribution of this particular data set to helioseismology is really important in the sense that many of the basic results previously obtained by other groups using one-site instrument were confirmed and, in many cases, extended (*i.e.* the solar cycle changes in the p-mode spectrum, the spacings of the p-modes and their relation with solar models, the background solar velocity spectrum, etc.). The BiSON

Project has set up a WWW page (http://bison.ph.bham.ac.uk) where much
more information can be found not only about the project but also on the
scientific results and publications.

2.2. GONG

The GONG (**G**lobal **O**scillations **N**etwork **G**roup) is a "community based
project to conduct a detailed study of the solar internal structure and dynamics, using Helioseismology" (Leibacher 1995). This project was developed in 1985 at the National Solar Observatory and funded by the NSF
(National Science Foundation) to take observations over at least three years.
The Network completed deployment in October 1995 and the first year's
data has been obtained recently. The Network consists of 6 nodes which,
from a 5 year site survey, have proven to be able to produce a duty cycle as
large as 93% over a full year. The basic instrument, a modified Michelson
interferometer, produces radial velocity, intensity and modulation images
of the whole Sun on a 256 x 243 pixel^2 CCD camera at the Ni 676.8 nm
spectral line. With the given geometry, GONG instruments will cover a
wide range of mode degree (ℓ up to 250), although the sensitivity to lowest
degree modes ($\ell < 3$) is still unclear. Due to the imaging capabilities of the
instruments, the network is producing a huge amount of data ($\sim$ 1 Gbyte
per day) which needs to the reduced, calibrated, combined and analysed. To
undertake this enormous task, GONG has not only supplied the required
computational and software tools but also, and more importantly, invited
scientists world wide to contribute significantly in the project's success. In
this spirit, the project has also tried, whenever possible, to involve the local
people and institutions that host the GONG nodes, thus converting GONG
into a really community-based research project.
The extensive analysis of the first operational month has been published
recently (GONG team 1996), showing not only the exceptional data quality but also new results on the Sun's internal structure (latitudinal dependence of the temperature distribution, the fast rotation just below its
surface, detection of poleward flows at the surface, etc.). A complete description of the project and related issues, can be also found at the WWW
(http://www.gong.noao.edu).

2.3. HDHN

The third network currently operational is so recent that no official acronym
yet exists. We have named it HDHN (**H**igh **D**egree **H**elioseismology **N**etwork). It has been concieved as a "three stations network capable of studying high degree p-modes on a nearly continuous basis for several month
a year" (Rhodes *et al.* 1995). The background of this joint international

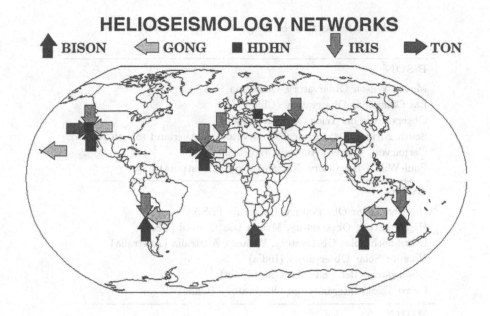

Figure 2.   Geographical location of the currently operational Helioseismology Network nodes.

research project was the helioseismic programme operated, since 1984, at Mount Wilson 60-Foot Solar Tower. The second node, located at the Crimean Astrophysical Observatory, came into operation in June 1996, thus converting this programme into a network (Rhodes 1996). Plans for a third node (possibly at the Solar Observatory of the V. G. Fesenkov Astrophysical Institute, Kazakhstan) already exists although negotiations have not yet been concluded. Even with the first two sites, it is expected to obtain up to 22 hours per day solar coverage 5 month per year. The basic instrument of the network is a sodium magneto-optical filter (MOF), originally developed by A. Cacciani (Cacciani *et al.* 1978), which will produce radial velocity maps of the whole Sun. The nominal detector, a 1024 x 1024 pixel^2 CCD camera, will allow angular mode degrees up to $\ell \sim 1000$ to be reached; the currently installed ones (490 x 490), covers only up to $\ell = 600$. One of the main interests of this network is to provide the Dopplergrams for coordinate space and ground-based studies of the high-degree oscillations during the flight of the SOI experiment on board *SOHO*. Of course, the capability of extending the observations over much longer time scales, results in a great interest in the investigation of temporal variability in the internal and atmospheric solar structure. The financial support for this project is provided by NASA and the NSF.

TABLE 1. Location of the ground-based network nodes.

BiSON

Mount Wilson Observatory, California (USA)
Las Campanas Observatory (Chile)
Observatorio del Teide, Tenerife (Spain)
South African Astronomical Observatory, Sutherland (South Africa)
Carnarvon, Western Australia (Australia)
Paul Wild Observatory, Narrabri, NSW (Australia)

GONG

Big Bear Solar Observatory, California (USA)
High Altitude Observatory, Mauna Loa, Hawaii (USA)
Learmonth Solar Observatory, Western Australia (Australia)
Udaipur Solar Observatory (India)
Observatorio del Teide, Tenerife (Spain)
Cerro Tololo Interamerican Observatory (Chile)

HDHN

Mount Wilson Observatory, California (USA)
Crimean Astrophysical Observatory, Nauchny (Crimea)

IRIS

IPS Solar Observatory, Narrabri, NSW (Australia)
Kumbel, Tashkent (Uzbekistan)
Oukaimeden (Morroco)
Observatorio del Teide, Tenerife (Spain)
European Southern Observatory, La Silla (Chile)
John Wilcox Solar Observatory, Stanford, California (USA)

TON

Huariou Solar Observing Station, Beijing (P. R. China)
Tashkent Astrophysical Institute, Tashkent (Uzbekistan)
Observatorio del Teide, Tenerife (Spain)
Big Bear Solar Observatory, California (USA)

2.4. IRIS

The IRIS (International Research of the Interior of the Sun) network is an "International Cooperative Project to deploy and operate a 6-site network, to perform continuous full disk measurements to give access to the solar modes of very low degree, along, at least, one complete solar cycle" (Fossat 1995). The project was conceived by the Nice University group in the mid eighties; their experience with the South Pole observations led the project to be funded by the INSU (Intitute National des Sciences de l'Univers), which allowed the development and construction of better instrumentation. Deployment started in 1989 and ended in summer 1994 with 6 nodes fully operational. The basic instrument is similar in concept to BiSON: by using the resonant scattering technique the radial velocity of the Sun is precisely measured, thus allowing a good determination of the low-degree solar oscillation modes. Unlike BiSON, IRIS uses the D1 sodium spectral line instead of potassium; some other technical differences also exist. Due to the limited resources of the project, in terms of money, spares and manpower and also to the goal of running over more than one solar cycle, the strategy has been to involve, at all levels, local institutions hosting the nodes. In the IRIS framework, collaboration between the different institutions has been exploited to the full, thereby resulting in a great benefit not only to the project but also to local groups. Furthermore, IRIS is an open scientific community in which any one interested in the research can access the data and the existing project facilities. Like GONG, IRIS has a clear structure and policy in order to give adequate credit to all participants according to the importance of their contributions. More extensive information about the project, scientific achievements and future plans can be found at the WWW address: http://boulega.unice.fr.

2.5. TON

The TON (Taiwan Oscillations Network) is a "ground-based network to measure solar K-line intensity oscillations to study the internal structure of the sun" (Chou et al. 1995). Unlike the other experiments, the oscillations are measured only as spectral intensity variations of the whole Sun and with a very high spatial resolution (1080×1080 pixel^2), allowing mode degrees up to $\ell \sim 1000$ to be reached. The whole network will consist of 6 nodes, 4 of which are already operational (see Figure 2). It is surprising how fast the construction, deployment and operation has proceeded, when compared with the other projects: TON has spent only 3 years in reaching its present state. The project, although funded by the National Research Council of the ROC, has very limited manpower resources compared with the enormous tasks that the data require: at present TON is producing ~ 6 Gbyte per day and once completed will go up to 9 Gbyte. Because of the high resolution and quality of these data, they have become very useful for sunspot and local helioseismology (Chen et al. 1996) and ring diagram analysis (González et al. 1995).

3. Discussion

Now the main characteristics of the five existing networks have been presented, it would be convenient to summarise, compare them and comment
on some aspects. In Table 3 some of the networks' characteristics are presented. First of all, one can appreciate that two of the networks are still not
fully deployed as planned. Furthermore, the BiSON group is pushing for a
large number of nodes (eight). The reasons given are twofold. First, after
many years of experience, they concluded that with 6 nodes a duty cycle
of not more than 80% over a full year could be achieved. Secondly, increasing the number of modes implies more overlap between different stations'
data, which can be used for many scientific/instrumental purposes, such as
checking the local clock timing, proving the validity of the data calibration,
studying the solar background noise by cross-correlation techniques, etc.
The reduced number of solar spectral lines used to measure the solar oscillations as well as the observed magnitudes (spectral intensity and Doppler
velocity), is also surprising. Concerning the observing strategies that the
different networks follow, we can separate these in three groups. The first
aims at the full involvement of the local people-institution performing the
observations, into the project. In this case, the degree of interest and care
taken over the observations is very high and is reflected in the degree of automatisation required in the instruments and also in the high operational
duty cycle. This is the case for IRIS and TON. For the second group, a
complementary strategy is used. By working really hard, it is possible to
get a rather simple instrumentation with a high degree of automatisation
and even remote control. Local people and institutions become irrelevant
as far as operation is concerned, and only minor services are required from
them. This is the way in which the BiSON project proceeds with a very
high level of performance from 1993 onwards. The third group is a mixture
of the other two and is probably the ideal one: instrumentation highly au-
tomated and local people deeply involved in the project. This is the case for
GONG. Of course in the first and last options, the network and local data
gathered (as well as the derived products) are shared among the community
involved, which results in a clear scientific benefit.

Finally, we should comment on the data duty cycles already obtained by
the networks. GONG had the expectation, after many years of site survey
data, of obtaining a 93% duty cycle over a full year with the selected nodes.
At present, and after the first 8 month of six-site operation (November 95 to
June 96), the achieved mean value is 90%, the best and the worst monthly
values being 94 and 82%, respectively. It looks, then, as if the claim by
the BiSON group of a maximum of 80% over a full year with six sites can
no longer be supported. The BiSON network has reached monthly peak

values up to 94% and the values for a complete year are 68 and 78% for 1993 and 1994, respectively. On the other hand, the IRIS network has not yet reached comparable figures. The best duty cycle for a full year's data is ~30%; the highest values for a two-month series between 1989 and 1994 are ~ 60%. This situation can radically change in the near future: thanks to the collaborative IRIS spirit, it is planned to use other similar data to be merged and combined with the IRIS ones in order to increase the observational duty cycle. In particular, integrated data provided by the MOF instrument (Cacciani *et al.* 1988), GONG and others is planned to be used.

TABLE 2. Main characteristics of the ground-based helioseismology networks: present and expected number of nodes, date of completion, observed physical magnitude (Doppler velocity, intensity and modulation), wavelength used, mode angular degree range and sampling time of the basic measurement.

	Present Nodes	Total Nodes	Fully Oper.	Obser. Param.	λ (nm)	$\Delta\ell$	Δt (s)
BiSON	6	8?	1993	V_D	KI 769.9	$\ell \leq 4$	40
GONG	6	6	1995	$V_D IM$	Ni 676.8	$\ell \leq 250$	60
HDHN	2	3	-	V_D	Na D1,D2	$\ell \leq 600^\dagger$	60
IRIS	6	$6^\ddagger$	1994	V_D	NaD1 589.6	$\ell \leq 4$	15
TON	4	6	-	I	CaK 393.4	$\ell \leq 1000$	60

($\dagger$): will go up to $\ell \sim 1000$. ($\ddagger$): uses data from other instruments.

4. Single-site experiments

Having seen the previous projects one can ask if still there is a role to play for the single-site helioseismology experiments. Apparently, they cannot compete in data quality, as they are limited to data duty cycles below 30% over a full year. There is, however, a positive answer: these experiments have an extremely important role in the future of helioseismology. In a recent review (Duval 1995) some of the important issues that small ground-based experiments can face have been pointed out: long-term (solar cycle) and high temporal frequency studies (for which high duty cycles are not so important), testing new techniques and developing new instrumentation, to compare their results with other experiments or with network's results in order to understand systematic errors, etc. Let us briefly describe each one of the operational experiments.

4.1. CRAO-WSO

This acronym stands for the **Cr**imea **A**strophysical **O**bservatory and the **W**ilcox **S**olar **O**bservatory, whose observational programmes have run in parallel since ~1977, (Kotov, Haneychuk & Tsap 1995; Scherrer, Hoeksema & Kotov 1993). Both instruments are modified magnetographs to measure the differential Doppler velocity of the central part of the solar disc, and thus have sensitivity to mode degrees $2 \leq \ell \leq 6$. Although observations at both places have not been continuous on a daily basis, they have significant coverage for most of the years. Their main interest has been focused on studying and proving the solar origin of the 160-minute oscillation: how it appears in both data sets separately and in the combined data; the frequency, phase and power stability over the last twenty years, etc. As everyone is aware today, the results claimed (Kotov *et al.* 1995) are not generally accepted and many opposite claims exist too. The Stanford data have also been used in the past (Henning *et al.* 1986) for p-mode studies to cover the lack between low and intermediate-degree mode determinations and for low-degree gravity modes detection (Delache & Scherrer 1983).

4.2. HLH

The **High-ℓ H**eliosismometer is an NSO project whose primary goal is the "long term observations of high degree frequencies and to study local helioseismology" (Bachmann *et al.* 1995). The instrument is a modification of the device used for the South Pole campaigns and is located at the Kitt Peak Observatory, measuring the brightness signal at the Ca-K line with a high spatial resolution (ℓ up to 1200). The operational plan, started in March 1993, is to run in a systematic way four days every four weeks; some additional and different plans are also foreseen (Hill 1996). These observations could be very useful in the framework of the new research fields arising in helioseismology, such as time-distance analysis, the ring diagrams, etc.

4.3. LOI-T

The **L**uminosity **O**scillations **I**mager installed at the Observatorio del **T**eide is the Qualification Model of the instrument flying in the VIRGO package on board *SOHO*. Its primary goal is to perform "measurements of the low degree p-modes and to test the performances and develop new techniques applied to the space model." It has been running continuously on a daily basis since 1994, measuring the brightness signal at 550 nm. Its special pixel design allows it to measure modes of degrees up to $\ell = 7$ and scientific exploitation of the data is now taking place: the rotational splittings and mean frequencies have been determined and inversions performed (Rabello-

Soares *et al.* 1996). Also, the solar radius signal provided by the instrument has been analysed (Rabello-Soares 1996). Overall, the great success of this instrument raises expectations of a good scientific output from its space counterpart, free from the earth's atmosphere and without the day-night interruption.

4.4. LOWL

The Low-ℓ is an HAO project whose primary goal is to "measure frequency splittings of low-degree modes in order to infer rotation rate and other properties of the solar core" (Tomczyk *et al.* 1995). By using the MOF technique at the potassium line with a moderate spatial resolution (25 arcsec), it produces full-Sun Doppler velocity maps. Installed at Mauna Loa Solar Observatory in Hawaii at the beginning of 1994, it has been continuously running since then on a daily basis. The first two years of data have already been analysed for p-mode frequency determinations, rotational splitting and the corresponding inversions (Schou, Tomczyk & Thompson 1996). The good point of this instrument is precisely the combination of the advantages of an imaging instrument (allowing separation of m-components), with the extreme stability provided by the resonant scattering technique. In addition, it provides a homogeneous set of p-mode frequencies ranging from $\ell=1$ up to 80, which is ideal for performing inversions of the deeper layers of the Sun. In principle, because of the spatial resolution, it should be able to cover the whole range from $\ell = 0$ up to $\ell = 100$, but some work should still be done on the instrumental and on the data analysis areas. Recently, the project has been funded to construct and deploy a second instrument, which will be operating possibly by the end of next summer, the LOWL thus becoming the sixth helioseismology network.

4.5. MKI

The MKI instrument corresponds to the first resonant scattering spectrometer made by the University of Birmingham group, back in 1974. It was installed in 1977 at the Observatorio del Teide where it ran, mainly over summer seasons, until 1984. At that point it was upgraded and since then it has been operated continuously, by the IAC group, on a daily basis. The extremely high temporal stability of this instrument, which unlike the rest of the BiSON network ones works in photon counting mode, makes it ideal for long-term study of solar oscillation behaviour. The scientific exploitation of the data provided by MKI has been very successful: it is at present the single ground-based instrument that produces the greatest quantity of scientific papers (Harvey 1995). Among the many issues studied with this data set, we should mention the detailed study of the p-mode spectrum (Anguera *et al.* 1992), its changes related with the solar activity cycle (Régulo *et al.* 1994), the study of the background solar velocity spectrum (Pallé *et al.* 1994), the gravity mode search, etc.

4.6. POI

The **P**-mode **O**scillation **I**mager is an instrument devoted to "*p*-mode measurements at intermediate degree and high temporal frequencies" (Ronan & LaBonte 1994). The instrument measures the Ca-K line brightness signal with a spatial resolution that allows mode degree up to $\ell \sim 300$ to be reached. It was installed in 1991 at the Mees Solar Observatory in Hawaii and the possibility of a second site at Fort Yukon (Alaska) is envisaged (LaBonte, Ronan & Kupke 1995).

TABLE 3. Characteristics of the single-site helioseismology projects

	Operated since	Obser. Param.	λ (nm)	$\Delta\ell$	$\Delta t(s)$	Institution
CrAO	1974	V_D	FeI 512.4	$2 \le \ell \le 4$	300	CrAO
HLH	1993	I	CaII 393.4	$\ell \sim 1200$	60	NS-NOAO
LOI-T	1994	I	500 ± 5	$2 \le \ell \le 5$	~ 53	SSD-IAC
LOWL	1994	V_D	KI 769.9	$\ell \le 80$	15	HAO-NCAR
MKI	1976	V_D	KI 769.9	$\ell \le 4$	2	IAC
WSO	1977	V_D	FeI 512.4	$2 \le \ell \le 5$	15	U. Stanford
POI	1991	I	CaII 393.4	$\ell \le 300$	30	U. Hawaii

5. Conclusions

From the above description of all operational ground-based helioseismology projects, the perspectives seem really exceptional. Although the various projects provide a good coverage of mode-degree sensitivities (some of them overlapping), the variety of spectral lines used, techniques and physical parameters measured, is rather poorer (see Tables 2 and 3). This drawback has already been pointed out (Harvey 1995) and can be overcome by simultaneous comparison of data from different instruments in order to minimise or even to identify the possible sources of systematic errors. In addition, data provided by currently operational space instruments, which are similar in type to those obtained at ground-based observatories, will really help in this task. It seems that at present the sources of discrepancy between results obtained by different experiments are more in the realm of data analysis techniques than in the instrumental part, although this is not conclusive. As a good example, we could mention the measurement of the frequency splittings of p-modes: different data sets produce different rotation profiles derived from inversions. In particular, the basic splitting measurement, that corresponding to the $\ell = 1$ p-modes, is very contradictory, as can be seen in Figure 3.From this figure it would seem that the state of the art of helioseismology is very crude, in that there is no agreement in one of the most basic measurements. We do not think that this is so; these measurements are extremely delicate and the physics of the solar oscillations is not yet well understood. This is the basic reason for running many experiments simultaneously: to get rid of systematics and also to establish the obtained results beyond doubt.

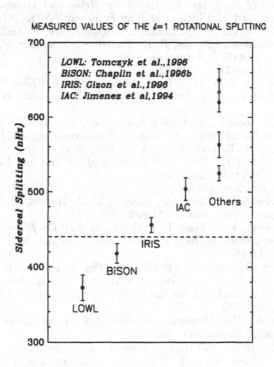

Figure 3. Some of the most recent observational determinations of the $\ell = 1$ rotational splitting. Under the label *Others*, values not yet published are included. The dashed line corresponds to the Sun's equatorial surface rotation.

References

Anguera Gubau, M., Pallé, P. L., Pérez Hernández, F., Régulo, C., Roca Cortés, T. (1992), *Astron. Astrophys.* **255**, 363–372

Bachmann, K. T., Duvall, T. L. Jr., Harvey, J. W., Hill, F. (1995), in *ASP Conf. Ser. 76, GONG '94: Helio- and Astero-Seismology from the Earth and Space*, ed. K. Ulrich et al. (San Francisco: ASP), 156–159

Cacciani, A., Rosatti, P., Ricci, D., Marquedant, R., Smith, E. (1988) in *Seismology of the Sun and Sun-Like Stars*, ed. E.J. Rofe (ESA SP-286), 181-184

Cacciani, A., Fofi, M. (1978), *Solar Physics* **59**, 179–189

Chaplin, W. J., Elsworth, Y., Howe, R., Isaak, G. R., McLeod, C. P., Miller, B. A., van der Raay, H. B., Wheeler, S. J., New, R. (1996a), *Solar Physics* **168**, 1–18

Chaplin, W. J., Elsworth, Y., Howe, R., Isaak, G. R., McLeod, C. P., Miller, B. A., New, R. (1996) submitted to *MNRAS*

Chen, K, Chou, D. and the TON Team (1996), *Astrophysical Journal* **465**, 985-993

Chou, D, and the TON Team (1995),*Solar Physics* **160**, 237–243

Claverie, A., Isaak, G. R., McLeod, C. P., van der Raay, H. B., Roca Cortés, T. (1979), *Nature* **282**, 591-594

Delache, P., Scherrer, P. H. (1983), *Nature* **306**, 651–653

Duvall, T. J. Jr. (1995) in *Fourth SOHO Workshop: Helioseismology*, ed. Hoeksema, J.T., Domingo V., Fleck, B., Battrick, B., (ESA SP-376), 107–111

Fossat, E. (1995), in *ASP Conf. Ser. 76, GONG '94: Helio- and Astero-Seismology from the Earth and Space*, ed. K. Ulrich et al. (San Francisco: ASP), 387–391

Gizon, L., Fossat, E., Lazrek, M., Cacciani, C., Ehgamberdiev, S., Gelly, B., Grec, G., Hoeksema, T., Khalikov, S., Pallé, P. L., Pantel, A., Schmider, F.-X., Wilson, P. (1996), submitted to *Astron. Astrophys.*

Gong Team (1996), *Science* **272**, 1233–1388

González, I., Patrón, J. and the TON Team (1995) in *Fourth SOHO Workshop: Helioseismology*, ed. Hoeksema, J.T., Domingo V., Fleck, B., Battrick, B. (ESA SP-376), 137–140

Grec, G., Fossat, E., Pomerantz, M. (1980), *Nature* **288**, 541–544

Harvey, J. (1995), in *Fourth SOHO Workshop: Helioseismology*, ed. Hoeksema, J.T., Domingo V., Fleck, B., Battrick, B. (ESA SP-376), 9–18

Henning, H.M., Scherrer, P.H. (1986) in *Seismology of the Sun and Distant Stars*, ed. D.O. Gough (Dordrecht: Reidel), 55–62

Hill, F. (1996), private communication

Jiménez, A., Pérez Hernández, Claret, A., Pallé, P. L., F., Régulo, C., Roca Cortés, T. (1994), *Astrophysical Journal* **435**, 874–880

Kotov, V. A., Haneychuk, V. I., Tsap, T. T. (1995), in *ASP Conf. Ser. 76, GONG '94: Helio- and Astero-Seismology from the Earth and Space*, ed. K. Ulrich et al. (San Francisco: ASP), 82–88

LaBonte, B. J., Ronan, R., Kupke, R. (1995), *Solar Physics* **158**, 1–10

Leibacher, J. and the GONG Project Team (1995), in *ASP Conf. Ser. 76, GONG '94: Helio- and Astero-Seismology from the Earth and Space*, ed. K. Ulrich et al. (San Francisco: ASP), 381–386

Pallé, P. L., Jiménez, A., Pérez Hernández, F., Régulo, C., Roca Cortés, T., Sánchez, L. (1994), *Astrophysical Journal* **441**, 952–959

Rabello-Soares, M. C. (1996). PhD thesis, University of La Laguna

Rabello-Soares, M. C., Roca Cortés, T., Jiménez, A., Appourchaux, T., Eff-Darwich, A. (1996), submitted to *Astrophysical Journal*

Régulo, C., Jiménez, A., Pallé. P. L., Pérez Hernández, F., Roca Cortés, T. (1994), *Astrophysical Journal* **434**, 384–388

Rhodes, E.J. Jr. (1996), private communication

Rhodes, E. J. Jr., Didkovsky, L., Chumak, O. V., Scherrer, P. H. (1995), in *ASP Conf. Ser. 76, GONG '94: Helio- and Astero-Seismology from the Earth and Space*, ed. K.

Ulrich *et al.* (San Francisco: ASP),   398–401

Ronan, R. S., LaBonte, B. J. (1994), *Solar Physics* **149**,   1–21

Scherrer, P. H., Hoeksema, J. T., Kotov, V. A. (1993), in *ASP Conf. Ser. 42, GONG 1992: Seismic Investigation of the Sun and Stars*, ed. T. M. Brown (San Francisco: ASP),   281–284

Schou, J., Tomczyk, S., Thompson, M. J. (1996), *Bull. Astr. Soc. India* **24**,   375–378

Tomczyk, S., Streander, K., Card, G., Elmore, D., Hull, H., Cacciani, A. (1995), *Solar Physics* **159**,   1–21

Tomczyk, S., Schou, J., Thompson, M. J. (1996), *Bull. Astr. Soc. India* **24**,   245–250

PRESENT STATUS OF THE TAIWAN OSCILLATION NETWORK

DEAN-YI CHOU AND THE TON TEAM[1]

Physics Department, Tsing Hua University

Hsinchu, 30043, Taiwan, R.O.C.

Abstract. The Taiwan Oscillation Network (TON) is a ground-based network measuring solar intensity oscillations for the study of the internal structure of the Sun. So far, four telescopes have been installed at Teide Observatory (Tenerife), Huairou Solar Observing Station (near Beijing), Big Bear Solar Observatory (California), and Tashkent (Uzbekistan). The TON telescopes take K-line full-disk solar images of diameter 1000 pixels at a rate of one image per minute. The TON high-resolution data is specially suitable to study local helioseismology. Here, we present recent results of three topics on local helioseismology from TON data: (1) Inference of Subsurface Magnetic Field From Absorption Coefficients of p-modes in Active Regions, (2) Subsurface Structure of Emerging Flux Regions From Helioseismology, and (3) Flow Around Sunspots From Measurements of Frequency Shift.

1. Project of Taiwan Oscillation Network

The Taiwan Oscillation Network[1] (TON) is a ground-based network measuring solar K-line intensity oscillations for the study of the internal structure of the Sun. The TON project has been funded by the National Research Council of ROC since the summer of 1991. The headquarters of the TON is located at Physics Department of Tsing Hua University, Hsinchu, Taiwan, where the telescope systems are designed, built, and tested. So far, four telescopes have been operated. The first telescope was installed at the

---

[1]The TON Team includes: M.-T. Sun, T.-M. Mu, S. Loudagh, Bala B., Y.-P. Chou, C.-H. Lin, I.-J. Huang (Taiwan), A. Jimenez, M. C. Rabello-Soares (IAC, Spain), G. Ai, G.-P. Wang (Beijing), H. Zirin, W. Marquette (Big Bear), S. Ehgamberdiev and S. Khalikov (Uzbekistan).

J. Provost and F.-X. Schmider (eds.), Sounding Solar and Stellar Interiors, 31-38.

Teide Observatory, Canary Islands, in August of 1993. The second and third
telescopes were installed at the Huairou Solar Observing Station near Beijing and the Big Bear Solar Observatory, California in 1994. The first three
telescopes have been taking data simultaneously since October of 1994. The
fourth telescope was installed in Tashkent, Uzbekistan in July of 1996, and
started taking data since then. The site selection and arrangement of the
fifth telescope in South America is underway.

The TON is designed to obtain informations on high-degree solar p-
mode oscillations, along with intermediate-degree modes. That is complementary to other ground-based networks, such as BISON, IRIS, and
GONG. A discussion of the TON project and its instrument is given by
Chou et al. (1995). Here we give a brief description. The TON telescope
system uses a 3.5-inch Maksutov-type telescope. The annual average diameter of the Sun is set 1000 pixels. A K-line filter, centered at 3934, of
FWHM = 10 and a prefilter of FWHM = 100 are placed near the focal
plane. The measured amplitude of intensity oscillation is of about 2.5%. A
16-bit 1242 by 1152 CCD is used to take images, but only 1080 by 1080
pixels are read out. The CCD is water-cooled, and the CCD chip is operated at a temperature of of about 223–233 °K. The exposure time is set to
800 ms. The photon noise, about 0.2%, is greater than the thermal noise
of the CCD and its circuit. The size of each image is 2.33MB. The image
data are recorded by two 8-mm Exabyte tape drives.

The TON full-disk images have a spatial sampling window of 1.8 arcsec-
onds per pixel, and they can provide information of modes up to $l \approx 1000$.
The TON high-resolution data are specially suitable to study local helioseismology. In this paper, we present results of three topics on local helioseismology from TON data. For a more detailed description of these three
topics, see the proceedings for posters of this conference.

2. Inference of Subsurface Magnetic Field From Absorption Co-efficients of p-modes in Active Regions

Local analyses of helioseismic data have shown that a sunspot would modify the amplitude and phase of p-mode waves (Braun et al. 1987; Braun
et al. 1992; Braun 1995) as the waves are passing through the sunspot.
The interaction between the waves and the magnetic field provides a tool
to probe the properties of sunspots, such as magnetic field, density, and
temperature, below the surface. In this study we will use a phenomeno-
logical model proposed by Chou et al. (1996) and measured absorption
coefficients to infer the subsurface magnetic structure of sunspots. In this
model, the interaction of p-mode waves with a magnetic region is described

by introducing a complex sound speed

$$c^2 = \frac{c_0^2}{1 - i\sigma(\vec{x})} ,$$ (1)

where c_0 is the sound speed in non-magnetic regions. The dimensionless

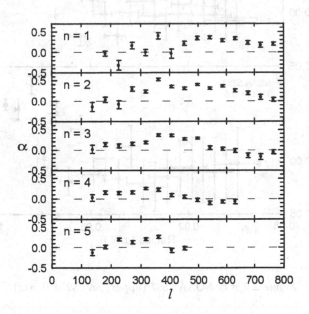

Figure 1. Absorption coefficient α of NOAA 7887 measured from TON data.

parameter σ is called the interaction parameter. σ relates to the distribution of the magnetic field that requires the theory of interaction mechanism. In general, σ is a complex. If σ is small compared with unit, to the lowest order, the real part of σ corresponds to the dissipation of waves, and the imaginary part describes the change of the phase velocity due to the change of physical conditions in magnetic regions. In this paper, we assume that σ is small, and use only measured absorption coefficients to invert the real part of σ. Our result shows that σ is small, and this assumption is a good approximation.

To make the problem tractable, we make the following assumptions about the magnetic region. (1) The distribution of σ is axisymmetric. (2) σ is constant in the horizontal direction within the spot, and zero outside the spot. (3) The spot radius is constant in depth. Then, the absorption coefficient α_{nl} is approximately related to σ by

$$\alpha_{nl} \approx \int K_{nl}(z)\sigma(z)dz ,$$ (2)

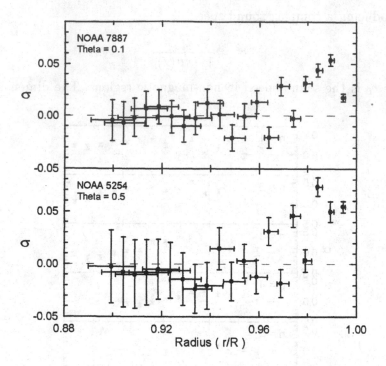

Figure 2.   σ of NOAA 7887 (upper) and 5254 (lower).

where the kernel K_{nl}, given in Chou et al. (1996), is a functional of the wavefunction, which can be computed from the standard solar model. α_{nl} can be measured by decomposing the p-mode oscillations around a sunspot into the modes propagating toward and away from the sunspot (Braun et al. 1987; Chen et al. 1996). We use the Optimal Averaging Kernel method (Christensen-Dalsgaard et al. 1990; Pijpers and Thompson 1992) to invert the measured α_{nl} to obtain the distribution of σ below the surface.

In this study, we invert α_{nl} of two active regions. α_{nl} of NOAA 5254 are measured by Braun (1995) from 67.7-hour data taken at the South Pole. α_{nl} of NOAA 7887, shown in Figure 1, are measured by us from five-day data taken by TON. σ inverted from NOAA 7887 and 5254 is shown in Figure 2. With the above assumptions, the curve of σ vs. depth will be scaled up and down by changing the spot radius. In this study, we use the radius of penumbra as the spot radius, which is 16.4 Mm for NOAA 7887 and 18 Mm for NOAA 5254. Figure 2 shows that distributions of σ for two regions are similar. It drops from about 0.05 near the surface to zero at a depth of about 0.04 solar radius (28 Mm), which is consistent with the result of Chou et al. (1996). Our method can also apply to invert measured phase shifts.

3. Subsurface Structure of Emerging Flux Regions From Helioseismology

The helioseismic data have shown that magnetic regions absorb energy of p-mode waves. The motivation of this study is as follows. Since the different p-modes sample different regions in the solar interior, the magnetic field before emergence would not interact with higher-l p-modes whose cavities are shallow. The comparison of p-mode absorption coefficients before and after emergence can be used to infer the depth of the magnetic field before emergence.

We have studied two emerging flux regions (EFRs), NOAA 7754 and 7874 with TON data taken at Big Bear. These two EFRs are selected based on the following criterions. (1) The EFR is isolated; namely, the surrounding region, used to compute the absorption coefficient α, is flux-free judging from Kitt Peak magnetograms. (2) At the location of interest, it is flux-free on the day before the flux appears on the surface. For NOAA 7754, the magnetic flux appears on July 12, 1994, and we measured α with data of July 11 and July 12, respectively. For NOAA 7874, the magnetic flux appears on June 2, 1995, and we measured α with data of June 1 and June 2, respectively. The measured α are shown in Figure 3. Apparently, α after emergence is greater than before emergence for most modes. An important feature of Figure 3 is that the absorption is detectable before the flux appears on the surface for some modes; for example, in $n = 2$ of NOAA 7754, and $n = 1$ of NOAA 7874. Although the error bars are not small in our measurements, from our experiences on the quiet Sun analyses, we believe that here we detect the absorption of p-modes before the flux appears on the surface. For higher n, α is slightly negative in all our quiet Sun analyses. Thus it is difficult to conclude from Figure 3 that there exists absorption prior to emergence for the modes with $n > 2$.

In principle, the difference of α measured before and after emergence could provide information about the depth of the top of flux at the time α is measured prior to emergence. We expect that for the case prior to emergence, for a fixed n, α decreases with l and becomes zero when the mode cavity is too shallow to interact with the flux. Because of the low S/N, the present results would not allow us to identify the l value for a fixed n where α drops to zero. But, we can estimate the depth of the top of the flux prior to emergence from the difference of α measured before and after emergence, which is shown in Figure 4. For $n=1$, the difference of absorption coefficients peaks around $l = 600$ for NOAA 7754 and $l = 500$ for NOAA 7874. If the peak approximately corresponds to the mode which does not interact with the flux, the top of flux is located at a depth of about 4000 km for NOAA 7754 and 4700 km for NOAA 7874. The low S/N of α is due to the limitations: (1) the time series used to compute α has to be short (512 minutes in this study) such that the measured α represents the information of the EFR at a certain depth; and (2) the absorption is weak for EFRs.

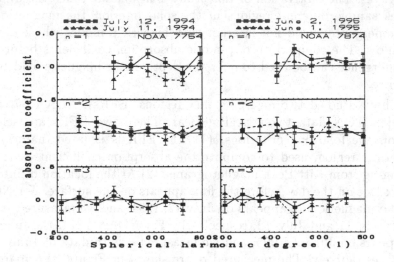

Figure 3. Absorption coefficients before and after emergence vs. l for different n.

4. Flow Around Sunspots From Measurements of Frequency Shift

We use five-day helioseismic data from TON to study the flow around a sunspot group, NOAA 7887, whose penumbra size is 16.4 Mm.

As in the analyses discussed in Section 2, the p-mode oscillations in an annular region centered at the sunspot are decomposed into modes propagating toward the sunspot and modes propagating away from the sunspot. The range of the annular region is $2° - 10°$. The frequencies of incoming modes and outgoing modes are determined by fitting the power spectra with a Gaussian profile (Libbrecht and Kaufman 1988). We find that for most modes the frequency of the outgoing mode is greater than that of the incoming mode. This indicates that the plasma is flowing outward from the sunspot. The frequency difference $\Delta \nu = \nu_{out} - \nu_{in}$ is shown in Figure 5. A region of quiet Sun in the same data set is studied to act as a control measurement. The corresponding $\Delta \nu$, also shown in Figure 5, is of about zero. We also measure $\Delta \nu$ with an annular region ($6° - 21°$) farther away from the sunspot. For most modes, $\Delta \nu$ is still positive but

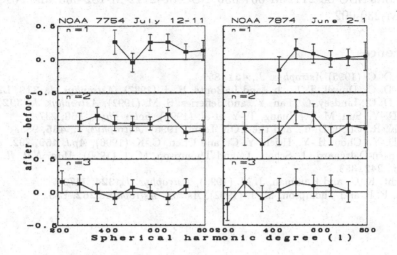

Figure 4. α(after emergence) - α(before emergence)

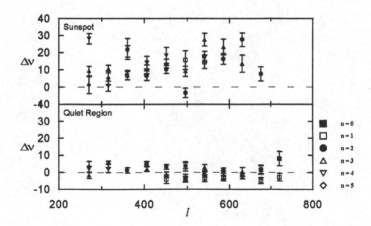

Figure 5. $\Delta\nu$ ($\nu_{out} - \nu_{in}$) vs. l for sunspot (upper) and quiet Sun (lower).

smaller than that of the annulus of $2° - 10°$. This is consistent with the picture that the outward flow is significant only near the sunspot. If we use $\Delta\nu \sim 2v\nu/c \sim vk/\pi$, where k is the wavenumber, to estimate the flow velocity, we obtain $v \sim 50 - 100m/s$.

Aknowledgement The TON project is supported by NSC of ROC under grants NSC-86-2112-M-007-036, NSC-86-2112-M-182-003, and NSC-86-2112-M-239-002.

References

Braun, D. C. (1995) *Astrophys. J.*, **451**, 859.
Braun, D. C., Duvall, T. L., Jr., and LaBonte, B. J. (1987), *Astrophys. J.*, **319**, L27.
Braun, D. C., Lindsey, C., Fan, Y., and Jefferies, S. M. (1992), *Astrophys. J.*, **392**, 745.
Chou, D.-Y., Sun, M.-T., Huang, T.-Y. et al. (1995), *Solar Phys.*, **160**, 237.
Chen, K.-R., Chou, D.-Y., and the TON Team (1996), *Astrophys. J.*, **465**, 985.
Chou, D.-Y., Chou, H.-Y., Hsieh, Y.-C., and Chen, C.-K. (1996), *ApJ*, **459**, 792.
Christensen-Dalsgaard, J., Schou, J., and Thompson, M. J. (1990), *Mon. Not. R. Astr. Soc.*, **242**, 353.
Libbrecht, K. G. and Kaufman, J. M. (1996), *Astrophys. J.*, **324**, 1172.
Pijpers, F. P. and Thompson, M. J. (1992), *Astron. Astrophys.*, **262**, L33.

A REVIEW OF SPACE HELIOSEISMOLOGY

T. TOUTAIN

Observatoire de la Côte d'Azur
BP 4229, F-06304 Nice cedex 04

1. Introduction

The helioseismology started in 1960 with the discovery of oscillations of the velocity field on the solar surface by Leighton, 1960. The mechanism producing these oscillations remained unexplained for the next ten years until Ulrich, 1970, and independently Leibacher and Stein, 1971, demonstrated that acoustic waves trapped beneath the solar surface could generate such oscillations. According to that theory the Sun should oscillate globally and not only locally as seen in the pionnering observations. First Deubner, 1975, and then Claverie et al., 1979, and Grec et al., 1980, demonstrated the existence of global oscillations of intermediate and low degrees. These observations analysed in terms of k-ω diagram and power spectrum led to the famous p-modes ridges and spectrum lines which are at the base of any inference of the solar structure. All these observations where carried out from the ground. Nevertheless ground-based observations at that time (early eighties) were suffering some limitations which could be over taken from space. This is the reason why the eighties saw the intensive development both of ground-based networks and space missions for helioseismology resulting in 1995 in the existence of four networks: BISON, IRIS, GONG and TON (see Pallé, 1996) and the launch of the SOHO satellite (Domingo et al., 1995). In the next section I review the limitations of ground-based observations and those of space missions. In section 3, I describe in a chronological order the past space helioseismology missions (from 1980 to 1994). Furthermore I discuss, in section 4, the present status of space helioseismology with SOHO and finally in section 5 the co-ordination of space missions and ground-based solar observations. In the last section the future and the prospects of space helioseismology are reviewed.

J. Provost and F.-X. Schmider (eds.), Sounding Solar and Stellar Interiors, 39-52.
© 1997 *IAU. Printed in the Netherlands.*

2. Limitations in the early eighties

At the end of the seventies and the beginning of the eighties the observations of the solar oscillations were suffering from mainly three limitations.

2.1. TIME COVERAGE OF OBSERVATIONS

At that time, the ground-based instruments observed the Sun separately with interruption during night-time and bad weather conditions. In terms of Fourier power spectrum the regular gaps in the time series due to night and day cycles introduce side bands at 11.57 μHz and other spurious peaks in the case of irregular interruptions. All these peaks make the identification of modes more difficult and the determination of their frequencies less accurate. From simulations (Noyes and Rhodes, 1984), we know that gap filling methods are able to recover the original power spectrum for duty cycles better than 80%, especially if the gaps are irregularly spaced in time. There was an attempt to bypass this limitation by Grec et al., 1980) who were able to observe the Sun continuously during five days from the South Pole. Though it was a great improvement in term of continuity, it showed the second limitation of experiments in helioseismology; namely the short-time duration of observations.

2.2. DURATION OF OBSERVATIONS

Long-uninterrupted observations can improve the determination of frequencies in two different ways. First by increasing the frequency resolution and second by increasing the signal-to-noise ratio for modes having lifetime longer than the observation duration. For example the 5 days of continuous observations obtained from the South Pole allow a frequency resolution of only 2.3 μHz which is not enough to resolve neither the rotational splitting of the modes (0.45 μHz) nor the lines of modes below 2 mHz which have lifetime of the order of a month. In addition the increase of solar noise with decreasing frequency makes detection of low frequency modes more difficult. This effect is due to granulation and supergranulation noise and is emphasized with ground-based observations because of the influence of the Earth's atmosphere.

2.3. INFLUENCE OF THE EARTH'S ATMOSPHERE

It is obvious that the motion of the Earth's atmosphere due to turbulence strongly affect the observations of solar oscillations. Low-degree mode observations are affected by an extra noise at low-frequencies whereas high-degree mode observations are affected by the loss of coherency on the solar disk. This produces a leakage of the high-degree modes and a conversion of the steady velocity flows in apparent time-varying signals. Simulations of seeing effect on p modes ridges (Noyes and Rhodes, 1984) show that modes for l >300 are embedded in the noise already with a 4" seeing which is a typical seeing under turbulent conditions.

The present status of observations in helioseismology is different. Ground-based networks do not suffer from the two first limitations though the last one is still there. The only way to avoid the influence of the Earth's atmosphere is to fly space missions. Nevertheless space missions have also some disadvantages like their cost and the impossibility to repair any instrumental defect. For these reasons the first observations of solar oscillations from space were not made with a dedicated instrument.

3. Past of space helioseismology (1980-1994)

From an historical point of view and though it was designed for solar constant measurements and not for helioseismology, ACRIM was the first instrument to observe solar oscillations from space.

3.1. ACRIM

The ACRIM (Active Cavity Radiometer Irradiance Monitor) instrument was launched the 14th of February 1980 aboard the SMM (Solar Maximum Mission) satellite of NASA. The orbit consisted in a 550-km, low-inclination Earth orbit. The corresponding orbital period was 95 mn with a night of 35 mn. This kind of regular data gaps due to night-time generate in power spectrum side bands at +/- 170 μHz. The instrument was composed of 3 independent cavity radiometers operated differentially against a common thermally massive heat sink (see Willson, 1979 for more details). As explained before this instrument was designed for total irradiance measurements and therefore sensitive to UV, visible and IR radiation. Its relative accuracy was far better then 0.1%. That is to say good enough to detect solar p modes in integrated sunlight measurements. At that time, none of the attempts from ground to detect p-mode signature in integrated sunlight were successful. Therefore ACRIM is both the first space experiment to detect p modes and also the first experiment to detect p modes in the integrated sunlight.

3.1.1. *Scientific outcomes*
The main results from ACRIM as far as helioseismology is concerned (Woodard and Hudson, 1983a,b, Woodard, 1984) were obtained with a time series of 10 months gathered during the first year of mission and before the loss of spacecraft fine-pointing control. The duty cycle of ACRIM during these

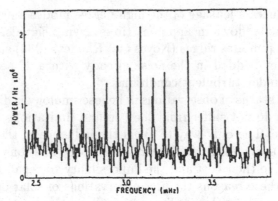

Figure 1.   10-month ACRIM spectrum Woodard, 1984

10 months was about 30%. With this time series was obtained the famous ACRIM p-mode spectrum (fig.1), where p modes of degrees $l=0,1$ and 2 are visible with a signal-to-noise ratio between 1 and 4 in the frequency range of 2.5 to 3.8 mHz. The frequency range was upper-limited by the 131 sec shutter-cycle, leading to a Nyquist frequency of 3.815 mHz. The quite good signal-to-noise ratio allowed measurement of the p-mode frequencies (with an accuracy of 0.4 μHz), their linewidths as well as their amplitudes.

3.2. STRATOSPHERIC BALLOONS

In 1980 and 1983 were launched two stratospheric balloons by the french space agency (CNES) with aboard a set of three sunphotometers dedicated to the measurements of the spectral solar irradiance (Fröhlich, 1984). As in the case of ACRIM, the instrument was not designed for helioseismology, but its good stability could allow detection of p modes. The concept was quite simple: three interference filters (368, 500 and 778 nm) at the entrance of the instrument to select thin wavelength bands of the visible continuum and a silicon-diode as detector. The observing times were rather short: 4 and 7 hours respectively.

3.2.1. *Scientific outcomes*

Because of the short duration of the observations it was not possible to identify separate peaks in the power spectra. Nevertheless it was the first time that p modes were observed at three different wavelengths (red, green and blue). It was therefore possible from the time series (fig.2) to determine an amplitude ratio and a phase shift between colors as well as their frequency-dependence (Fröhlich and van der Raay, 1984).

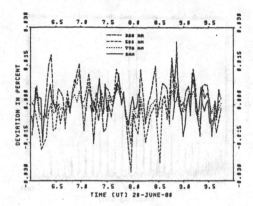

Figure 2. Time series of the three sunphotometer channels and of ACRIM total irradiance (Fröhlich, 1984)

3.3. IPHIR

The IPHIR (InterPlanetary Helioseismology by IRradiance measurements) instrument was launched on July 7 and July 12, 1988 aboard the interplanetary russian probes PHOBOS I and PHOBOS II, respectively (Fröhlich et al., 1988). IPHIR can be considered as the first space experiment dedicated to helioseismology. The idea was to take advantage of the long interplanetary flight between the Earth and Mars to observe continuously the Sun. In fact PHOBOS I was lost on September 2, 1988 only 2 months after the launch because of a wrong telecommand sent to the probe. The lifetime of the second probe was quite longer, PHOBOS II was lost in March 27, 1989, far later after the encounter of the probe with the martian satellite, because of a breakdown of the onboard computer. Thus, it allowed the IPHIR experiment aboard PHOBOS II to observe the Sun continuously for more than 160 days. The IPHIR instrument was composed of a triple sunphotometer with 3 interference filters(335, 500 and 862 nm) and a silicon-diode detector allowing observation of the integrated sunlight at 3 wavelengths of the visible continuum. The relative accuracy of the instrument (better than 1 ppm) and the fine pointing provided by a two-axis sun sensor have led to data of good quality though slightly polluted by light diffused inside the instrument. Moreover the blue channel of the photometers never provided good data because of the strong and quick degradation of the blue filter.

3.3.1. *Scientific outcomes*

The long continuous duration of observations with IPHIR yielded to a rather clean power spectrum (fig. 3, Toutain and Fröhlich, 1992) compared to the ACRIM one. The improvement in terms of signal-to-noise ratio was quite significant. This ratio was about 20 for p modes around 3 mHz. P-

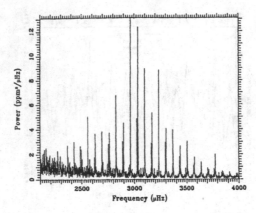

Figure 3. IPHIR green power spectrum (Toutain and Fröhlich, 1992)

mode parameters of l=0,1 and 2 modes were measured in a frequency range spanning from 2.4 to 3.8 mHz. Frequencies were measured with an accuracy of 0.1 μHz. Thanks to the continuous observations it has been demonstrated with IPHIR that p-mode amplitudes change quite strongly with time and in a way which agrees quite well with a model of stochatic excitation (Toutain and Fröhlich, 1992, Baudin et al., 1996). Nowadays it is widely admitted that p-modes are stochastically excited by turbulent convection. With IPHIR, was also measured a mean splitting of 0.56 + 0.02 μHz for modes l=1,2. It has been shown (Toutain and Kosovichev, 1994) that this can be considerably reduced (0.47 + 0.04 μHz) if a constant splitting is assumed throughout the mode frequency range, thus showing how difficult it is to measure accurately the splitting of low-degree p-modes. Up to now there is still no a complete agreement on the value of the mean splittings of low-degree p modes between the various experiments which are able to measure these quantities (Appourchaux et al, 1995).

3.4. SOVA

Four years after IPHIR was launched on July 31, 1992 aboard the EURECA (EUropean REtrievable CArrier) platform the SOVA (SOlar constant and VAriability) instrument (Crommelynck et al, 1995). The platform was retrieved 1 year later on June 24, 1993 by the space shuttle. This platform was originally planned to be launched in the mid-eigthies but was delayed because of the problems the space shuttle programme encountered. The orbit of EURECA was a 508-km Earth orbit with a period of about 95 mn and a night of 40 mn (Innocenti 1993). The SOVA instrument was composed of radiometers measuring the solar constant and two sets of IPHIR-like sunphotometers looking at the solar oscillations at wavelengths of 330,

480, 500, 545 and 865 nm. The photometers have been observing the Sun for 9 months with a duty cycle of 60%. That is to say with about the same duration as ACRIM but with a better duty cycle. Unfortunately, the temperature stabilization of the SOVA instrument introduced in the photometer signals spurious periods of the order of several minutes avoiding the detection of many p modes.

3.4.1. *Scientific outcomes*
With a time series of 7 months it was possible to detect only the strongest p modes of degree $l=0,1$ in the frequency range from 2.9 to 3.5 mHz with a signal-to-noise ratio close to what it was obtained with ACRIM.

3.5. DIFOS

The last experiment of this history of space helioseismology to be launched was DIFOS on March 2,1994 aboard the CORONAS-I satellite (Gurtovenko et al., 1994). This satellite is still flying, its orbit is a circular quasi-synchronous Earth orbit with an altitude of 500 km and an inclination of 82.5 degrees. This kind of orbit allows continuous observation of the Sun during 20 days each 3 months, the remaining time consisting in orbit of 95 mn with 35 mn night. The DIFOS instrument is composed of a silicon photodiode detector with 2 glass filters at 550 and 750 nm and also a neutral glass leading to a bandwith of 400-1100nm which is the spectral sensitivity range of the detector. DIFOS has been working until the 7th of May 1994 when it broke down just before the satellite entered the period of 20 days of continuous observations. The time series obtained is therefore long of 50 days with a duty cycle of 65% (Lebedev et al, 1995).

3.5.1. *Scientific outcomes*
In a frequency range spanning from 2.2 to 3.8 mHz were detected p modes of degrees $l=0,1$ and 2. The signal-to-noise ratio better than 10 yielded to measurement of frequencies with an accuracy of 0.4 μHz. As with IPHIR, it was also possible to follow the time-dependence of the modes and see the large variations of their amplitudes with time (Hasler et al., 1996).

4. Present of space helioseismology

The present of space helioseismology is fruitful because at the same time are flying three complementary experiments of helioseismology aboard the SOHO satellite (Domingo et al., 1995). SOHO (SOlar and Heliospheric Observatory), the solar mission of ESA and NASA, was successfully launched on December 2, 1995 and placed into halo orbit around the Earth-Sun lagrangian point L1. This mission should last three years at least and there is enough fuel in the spacecraft for many years. It is useful to briefly remind here the historical development of SOHO

4.1. HISTORY OF SOHO

In the context of helioseismology at the beginning of the eighties (see section 2) it was obvious that flying an experiment of helioseismology could greatly improve our knowledge of the solar structure. For this reason, in 1981 P. Delache proposed helioseismology as one of the science objectives of the potential ESA space mission DISCO (Dual Irradiance and Solar Constant Observations) during its assessment study.

One year later in November 1982 was proposed the SOHO mission in response to a call for mission proposals by ESA. Originally SOHO meant SOlar High-resolution Observatory and was dedicated to high-resolution spectroscopic investigation of the upper solar atmosphere. But, during the assessment study of SOHO in 1983 helioseismology was included in the scientific objectives of SOHO because DISCO was not being implemented.

Meanwhile in the United States was commissioned two SWG (science working group) one in 1978 to discuss what will be the best mission to follow the SMM mission and the other one in 1983 to consider the possibility of a space mission in helioseismology. Therefore already in 1983 thanks to their common goals, ESA and NASA envisaged to collaborate in SOHO.

For this reason the Science Study Team responsible of the Phase A study of SOHO was composed both of European and US scientists supported by ESA and NASA. Finally in February 1986 the Science programme Committee of ESA approved the STSP (Solar Terrestrial Science Programme) as the first cornerstone of the ESA's "Horizon 2000 programme". SOHO being one of the two components of the STSP with CLUSTER.

In March 1987 ESA and NASA made an Announcement of Opportunity for the STSP missions. This call for proposals led one year later to the selection of 12 experiments for SOHO which are shown here in a sketch of the satellite (see fig. 4) . Three of these experiments are dedicated to helioseismology : VIRGO, GOLF and SOI-MDI.

4.2. VIRGO

VIRGO (Variability of solar IRradiance and Gravity Oscillations) is an experiment which aims at determine the characteristics of low-degree solar p modes and g modes measuring spectral and total irradiance changes (Fröhlich et al.,1995 , Fröhlich et al., these proceedings). VIRGO is composed of four different instruments :
- two different kinds of radiometers (PMO6-V and DIARAD),
- an IPHIR-like triple sunphotometer with interference filters (402, 500 and

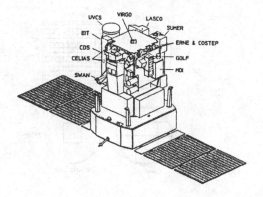

Figure 4. SOHO satellite (Domingo et al., 1995)

862 nm) and a Si-diode detector allowing detection of modes up to degree $l=3$,
- a Luminosity Oscillation Imager (LOI) based on a small Ritchey-Chretien telescope with a 16-pixel deep diffused silicon photodiode allowing detection of modes up to degree $l=7$.

4.3. GOLF

GOLF (Global Oscillations at Low Frequency) has also for scientific objectives to determine the characteristics of low-degree solar p modes and g modes. The complementarity with VIRGO is that GOLF measures variations of the integrated Doppler velocity on the solar disk (Gabriel et al., 1995, Gabriel et al., these proceedings) instead of integrated sunlight variations. The GOLF instrument is based on an optical resonance technique using a sodium-vapour resonant scattering cell which allows to measure small Doppler shifts of the solar Na D1 and D2 lines. This technique is similar to the technique carried out in the IRIS network. (Grec et al., 1991)

4.4. SOI-MDI

SOI (Solar Oscillations Investigation) is an experiment which aims at observe intermediate and high degree modes (Scherrer et al., 1995, Scherrer et al. these proceedings) and is therefore complementary to VIRGO and GOLF. The scientific objectives are mainly to determine the characteristics of intermediate and high-degree solar p modes, detection and identification of g modes, study of the convective zone dynamics and of magnetic structures. The instrument is a MDI (Michelson Doppler Imager) which is a kind of modified Fourier Tachometer tuned across the Ni line (676.8 nm). The detector is a 1024x1024 CCD camera with 2 resolutions: a low resolution of

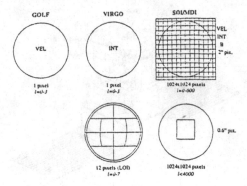

Figure 5. Comparison of GOLF-VIRGO and SOI-MDI kind of observations and degree coverage (Domingo et al., 1995)

2" per pixel and a high resolution of 0.6" per pixel. SOI produces various observables: Doppler velocity images, Ni line and continuum intensity,line-of-sight magnetic field, transverse velocity as well as limb position.

In Fig.5 is summarized the type of observations done by each instruments and the corresponding degree coverage showing the complementarity of VIRGO-GOLF and SOI-MDI. These three instruments are now working well since more than 6 months. Their on-flight performances as well as the first results obtained with these ones can be found in these proceedings.

5. Co-ordination of SOHO experiments with networks and other solar observations

As shown in the previous section the three experiments of helioseismology aboard SOHO are complementary. In terms of mode degree the coverage of VIRGO, GOLF and SOI-MDI is from $l=0$ to about 4000 though above 1500 the overlap of the lines makes determination of frequencies difficult. The degree coverage is important because the more large it is the more accurate is the inference of the solar structure. In addition to degree coverage, time coverage is also very important. We know that p-mode frequencies are changing with the solar cycle (Pallé et al (1995)). It is therefore necessary to observe solar oscillations on timescale of the order of the solar cycle to better understand the physics of p modes. This has already been done with the BISON and the IRIS networks which are working since the beginning of the eighties. In figure 5 is a plot of the degree coverage vs time coverage of the SOHO experiments and the ground-based networks. It shows how SOHO and the networks are complementary from the point of view of time coverage and degree coverage. Moreover, both velocity and irradiance observations are represented among these experiments. Helioseismology has

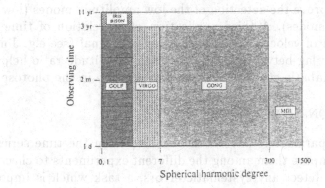

Figure 6. time coverage vs degree coverage of ground-based networks and SOHO experiments

never been so strong in terms of instruments observing at the same time the same oscillations from ground and from space in velocity and in intensity. It is a good opportunity to combine these different data sets together or with other solar observations in order to reinforce their impact on the inference of the solar structure and rotation.

5.1. CO-ORDINATION WITH SYNOPTIC PROGRAMMES

It well-known that experiments looking at oscillations in the Doppler shift of lines (like GOLF) are very sensitive to the magnetic activity on the solar surface (Ulrich et al.,1991). In this case it is useful to get informations from the synoptic programmes on ground to correct the line shape deformation due to magnetic effects. This will tend to reduce the solar noise in the power spectra and consequently to improve the detection of modes of low amplitude. These synoptic programmes are also useful because they provide helioseismologists with various indexes of activity which can be compared to the time-dependence of the solar oscillations parameters (frequency, amplitude) in order to better understand how p modes are sensitive to solar activity.

5.2. CORRELATION OF TIME SERIES

It is very unlikely that any of the experiments of helioseismology is free of non solar noise. With a time series of only one instrument it is difficult to distinguish between noise of solar origine and other noises. Therefore it is important to compare equivalent time series of different experiments doing the same kind of observations to detect any differences which could be explained in terms of non solar noise. Moreover correlation of time series

will certainly improve the detection of the low amplitude modes (low order
p modes and g modes). A third application of correlation of time series
is the correlation of velocity signal with intensity signal (see e.g. Jiménez,
1994). The phase lag between signals and their amplitude ratio help us to
study the nonadiabatic effects which are taking place in the photosphere.

5.3. COMPARISON OF MODE PARAMETERS

Once the mode parameters have been obtained from the time series it is
also useful to compare them among the different experiments to check their
consistency and detect any systematic errors, a task which is impossible
with only one set of parameters. For example, from the existing sets of
low-degree p-mode frequencies we can see that for some modes the values
obtained with various experiments do not agree within the given error bars.
Thus, either there are systematic errors or frequencies are changing on time
scale quite smaller than the solar cycle. In such a case it is important to
work on similar time series from different sources to understand where
the discrepancies come from: systematic errors or unaccurate model of the
physics of p modes? These differences are even larger in the case of low-
degree p-mode splittings, showing how much work still remains to do as far
as comparisons are concerned.

5.4. MERGING OF FREQUENCY OR SPLITTING SETS

With SOHO experiments and ground-based networks, the number of datasets
which are or will be soon available has never been so large as well as the
degree coverage. Assuming that frequency or splitting sets have been corrected from any systematic errors or that our model of mode physics has
been improved their merging and their inversion will greatly extend our
knowledge of the Sun's structure and its rotation.

6. Future and prospects

The next future of space helioseismology missions is of course still based
on SOHO which is maintained until 1998 at least. Nevertheless, keeping
SOHO observing all along the coming solar cycle could greatly improve the
scientific outputs of the 12 embarked experiments.

    It is also planned by Russia to fly in the next years two more CORONAS
satellites (Oraevsky and Fomichev, 1996). On the first one (CORONAS-F)
will be embarked a new DIFOS experiment. This one is slightly different
from the original one described in section 2.5 , the reason is that the new
instrument is based on a 6-channel sunphotometer instead of a 3-channel
sunphotometer.

References

Appourchaux, T., Toutain, T., Jiménez, A., Rabello Soares M.C., Andersen, B.N., Jones, A.R., (1995) Results from the luminosity oscillations imager, *ESA-SP publications*,Vol. **376**,pp. 265

Baudin, F., Gabriel, A., Gibert, D., Pallé, P.L., Régulo, C., (1996) Temporal characteristics of solar p-modes, *Astron. Astrophys.*,Vol. **311**,pp. 1024

Claverie, A., Isaak, G.R., McLeod, C.P., van der Raay, H.B., Roca Cortes, T., (1979) Solar structure form global studies of the 5-minute oscillation, *Nature*,Vol. **282**,pp. 591

Crommelynck, D., Fichot, A., Lee, R.B., III; Romero, J.,(1995) First realisation of the space absolute radiometric reference (SARR) during the ATLAS-2 flight period, *Advances in Space Research*, Vol. **16**, pp 17

Deubner, F.L., (1975) Observations of low wavenumber nonradial eigenmodes of the Sun,*Astron. Astrophys.*,Vol. **44**,pp. 371

Domingo, V., Fleck, B., Poland, A.I., (1995) The SOHO mission: an overview, *Sol.Phys.*,Vol. **162**,pp. 1

Fröhlich, C., (1984) Wavelength dependence of solar luminosity fluctuations in the five minutes range, *Mem. S.A.It.*, Vol. **55**,pp. 237

Fröhlich, C., van der Raay, H.B., (1984) Global solar oscillations in irradiance and velocity: A comparison,*ESA The Hydromagnetics of the Sun*,pp. 17

Fröhlich, C., Bonnet, R.M., Bruns, A.V., Delaboudinière, J.P., Domingo, V., Kotov, V.A.; Kollath, Z.; Rashkovsky, D.N., Toutain, T., Vial, J.C., (1988) IPHIR: the helioseismology experiment on the PHOBOS mission, *ESA Publications*, SP-286,pp. 359

Fröhlich, C., Romero, J., Roth, H., Wehrli, C., Andersen, B.N., Appourchaux, T., Domingo, V., Telljohann, U., Berthomieu, G., Delache, P., Provost, J., Toutain, T., Crommelynck, D., Chevalier, A., Fichot, A., Däppen, W., Gough, D., Hoeksema, T., Jiménez, A., Gómez, M.F., Roca cortés, T., Jones, A.R., Pap, J.M., Willson, R.C., (1995) VIRGO: Experiment for helioseismology and solar irradiance monitoring,*Sol.Phys.*, Vol. **162**,pp. 101

Gabriel, A.H., Grec, G., Charra, J., Robillot, J.-M., Roca cortés, T., Turck-chièze, S., Bocchia, R., Boumier, P., Cantin, M., Cespédes, E., Cougrand, B., Crétolle, J., Damé, L., Decaudin, M., Delache, P., Denis, N., Duc, R., Dzitko, H., Fossat, E., Fourmond, J.-J., Garcia, R.A., Gough, D., Grivel, C., Herreros, J.M., Lagardère, H., Moalic, J.-P., Pallé, P.L., Pétrou, N., Sanchez, M., Ulrich, R., van der Raay, H.B., (1995) Global oscillations at low frequency from the SOHO mission (GOLF),*Sol.Phys.*, Vol. **162**,pp. 61

Grec, G., Fossat, E., Pomerantz, M., (1980) Solar oscillations: Full disk observations from the geographic south pole,*Nature*, Vol. **288**,pp. 541

Grec, G., Fossat, E., Gelly, B., Schmider, F.X., (1991) ,*Sol. Phys.*, Vol. **133**,pp. 13

Gurtovenko, E.A., Kesel'man, I.G., Kostyk, R.I., Osipov, S.N., Lebedev, N.I., Kopayev, I.M., Oraevsky, V.N., Zhugzhda, Yu.D., (1994) Photometer "DIFOS" for the study of solar brightness variations, *Sol.Phys.*,Vol. **152**, pp. 43

Hasler, K.-H., Zhugzhda, Yu.D., Lebedev, N.I., Arlt, R., (1996) Observation of solar low-l p-modes by the CORONAS-DIFOS experiment, *to appear in Astron. Astrophys.*

Innocenti, L., (1993), EURECA-1 Science operations,*ESA mission report*

Jiménez, A., (1994) Phase differences between irradiance and velocity low degree solar acoustic modes revisited, *Sol. Phys.*, Vol. **152 no. 1**, pp. 319

Lebedev, N.I., Oraevsky, V.N., Zhugzhda, Yu.D., Kopayev, I.M., Kostyk, R.I., Pflug, K., Rüdiger, G., Staude, J., Bettac, H.-D., (1995) First results of the CORONAS-DIFOS experiment,*Astron. Astrophys.*, Vol. **296**,pp. L25

Leibacher, J.W., Stein, R.F., (1971) A new description of the five-minute solar oscillation,*Astrophys. lett.*,Vol. **162**,pp. 191

Leighton, R.B., Noyes, R.W., Simon,G.W., (1960) Aerodynamic phenomena in stellar atmosphere, *Suppl. Nuovo Cimento*, Vol. **22**, pp. 321

Noyes, R.W., Rhodes Jr., E.J.,(1984) Probing the depths of a star: the study of solar oscillations from space, *NASA report*

Oraevsky, V.N., Fomichev, V.V., (1996) Russian future projects,*AdSpR*, **Vol. 17 no. 4,5**,pp. 359

Pallé, P.L.,(1996) The state of art on helioseismic ground-based experiments, *These proceedings*

Pallé, P.L.,(1995) Solar cycle frequency shifts at low l, *ASP Conf. series*, **Vol. 76**, pp. 239

Scherrer, P.H., Bogart, R.S., Bush, R.I., Hoeksema, J.T., Kosovichev, A.G., Schou, J., Rosenberg, W., Springer, L., Tarbell, T.D., Title, A., Wolfson, C.J., Zayer, I., MDI engineering team, (1995) The solar oscillations investigation-Michelson Doppler imager, *Sol.Phys.*, **Vol. 162**,pp. 129

Toutain, T., Fröhlich, C.,(1992) Characteristics of solar p-modes: results from the IPHIR experiment, *Astron. Astrophys.*,**Vol. 257**, pp. 287

Toutain, T., Kosovichev, A.G., (1994) Anew estimate of the solar core rotation from IPHIR, *Astron. Astrophys.*,**Vol. 284**,pp. 265

Ulrich, R.K. (1970) The five-minute oscillations on the solar surface, *Astrophys. J.*,**Vol. 162**,pp. 993

Ulrich, R.K. (1991) A co-ordinated and synergistic analysis strategy for future gorund-based and space helioseismology, *AdSpR*,**Vol. 11 no. 4**,pp. 217

Willson, R.C., (1979) Active cavity radiometer type IV, *Appl. Opt.*, **Vol. 18**,pp. 179

Woodard, M., (1984) Short-period oscillations in the total solar irradiance, *Phd Thesis*,**University of California**

Woodard, M., Hudson, H., (1983) Solar oscillations observed in the total irradiance, *Sol. Phys.*, **Vol. 82**,pp. 67

Woodard, M., Hudson, H., (1983),Frequencies, amplitudes and linewidths of solar oscillations from total irradiance observations, *Nature*, **Vol. 305**,pp. 589

PERFORMANCE AND FIRST RESULTS FROM THE GOLF INSTRUMENT ON SOHO

A. H. GABRIEL[1], J. CHARRA[1], G. GREC[2] , J.-M. ROBILLOT[3]
T. ROCA CORTÉS[4], S. TURCK-CHIÈZE[5] , R. ULRICH[6]
F. BAUDIN[4], L. BERTELLO[6], P. BOUMIER[1] , M. DECAUDIN[1]
H. DZITKO[5], T. FOGLIZZO[5], E. FOSSAT[7] , R. A. GARCÍA[5]
J. M. HERREROS[4], M. LAZREK[2], P. L. PALLÉ[4] , N. PÉTROU[5]
C. RENAUD[2] , C. RÉGULO[4]

[1] Institut d'Astrophysique Spatiale, CNRS et Université
Paris XI, 91405 Orsay, France
[2] Département Cassini URA 1362 du CNRS,
Observatoire de la Côte d'Azur, 06304 Nice, France
[3] Observatoire de l'Université Bordeaux 1,
BP 89, 33270 Floirac, France
[4] Instituto de Astrofísica de Canarias,
38205 La Laguna, Tenerife, Spain
[5] Service d'Astrophysique, DSM/DAPNIA, CE Saclay,
91191 Gif-sur-Yvette, France
[6] Astronomy Department, University of California
Los Angeles, U.S.A.
[7] Département d'Astrophysique, URA 709 du CNRS,
Université de Nice, 06034 Nice, France

1. Introduction

GOLF is designed to measure the Global Oscillations of the integrated solar disk, by determining the line-of-sight velocity of the photosphere as a function of time, over the frequency range 10^{-7} to 10^{-2} Hz.

For a full description of the GOLF instrument concept and design, the reader is referred to a detailed article published before launch (Gabriel *et al.*, 1995), together with a description of the SOHO platform and mission (Domingo *et al.*, 1995). The special L1 orbit of SOHO, with continuous sun-

J. Provost and F.-X. Schmider (eds.), Sounding Solar and Stellar Interiors, 53-66.

centre pointing, is ideally suited for the long-term stability requirements of GOLF.

GOLF is a development for the space environment of instruments operated for many years on the ground, in particular the BISON (Chaplin *et al.*, 1994) and IRIS (Grec *et al.*, 1991) networks. More closely related to IRIS, GOLF compares the frequency of the solar sodium D Fraunhöfer lines with that of an atomic standard sodium vapour cell carried in the instrument. The Doppler shift between these two gives the line-of-sight relative velocity and analysis of its time variation gives the global solar frequencies. Each sodium line is split into two Zeeman components by a strong permanent magnet on board, and these monitor the two wings of the broader solar absorption line. The actual configuration of GOLF differs somewhat from the IRIS concept; in the choice of the plane of polarisation used, in the use of both of the sodium D lines and in the system of polarising mechanisms adopted for switching between the two wings. In addition, GOLF carries small electrical coils, allowing the 5000 gauss permanent magnetic field to be modulated by a small field of about ±100 gauss, to enable also the mean slope of the operating points on the two wings of the solar line to be determined. GOLF carries an additional rotatable quarter-wave plate. This has the dual objective of measuring the intrinsic global solar magnetic field and providing a redundant back-up in the case of failure of the other GOLF polarising mechanism.

The pre-launch plan foresaw a 40 sec cycle of operations, in which at least one of the two rotating mechanisms and the magnetic field modulation were switched each 5 sec. In this way, ratios could be derived giving data points every 40 sec, relating to the line-of-sight velocity, the mean slope of the solar absorption line and the global magnetic field.

The SOHO mission was launched with outstanding precision on December 2 1995. GOLF was switched on some hours later and was subjected to some months of commissioning and calibration during the transfer to L1. In most respects GOLF has been shown to meet fully the scientific specification and is producing data of an outstanding quality, never before obtainable.

2. Commissioning and Calibration

GOLF was switched on for a period of 8 minutes on December 3 1995. This served to verify that the protective door was closed and enabled a check on the instrument temperatures, which were all nominal. GOLF Channel A was powered on on December 10 and put into an operating configuration, ecept for the cell, which was unheated and the door, which was kept closed. On December 11, the cell stem heater was turned on and the temperature allowed to stabilise. The functional testing of Channel A was then completed, as well as a temperature scan of the entrance filter. At this time, the detector background and cosmic ray counts were also established.

In January 1996, the door was opened, and the full operations commenced. During January, February and part of March, while the spacecraft was still on its way to the final L1 orbit, a range of tests were carried out of the sub-systems performance. This served to fully qualify the functioning and calibration of Channel A. It was decided to follow the policy adopted by the other SOHO instruments and not to fully commission the redundant Channel B, until it might be needed. However a brief 6 hour operation of Channel B served to show that it is fully functional, and can be used if required.

2.1. CELL TEMPERATURE

Critical calibrations were carried out on the two systems most sensitive to temperature variations. The cell temperature was scanned slowly, and the changes in the overall detector count rates were monitored, as well as the amplitude of the variations associaated with the 5 minute solar oscillations. As a result of these tests, the cell temperatures were set at optimum values of 170 deg C for the bulb and 188 deg C for the stem. It should be emphasised that there is an unknown but constant difference between these thermister readings and the real cell or vapour temperatures due to contact conductivity effects. At the temertures adopted, the counting rates were close to $5.5 \ 10^6$ per sec for each detector, with a sensitivity of 221000 c/s per one deg C change of stem temperature, in accordance with pre-launch calibration.

2.2. FILTER TEMPERATURE

The filter band-pass is directly influenced by its temperature, as observed and studied before launch. In addition, since the filter area is directly imaged in the vapour cell, any variations across its surface can introduce disturbances in the symmetry and uniformity of the scattering system. As the two photomultipliers cannot view precisely the same scattering volume, this can be tested by comparing their outputs. The filter temperature was scanned very slowly and automatically on January 23 and 24, to be completed on February 6 and 7. The range scanned was from 17 to 20 deg C. The operating temperature of 17.5 deg C was chosen as a result of these scans and has been adopted since February 8 for subsequent observations. At this temperature, the velocity signals from the two PMs are similar and do not vary rapidly with temperature. The filter heater current is maintained constant and has only once been adjusted since, in order to compensate for some very slow temperature drift due to variation in the performance of the various thermal coatings. Monitoring of filter temperatures is maintained at the level of 0.01 deg C, and to this precision no variations are observed at frequencies relevant to the GOLF measurements.

2.3. OFF-SET POINTING SENSITIVITY

Measurements of the sensitivity of GOLF to spacecraft pointing errors was considered to be of great importance, since it had been judged too difficult to carry out meaningful tests on the ground before launch. Moreover, such measurements as had been quoted from tests on IRIS instruments reported a high sensitivity to off-set pointing, at levels that could have produced serious problems for the spacecraft platform mounting of GOLF.

GOLF Sensitivity to off-set pointing was measured for Yaw and Pitch excursions on February 21 and 22 respectively. Calibration for Roll had to be postponed to March 15, after SOHO Star Sensor software patching. During 16 continuous hours of each of these days the spacecraft was oscillated around one of its three reference axis, with a period of 800 sec and a square-law amplitude of 30 arcsec in Pitch and Yaw and 90 arcsec in Roll. This is more than an order of magnitude larger than the worse case short term variations of SOHO observed pointing errors. Should the experiment have been sensitive to such movements, a peak would have appeared at 1.250 mHz in the Fourier spectrum of the detector signals. This frequency being in the area where the solar background "noise" reaches a minimum provides the best conditions to observe the created peak with the largest signal to noise ratio. For all three axes, no peak was detectable above the normal experiment noise- level. This confirmed the efficiency of the instrument optical system which has been designed specifically to minimise the sensitivity to spacecraft pointing off-sets.

This good performance was confirmed when large static off-set pointings were performed later for some hours for a specific South Pole Joint Observation Program. This showed no detectable signal perturbation for several arcminutes of pointing off-set. Depending on the off-set direction and duration, disturbance of the experiment thermal equilibrium could become an eventual source of detected signal. 360 deg spacecraft roll manoeuvres, carried out on March 19 and 20, established also that measured solar surface velocities were undisturbed by up to 90 deg rotation angles, in both positive and negative directions, a small measured velocity amplitude off-set being detected for larger angles.

3. Operating Sequence

The nominal pre-planned operating sequence of GOLF is fully described in the earlier instrument publication (Gabriel *et al.*, 1995). In this, switching the orientation of the two mechanisms (the quarter-wave plate QW and the polariser POL) as well as switching the direction of the magnetic field modulation are carried out in a sequence of 40 sec total, with at least one component switched each 5 sec. By regrouping the data, it is possible to construct signals characteristic of solar velocity, solar magnetic field and slope of the solar line wings each 40 sec, or even more frequently if desired.

This nominal sequence was successfully operated from the switch-on until early February 1996. At this point, the intermittent mal-functioning of the QW mechanism led to a decision to stop it in its optimum position. The operation continued in an approved back-up mode, in which the full data set is recovered by use of the POL mechanism alone, losing only the possibility to extract a secondary GOLF objective, the global magnetic field data.

The POL mechanism developed the same mal-funtion towards the end of March 1996. Early in April a decision was taken to stop the POL mechanism in its optimum position. This left the possibility of measuring the solar velocities using only the blue wing of the solar line, a mode of operation unforseen before launch. The limitations of this mode, which uses no mechanical motions were found to be less serious than anticipated, and will be discussed later. The advantage of freeing the observation of all further risk from mechanical failures persuaded us to leave GOLF in this mode for a substantial period of time and not to risk compromising a long coherent data set by using the flight instrument to attempt to diagnose the fault.

4. Technical Performance

With the exception of the operating sequence changes, already described, all of the GOLF sub-systems are functioning within their designed specification. The overall GOLF performance is thus at the level of sensitivity forseen and the analysis potential of the data remains limited only by the intrinsic solar problems of separating global modes from non-coherent solar velocity signals.

The measured stray diffused light in the cell has remained unaltered from the pre-launch value, as verified during switch-on, and again in March 1996 and September 1996, during brief cold cell periods. No change in overall counting rate has been observed which might be attributable to a deterioration in the cell performance or in the window transparency.

The detection system has been functioning nominally. The dark current is very close to pre-launch levels. Large pulses are observed, attributable to cosmic ray particles, but at a frequency which is negligible in comparison to the scientific count-rate. The system is equiped with a facility to record a pulse-height analysis from time to time, without interrupting the scientific data stream. These show that the pre-launch measurements and prediction of the decay of gain with time were unduly pessimistic. In reality this decay rate has been found to decrease with operating time. This can be extrapolated to predict many years of useful life from Channel A of GOLF, without needing to consider a switch to Channel B.

The thermal behaviour of the instrument confirms well the thermal model and the pre-launch tests. Spacecraft and GOLF sensor temperatures drift slowly throughout the observations. This is due in part to variations in the spacecraft/sun distance and in part to drift in the properties of the thermal coatings. It is difficult to separate these two effects without more than one year's observations. In general, the overall sensor drift is seen directly on the cell temperatures. However, the filter sub-system is more directly influenced by the solar flux and can show larger variations. Since the last thermal adjustment control on January 23 1996, a very slow positive drift has been observed and should reach about 1.6 deg C over the first year. This is due to the ageing of the thermal finishes on the GOLF sensor and on the spacecraft. Seasonal variation of solar flux adds to this drift an estimated annual modulation of about 1 deg C amplitude.

The data retrieval from GOLF has proved to be quite exceptional. The data reception provided by the NASA and DSN systems has been averaging better than 99 %, on the basis of CD roms, delivered some 3 weeks after the observations, a mean figure that conceals the fact that on some poor days the figure falls to 90 %. However, with the GOLF data buffer and retransmission after 10 hours and 16 hours, the total recovery rate is more than 99.99 %. With the sensor unit operating continuously between April 11 and early September 1996, this provides by far the most complete 5 months sequence of helioseismic data ever recorded.

5. Scientific Data

The most obvious and immediate method of analysing the data consists in first combining the signals from the two active PMs, and then, according to the sequence adopted, in choosing the appropriate algorithm to combine different intensities in order to derive a function which depends on line-of-sight velocity. For the pre-planned sequences of Mode A or B (see Section 4) we showed in the earlier paper (Gabriel et al., 1995) that the velocity is given by

$$v = v_a \frac{I_b - I_r}{I_b + I_r + 2s},$$

$$(1)$$

s is the signal due to stray solar light reaching the detectors without resonant scattering. v_a is the calibration factor, which can be taken approximately as $4kms^{-1}$, but is ultimately calibrated by using either the known slow variation in SOHO/sun distance, or the calculated shift due to the magnetic modulation of the instrument magnetic field. The function in Equation (1) is insensitive to first order to changes in the overall sensitivity of GOLF to the solar intensity. Such changes can arise from changes in the solar intensity or solar distance, or through changes in the instrument, such as the temperature of the cell stem.

For the sequence Mode C, using only the blue wing, the equation becomes

$$v = \alpha_b I_b. \tag{2}$$

The quantity α_b must also be calibrated and again the magnetic modulation provides one method of approach. However it is obvious that, without the denominator, equation (2) is also sensitive to first order changes in solar intensity or instrument sensitivity. With the exception of the intensity variations due to solar oscillations themselves, most of the variations are slow compared to the 5 minute oscillations. Such variations will not contaminate the spectrum of p modes, but they will lead to uncertainty and variation in the velocity calibration to be applied to the p modes. However, for low frequency oscillations, as when searching for g modes, it may become important to calibrate or to otherwise compensate for the variation of α_b, due to aging or instrumental drifts.

Several different techniques are being explored at present, in order to correct for these effects; both the first order variations in α_b and the higher order variations in v_a. Although these corrections are unimportant for p-mode analysis, and may not be important for the search for g modes, they will become necessary for later analysis, involving detailed comparison with other measurements, such as intensity oscillations from VIRGO, or spatially resolved measurements from MDI.

Using the early two-wing measurements from GOLF, a comparison has been made between its sensitivity and that of a good ground based solar oscillation instrument. The comparison was made with the Mark I instrument at Tenerife (Brookes, Isaak and van der Raay 1978). This is a potassium vapour instrument and a component of the BISON network.

In order to improve the validity of the comparison, the spectra were computed using 11 hour periods or "days" for each. Fig. 1 shows the average of 280 days of Mark-I, compared with 70 "days" of GOLF spectra obtained in Mode B. Frequency resolution is limited in this plot by the 11 hour periods used for the Fourier transforms. This approach has been used in order to demonstrate the level of the "continuous" spectrum, rather than the p modes, which are not here well resolved.

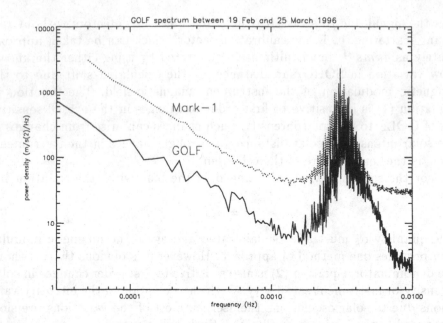

Figure 1. Low-resolution spectrum from GOLF compared with that from the ground-based instrument Mark 1, showing the lower background level obtained from space.

Examination of Fig. 1 shows several interesting features. The photon noise flat background instrumental spectrum can be seen in both cases at the high-frequency end. At the low frequency end the general (frequency)$^{-1}$ variation of the background signal is well demonstrated. This signal has normally been thought of as due to the random velocity fields of solar convection. Whereas this might be the case for the space measurements from GOLF, it is clear that for the Mark-I there must be an additional contribution, since its level is almost an order of magnitude higher. It seems reasonable to assume that this additional signal is due to noise from the earth's atmosphere. The much lower photon noise level of GOLF at the high-frequency end is in accordance with its design specification and the high PM counting rates adopted.

For the later GOLF data, based upon only the blue wing measurements, we have attempted to evaluate more precisely the real loss in sensitivity due to the loss of the red wing data. The technique adopted is to use the early two-wing data and to analyse it using the two techniques; two wings, or the blue wing only. This comparison is demonstrated in Fig. 2, where the two methods have been normalised for the same power in the p-mode region. Here we see that the increase in background noise level in the blue-wing only case is something between 1.4 and 2 mHz in the g-mode region. This

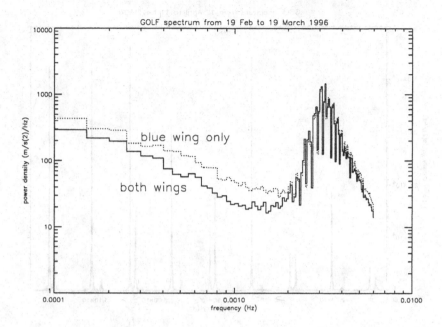

Figure 2. Low-resolution spectrum from blue wing only (dotted), compared with that from blue and red wings (full line).The curves are normalised for the same energy in the 2 to 4 mHz range.

would be a valid figure for the loss of signal to noise for the g modes, if we assume that their relative sensitivity on the two wings is the same as for the p modes. Taking account of uncertainties in these assumptions, we cannot say more than that the factor loss of detectability for g modes, due to the Mode C operation appears to be 2 or lower in units of power.

This small factor justifies the decision to continue operations in Mode C rather than endeavour to re-activate the mechanisms. Effort is instead being invested in improving the resultant difficulties in obtaining a reasonable precision in the oscillations amplitude calibration.

The lower background level of the GOLF spectrum is more clearly seen in a high-resolution spectrum of the p-mode region. Fig. 3 shows a section of the p-mode spectrum obtained by Fourier analysis of a 6-month period of GOLF data. Here again one can see the exceptionally clean background, allowing the identification of some very low amplitude modes. A first identification of GOLF p modes has been made, using a variety of techniques, but for the most part by the fitting of Lorenzian profiles by maximum likelihood technique to a Fourier spectrum. This is the subject of separate more detailed presentations (Grec *et al.*, 1997, Lazrek *et al.*, 1997).

At this time, around 100 p modes have been identified and their frequencies measured. This number will be increased and the precision of

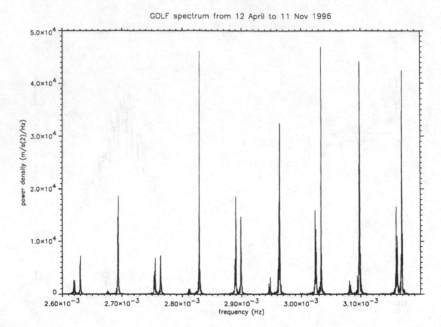

Figure 3. Section of the *p*-mode spectrum obtained from 9 months of data. Note the extremely low background level.

measurement improved with more sophisticated analysis techniques. In addition, significant improvements are expected from the analysis of a longer GOLF data set. A preliminary inversion of these *p*-mode frequencies has been carried out, using the LOWL data for the higher *l* modes (Tomczyk *et al.*, 1996). The precision obtained can be seen in the plot of sound speed in Fig. 4 (see also Turck-Chièze *et al.*, 1997).

6. Excitation Effects

A time-frequency analysis has been carried out on many of the observed *p* modes from GOLF (Baudin *et al.*, 1996, 1997). This uses the same techniques developed for analysis of the space observations in solar intensity oscillations from PHOBOS (Baudin *et al.*, 1994). An example of the time variation observed for an $l = 0$ mode (without fine structure or rotational splitting) is shown in Fig. 5. In the present case of the low-noise GOLF data, it is much easier to demonstrate that noise, from either solar velocity fields or photon statistics does not contribute significantly to the observed effects. We can deduce from the observed variations that there is an important fluctuation in both the amplitude of the mode and in its observed frequency, on time scales as short as a few days. This limit is imposed by the need for a trade- off between time and frequency resolution in the

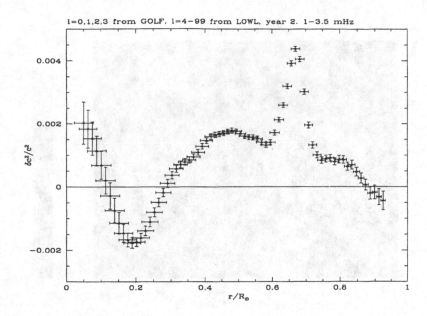

Figure 4. Difference in sound speed as derived from observations (GOLF and LOWL), compared with the model of Basu *et al.*

analysis and the importance of avoiding beating between adjacent modes. It is this time variation that produces the characteristic spiky line profile when analysed by Fourier techniques. A number of studies have aimed at interpreting these variations as due to the stochastic nature of the excitation process for the modes (Chang, 1995, Chang and Gough, 1995). García *et al.*, (1997), developing the theory of Woodward (1984) and Duvall (1990), have modelled the excitation of a damped oscillator by stochastic impacts, and are able to reproduce Fourier profiles with similar appearance to those of Fig. 5. According to García *et al.*, the observed frequency shifts can be explained by excitation effects, without requiring shifts in the frequency of the solar oscillator, although this remains an open question. What is clear is that the time frequency analysis, as well as the complex profile of the Fourier spectrum, contain important information on the basic excitation mechanism of the modes.

These manifestations of stochastic excitation impose important constraints on the precision obtainable in the frequency measurements of the modes. It is generally accepted that with a long enough time-series, a mean profile of a modified Lorenzian form can be accurately applied. However, it is not clear how rapidly we can converge to this limit, or whether real frequency shifts due, for example to solar cycle effects will intervene before we reach a high precision. With the present data set of some 10 months, there

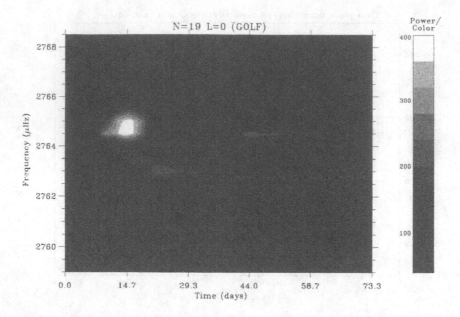

Figure 5. Time/frequency analysis of a single GOLF $l = 0$ mode, showing variations in amplitude and apparent changes in frequency.

is clearly a limitation in precision imposed by the stochastic excitation.

This difficulty is particularly troublesome when measuring the difference between two close resonances, as is the case in determining rotational splitting. Reported rotational splittings in the literature cover a very large range of values, much wider than the precision quoted in fitting the Lorenzian profiles. Values are now beginning to emerge from the GOLF data, but it is obvious that with less than one year of data, it is too soon to make an important contribution to the debate concerning solar internal rotation rates.

7. The search for *g* modes

The discovery of *g*-mode oscillations from GOLF could never be confidently anticipated, since there are no reliable predictions of their expected amplitudes. However, it was always clear that their detection would provide an inestimable advantage for the interpretation of conditions in the solar core. The search for *g* modes is therefore an important priority for GOLF, whose low-noise long-term stability offers the best chance for their detection. The solar spectrum in the expected *g*-mode range is dominated by the more random velocity fields produced by the various scales of convection. The approximate form of this continuous spectrum has been predicted by

Harvey (1985). When we examine the Fourier transform of GOLF data in this region, we see the background shape predicted by Harvey, with a number of superposed peaks, well above the background level. Although, this is somewhat what one might expect from g modes, this interpretation has not yet withstood critical tests concerning long-term invariance of the features. Since such tests depend strongly on the length of the available data base, it is with more months of data that we can expect to arbitrate with more confidence.

8. Conclusions

The GOLF instrument has been successfully launched and is operating for the most part very well. The exception concerns the rotating polariser mechanisms. After persistent malfunctioning, these have been stopped in optimum rotation orientations and an unforeseen mode of instrument operation has been adopted. This mode, which now involves no moving mechanisms, allows the full determination of frequencies to be performed, with little or no loss of precision. It does however pose difficulties for the absolute velocity calibration, which is currently of the order of 20 % or better. In all other respects, GOLF is maintaining its full design specification and is showing no degradation which would impede its continued operation for up to 6 years or more.

The p-mode oscillations from GOLF have a considerably improved signal to noise ratio, which has already allowed the identification of a some of new resonances. The precision today for the determination of p-mode frequencies must be regarded as preliminary. Significant advances are anticipated, both from improved analysis techniques on the existing data base as well as of course a substantial prolongation of this base. Continuity of the data set is so far quite exceptional, offering ideal opportunities for the study of excitation processes and solar cycle effects. The search for g-mode oscillations is continuing, but they have not so far been identified. All indications at this point are that GOLF is capable of continuing operations into the next solar maximum.

Acknowledgements

The GOLF programme is based upon a consortium of institutes involving a large number of scientists and engineers, as enumerated in the earlier instrument publication. The analysis work described here is also a team activity, involving many workers outside of the instrument teams.

The exceptional performance of the GOLF instrument is in large part due to the unique facilities offered by the SOHO platform, in terms of long-term stability of temperature and pointing and an extremely reliable and accurate clock. The continuous and dedicated attention of the ground system, including the Experiment Operations Facility at GSFC and the Deep Space Network have enabled an exceptional continuity in the data reception, essential for this project. Our thanks are due to these vital contributions from ESA, NASA and Matra Marconi.

References

Basu S., Christensen-Dalsgaard J., Schou J., Thomson M. J., Tomczyk S. (1996) *Bull. of Astron. Soc. of India*, **24-2**, p. 147.

Baudin F., Gabriel A.H. & Gibert D. (1994) *Astron. Astrophys.* **285**, L29.

Baudin F., Gabriel A.H., Gibert D., Pallé P. L. & Régulo C. (1996) *Astron. Astrophys.* **311**, p. 1024.

Baudin F., Régulo C., Gabriel A.H., Roca Cortés T. & the GOLF Team (1997) *Posters from the I.A.U. Symposium No. 181, Nice*, OCAN, in press.

Boumier P. & Damé L. (1993) *Experimental Astron.*, **4**, p. 87.

Brookes J.R., Isaak G.R. & van der Raay H.B. (1978) *Monthly Notices Roy. Astron. Soc.*, **185**, p. 19.

Chaplin W.J., Elsworth Y., Howe R., Isaak G.R., McLeod C.P., Miller B.A., van der Raay H.B. Wheeler S.J. and New R. (1996) *Solar Phys.*, in press.

Chang H.-Y. (1995) *PhD Thesis*, University of Cambridge.

Chang N.-Y. & Gough D. (1995) *GONG'94 Helio- and Astro-Seismology*, eds. R.K. Ulrich, E.J. Rhodes & W. Dappen, p. 512.

Domingo V., Fleck B. & Poland A. (1995) *Solar Phys.*, **162**, p. 1.

Duvall T.L. (1990) *Inside the Sun*, eds. G. Berthomieu & M. Cribier, Kluwer, p 253.

Gabriel A.H., Grec G., Charra J., Robillot J.M., Roca Cortés T., Turck-Chièze S., Bocchia R., Boumier P., Cantin M., Cespédes E., Cougrand B., Crétolle J., Damé L., Decaudin M., Delache P., Denis N., Duc R., Dzitko H., Fossat E., Fourmond J.J., García R.A., Gough D., Grivel C., Herreros J.M., Lagardère H., Moalic J.P., Pallé P.L., Pétrou N., Sanchez M., Ulrich R. & Van der Raay H.B. (1995) *Solar Phys.*, **162**, p. 61.

García R.A., Foglizzo S., Turck-Chièze S., Baudin F., Boumier P. & the GOLF Team (1997) *Posters from the I.A.U. Symposium No. 181, Nice*, OCAN, in press.

Grec G., Fossat E., Gelly B. & Schmider F. X. (1991) *Solar Phys.* , **133**, p. 13.

Grec G., Turck-Chièze S., Lazrek M., Roca Cortés T., Bertello L., Baudin F., Boumier P., Charra J., Fierry-Fraillon D., Fossat E., Gabriel A.H., García R.A., Gouiffes C., Gelly B., Gouiffes C., Régulo C., Renaud C., R.A., Robillot J.M. & Ulrich R. K. (1997) *Proceedings of the I.A.U. Symposium No. 181, Nice* , Kluwer, Dordrecht.

Harvey J. (1985) *Proceedings of the ESA Workshop on Future Missions in Solar Heliospheric and Space Plasma Physics*, eds. Rolfe & Battrick, ESA SP 235, p. 199.

Lazrek M., Régulo C., Baudin F., Bertello L. García R.A., Gouiffes C., Grec G., Roca Cortés T., Turck-Chièze S., Ulrich R. K., Robillot J.M., Gabriel A.H., Boumier P., Charra J. & the GOLF Team (1997) *Posters from the I.A.U. Symposium No. 181, Nice*, OCAN, in press.

Tomczyk, S., Streander, K., Card, G., Elmore, D., Hull, H., Cacciani, A. (1995) *Solar Phys.* ,**159-1**, p. 1.

Turck-Chièze S., Basu S., Brun S., Christensen-Dalsgaard J., Eff-Darwich A., Gabriel M., Henney C.J., Kosovichev A.G., Lopes I., Paternò L., Provost J., Ulrich R. K. & the GOLF Team (1997) *Posters from the I.A.U. Symposium No. 181, Nice*, OCAN.

Woodard M. (1984) *PhD Thesis*, University of California, San Diego.

FIRST RESULTS FROM VIRGO ON SOHO

C. FRÖHLICH[1], B.N. ANDERSEN[2] , T. APPOURCHAUX[3]
G. BERTHOMIEU[4], D.A. CROMMELYNCK[5] , V. DOMINGO[3]
A. FICHOT[5], W. FINSTERLE[1], M.F. GÓMEZ[6] , D. GOUGH[7,8]
A. JIMÉNEZ[6], T. LEIFSEN[9], M. LOMBAERTS[5] , J.M. PAP[10,11]
J. PROVOST[4], T. ROCA CORTÉS[6] , J. ROMERO[1]
H–J. ROTH[1], T. SEKII[7] , U. TELLJOHANN[3]
T. TOUTAIN[4] , C. WEHRLI[1]

[1] Physikalisch-Meteorologisches Observatorium Davos
World Radiation Center, CH-7260 Davos Dorf
[2] Norwegian Space Centre, N-0309 Oslo 3
[3] Space Science Department, ESTEC, NL-2200 AG Noordwijk
[4] Departement Cassini, URA CNRS 1362,
Observatoire de la Cote d'Azur, F-06304 Nice Cedex 4
[5] Institut Royal Météorologique de Belgique, B-1180 Bruxelles
[6] Instituto de Astrofísica de Canarias,
Universidad de La Laguna, E-38071 La Laguna, Tenerife
[7] Institute of Astronomy, University of Cambridge,
Cambridge CB3 0HA, UK
[8] Department of Applied Mathematics and Theoretical
Physics, University of Cambridge, Cambridge CB3 9EW, UK
[9] Institute of Thoreretical Astrophysics, University of Oslo,
N-0315 Oslo
[10] Department of Physics and Astronomy, Division of
Astronomy and Astrophysics, University of California,
Los Angeles, CA90095-1562 U.S.A.
[11] Jet Propulsion Laboratory, California Institute of
Technology, Pasadena, CA 91109 U.S.A.

Abstract. First results from the 4–6 months observations of the VIRGO experiment (Variability of solar IRradiance and Gravity Oscillations) on the ESA/NASA Mission SOHO (Solar and Heliospheric Observatory) are reported. The time series are evaluated in terms of solar irradiance vari-

J. Provost and F.-X. Schmider (eds.), Sounding Solar and Stellar Interiors, 67-82.

ability, solar background noise characteristics and p-mode oscillations. The solar irradiance is modulated by the passage of active regions across the disk, but not all of the modulation is straightforwardly explained in terms of sunspot flux blocking and facular enhancement. The observed p-mode frequencies are more-or-less in agreement with earlier measurements, but it is interesting to note that systematic differences seem to exist between the observations in different colours. There is also evidence that magnetic activity plays a significant role in the dynamics of the oscillations beyond its modulation of the resonant frequencies. Moreover, by comparing the amplitudes of different components of p-mode multiplets, each of which are influenced differently by spatial inhomogeneity, we have found that activity enhances excitation.

1. Introduction

The aim of VIRGO is to determine the characteristics of pressure and internal gravity oscillations by observing irradiance and radiance variations, to measure the solar total and spectral irradiance and to quantify their variability over periods of days to the duration of the mission as described by Fröhlich et al.(1996). VIRGO contains two different active-cavity radiometers (DIARAD and PMO6-V) for monitoring the solar 'constant', two three-channel sunphotometers (SPM) for the measurement of the spectral irradiance at 402, 500 and 862 nm with a bandwidth of 5 nm, and a low-resolution imager (Luminosity Oscillation Imager, LOI) with 12 'scientific' and 4 guiding pixels, for measuring the radiance distribution over the solar disk at 500 nm.

SOHO was successfully launched on 2 December, 1995. VIRGO was switched on a few days later. The covers were kept closed with the lock released in order to outgas the experiment thoroughly during the following 7 weeks. Real observations with the radiometers started mid-January, and observations with the sunphotometers at the end of January 1996. At that time it was realized that the LOI cover could not be opened. A concerted effort, as described by Appourchaux et al.(1997a), was needed, and finally on 27 March the cover was successfully opened. The LOI has been operating continuously since the end of March 1996. The performance of all the instruments is very good.

The observational data products are time series of the total and spectral irradiance and the LOI radiances which are corrected for all a priori known effects such as temperature, pointing, instrument–sun distance and relative velocity (for level 1), and inferred effects such as degradation (for level-2

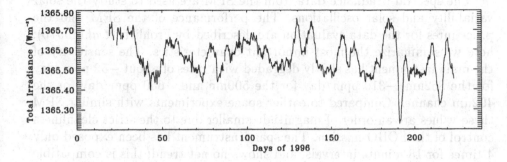

Figure 1. 6-hours averages of total solar irradiance in Wm^{-2} observed by VIRGO.

data products). The present scientific evaluation is carried out on preliminary level-2 data, which may change in the future owing to improvements
in the production of level-2 data from level-1 data. Such datasets from the
first 4–6 months of operation are the basis for the results we present here.

2. Total and spectral solar irradiance and radiance measurements

Observations of the total solar irradiance are a continuation of previous
measurements from satellites, which have been performed by several experiments since the end of 1978. The performance of the VIRGO radiometers and the procedures for the data evaluation are described by Fröhlich
et al.(1997b). The radiometers are operated in active mode, and have a
basic sampling cadence of 3 minutes for the DIARAD and of 1 minute for
the PMO6-V. The continuously exposed left DIARAD channel shows no
measurable degradation up to August 1996. This conclusion was obtained
from several comparisons with the right DIARAD channel which is operated
about every 60th day for 30 minutes. In contrast, the operational PMO6-V
radiometer degrades at a rate of $\approx$ 2.6 ppm/day relative to DIARAD. The
absolute irradiance values are based on the radiometric characterization of
each radiometer obtained before launch. In space the two type of radiometers differ by about 1 Wm^{-2}, namely 0.07%, well within the estimated
absolute accuracy of each radiometer ($\approx$ 0.15%). The VIRGO solar irradiance values as shown in Fig. 1 are DIARAD values decreased by 0.5 Wm^{-2}
to account for the difference between both radiometers. The absolute values
will be checked against measurements with similar radiometers on planned
stratospheric balloon in spring 1997 and later with the experiment SOL-
CON onboard the NASA Hitchhiker program (Crommelynck *et al.*1996),
and complemented by a detailed study of the time series of simultaneous
measurements of VIRGO and ACRIM-II on UARS.

The spectral irradiance data from the SPM are used to study the solar variability and solar oscillations. The performance of the SPM and the procedures for the data evaluation are described by Fröhlich *et al.*(1997b); here we summarize the most important characteristics. The sensitivity of the main instrument has slowly degraded with rates of about −52 ppm/day for the 862nm, −310 ppm/day for the 500nm and −650 ppm/day for the 402nm channel. Compared to earlier space experiments with similar SPM these values are an order of magnitude smaller due to the strict cleanliness control of the SOHO mission. The spare instrument has been exposed only 4 times for 18-minute intervals, and shows no net trend; this is compatible with the observations of the total solar irradiance. The absolute values are about 2.9% below the ones of Neckel & Labs (1984). While this is still within the combined errors of the NBS-1973 spectral irradiance scale of the lamps used and the reference spectrum of Neckel & Labs (1984), the lower values of all channels hint at a systematic error during calibration. The absolute value will be checked during the stratospheric balloon flight planned for 1997 by comparing with a similar SPM calibrated before and after the flight, by means of the recently developed, very accurate radiometric method.

The variations seen in the time series of spectral and total irradiance are caused by a superposition of random and periodic phenomena. Variability caused by individual phenomena at the photosphere is modulated further by the solar rotation, leaving a very complicated mixture of temporal and spatial effects. Although the main objective of the LOI is the study of solar oscillations (Section 5), its spatial resolution allows us to deconvolve the effects of different spatial and temporal influence of photospheric variations on the solar irradiance. The performance of the LOI and the procedures for the data evaluation are described by Appourchaux *et al.*(1997a). The signals from the 16 pixels are corrected to fluxes measured at 1 AU by taking into account the shape of the pixels, the limb darkening at 500 nm, and the actual size of the solar image. In addition, for comparing the LOI integrated flux with the SPM, conversion factors are derived for the four type of pixels. The comparison of the LOI integrated flux with the SPM green channel shows that the LOI sensitivity decreases much faster with time. At the beginning of the exposure it amounted to about 1000 ppm/day, decreasing to 500 ppm/day after 100 days. The stronger degradation is probably due partly to the fact that the LOI entrance filter sees more solar wind particles and is partly a result of having different coatings on the front surfaces.

3. Solar noise

The continuous measurements allow us to produce for the first time alias-free power spectra of the variance of total and spectral irradiance over a wide range of frequencies corresponding to periods from 2 minutes to the duration of observation. This not only permits studies of the Sun with helioseismology, but also allows us to study the variability itself. In Fig. 2 the power density spectra of the SPM green channel is shown as an example. The most important features are:

- the p-mode signal stands above the solar background noise in all channels with a signal–to–noise at the peak of about 120:1 (in power);
- there is no visible excess in the solar noise in the p-mode region;
- in and above the p modes, the solar noise decreases with frequency as about ν^{-4}; this value depends weakly on the wavelength of the observations, the blue and green channels being the steepest;
- from 10 to 100 μHz the solar noise decreases with frequency as ν^{-2};
- from 100 to 800 μHz the solar noise spectrum is quite flat; this was not expected, and it is a very encouraging finding because this is the range where g modes are expected;
- a distinct hump in the power spectrum is seen at about 1.2 μHz in all spectra;
- fitting the spectra with function types proposed by Harvey (1985) gives time constants of the three regimes of the solar noise to be about 66, 480 and 40000 seconds; this indicates that there are no structures visible with time constants in the mesogranulation regime; the shortest time constant could possibly be attributed to network bright points;
- at low frequencies peaks exist corresponding to periods of about 27, 9 and 7 days which are related to activity features crossing the visible disk.

Another new result is the frequency dependence of the ratio of the variance of the different spectral channels to that of total irradiance, which are quite complicated. In the frequency range of granulation and meso-granulation they are more or less constant; then they decrease towards lower frequencies with a broad minimum with about equal ratios for all colours at around 10 μHz, and then at even lower frequencies there is a broad hump with the blue 10 times higher than the total and about 6 times higher than the red and green at periods of a few days.

4. Influence of active regions on irradiance

The time scales associated with the lifetime of magnetically active regions have revealed extremely interesting variations in total solar irradiance.

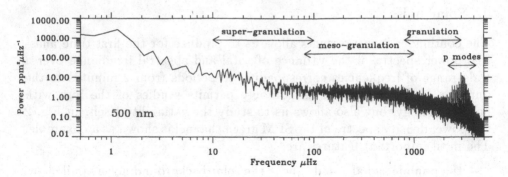

Figure 2. Power spectra of the green channel spectral irradiance time series for the period from January, 28 until August, 10 1996. The total, red and blue are very similar. The power ratios in the flat part of the spectrum (50–200 μHz) to the green are 0.47, 0.64 and 1.93 for the total, red and blue.

Analyses based on former observations of total solar irradiance showed that the most striking events in the short-term irradiance changes are the sunspot-related temporary dips (Willson *et al.*, 1981). The VIRGO experiment provides the first high-precision observations of solar spectral irradiance at 402, 500, and 862 nm in parallel with the total solar irradiance observations. These observations permit the study of simultaneous temporal changes in the near-UV, visible, and infrared ranges, and the determination of the spectral distribution of total solar irradiance variations, as is quite evident from Fig. 1. Here we discuss the occurrence of active region NOAA 7962 in early May 1996.

At the top of Fig. 3 we show magnetograms from Mt Wilson to illustrate the active region, and below we show time series of the total and spectral irradiance and of the four southern pixels of the LOI. The active region is causing significant changes in the spectral irradiance in the blue and green channels and the LOI, and much less in total irradiance and in the near IR. This particular active region appeared close to the limb in the southern part of the visible solar disk on 6 May (day 127) and contained a very small sunspot which vanished around 14 May. Very close to this region, a smaller active region without sunspots was also visible.

The LOI east pixel intensity value started to rise around 6 May with the appearance of the faculae in the active region rotating onto the east limb of the solar disk with a primary peak three days later. In parallel with the LOI east pixel intensity increase, both total solar and spectral irradiance in all the 3 wavelengths started to increase: by ≈300 ppm for the blue, ≈200 ppm for the green and ≈100 ppm in both the red channel and the total irradiance. It is interesting to note that at the central meridian transit only the blue and green irradiances reached the quiet sun level,

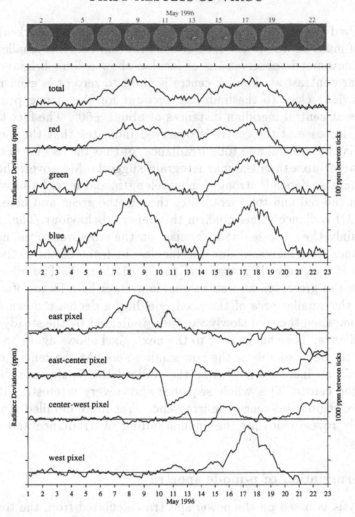

Figure 3. Passage of the active region NOAA 7962 in May 1996 shown by the Mt. Wilson magnetograms in the Fe I line (top panel, black and white correspond to ±5 Gauss), the time series of total and spectral irradiance (middle panel) and of the four LOI pixels south of the equator (bottom panel).

whereas the red channel and the total irradiance decreased by only about 40% and 60% of the maximal increase respectively. The photometric deficit of the spot group as determined by the San Fernando Observatory was around 50 ppm (Solar-Geophysical Data, 1996) at the maximum sunspot influence (around May 12–13). Taking the wavelength dependence of the spot contrast into account one would expect 45, 65 and 75 ppm for red, green and blue channels respectively. This accounts for only about half of the observed decrease in total irradiance by the sunspot darkening. The other half must be due to the facular contribution which is smaller at disk center. Photometric observations of faculae show a limb brightening and the facular contrast at the disk centre is close to zero or is even negative. The behaviour close to the limb can account for the strong peak of the irradiance at central meridian distances of about $\pm 60°$. The fact that only a partial decrease at the centre is observed indicates that the faculae are still seen in the spectral and total irradiance, or that the faculae extend over a much larger area than the magnetogram suggests. Moreover, the centre-to-limb increase depends strongly on wavelength, and must be substantially smaller in the red and total irradiance than in the green and blue.

The LOI radiance values confirm this general behaviour. The east pixel shows mainly the increase due to faculae; on the crossover to the next pixel the radiance first increases due the faculae in front of the active region and then it suddenly decreases due to the dark spot. The effect of the 65 ppm in the green sun-as-a-star signal is greater by a factor of about 12, owing to the smaller area of the pixel, yielding a decrease down to about -1100 ppm; then the spot slowly vanishes, indicated by the steady increase of the radiance. The change over to the next pixel shows again an increase and then the decrease due to the now smaller spot. At the central crossover there is still an increase, indicating that indeed the faculae are still bright close to the centre. This whole sequence shows very interesting details of how active regions influence the irradiance. Yet more detailed studies will definitively reveal clues for the understanding of irradiance and radiance variability.

5. Determination of p-mode spectra

The analysis is based on the power spectra calculated from the time series of the SPM channels and the 12 LOI pixels. The time series are detrended by piecewise fits of quadratic polynomials for the SPM and with a curve obtained by triangular smoothing with a full width of one day for the LOI. The power spectra are calculated from the Fourier transform of apodized time series by squaring the complex amplitude. As an example the p-mode spectrum from the SPM green channel is shown in Fig. 4. The amplitudes observed in the red, total and blue are ≈ 0.6, 0.7 and 1.4 times the green amplitude respectively.

6. Frequencies of p modes

From the SPM spectra the p-mode parameters are derived by fitting Lo-rentzians using a maximum-likelihood method as described in Toutain and Fröhlich (1992). The only difference is that the mode amplitudes in a given

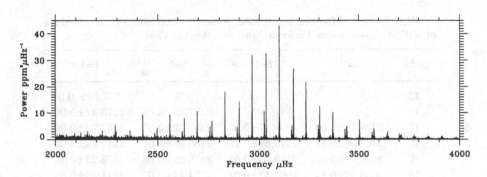

Figure 4. p-mode spectra from about 200 days of observation of the green irradiance. Note the high signal-to-noise ratio of about 120:1 (power).

(l, m) multiplet are no longer constrained to be the same. For the LOI a given (m, ν) diagram is fitted simultaneously using a maximum-likelihood method as described in Appourchaux *et al.*(1995). Moreover, the crosstalk between the $2l + 1$ components of a multiplet as well as the aliases from higher-degree modes are appropriately taken into account (Appourchaux *et al.*, 1997a).

The most important parameters of the p-mode characteristics are the frequencies. Five such sets are available in VIRGO. As the degeneracy splitting is not resolved in the low-degree modes determined from global observations of the sun the frequencies of the $l = 1, 2$ from the SPM or total spectra can be biased. The frequencies of the uninfluenced $l = 0$ modes determined from the SPM and the total channels are in good agreement among themselves and with the ones determined from the LOI. In order to have an internally consistent data set we present as VIRGO first-result frequencies those determined from LOI. These frequencies are listed in Table 1. They can be compared with those determined by the IPHIR experiment (Toutain & Fröhlich, 1992) during the second half of 1988. It is interesting to note that the frequencies of the $l = 1$ modes are systematically higher, with a slight increase towards higher frequencies, which could well be due to differences in solar activity in 1988 and 1996.

P-mode frequencies from Table 1 have been inverted to investigate the structure of the solar core as reported by Fröhlich *et al.*(1997a) and Appourchaux *et al.*(1997b). From the current results alone there is no compelling evidence that the sun is substantially different from the reference model.

7. Linewidths and Amplitudes of p modes

Linewidths are determined from the same fitting procedure from both the SPM green channel and the LOI. The results are shown in Fig. 5. Note the

C. FRÖHLICH ET AL.

TABLE 1. List of frequencies of solar p modes determined from 135 days of VIRGO observations between April and August 1996.

order	l=0	l=1	l=2	l=3
13	-	-	-	2137.79± 0.07
14	-	-	2217.63±0.08	2273.47±0.06
15	2228.34±0.09	2292.17±0.05	2352.29±0.07	2407.73±0.09
16	2362.96±0.12	2425.68±0.10	2486.05±0.07	2541.70±0.06
17	2496.14±0.09	2559.29±0.07	2619.69±0.08	2676.23±0.05
18	2630.03±0.15	2693.32±0.07	2754.44±0.07	2811.49±0.06
19	2764.43±0.16	2828.19±0.08	2889.59±0.05	2946.97±0.04
20	2898.95±0.10	2963.32±0.08	3024.75±0.04	3082.31±0.04
21	3033.75±0.09	3098.16±0.11	3159.83±0.07	3217.74±0.05
22	3168.67±0.09	3233.21±0.04	3295.25±0.06	3353.58±0.07
23	3303.21±0.12	3368.64±0.11	3430.76±0.11	3489.65±0.08
24	3438.95±0.16	3504.39±0.17	3567.02±0.15	3626.35±0.12
25	3574.62±0.29	3640.35±0.24	3703.41±0.22	3763.01±0.18
26	3711.45±0.98	3777.07±0.18	3840.84±0.31	-
27	3847.28±1.57	-	-	-

order	l=4	l=5	l=6	l=7
13	2188.43±0.35	2235.42±0.02	2280.08±0.07	2322.40±0.09
14	2324.31±0.05	2371.23±0.07	2415.59±0.06	2458.25±0.07
15	2458.53±0.05	2506.00±0.06	2551.12±0.05	2594.27±0.06
16	2593.16±0.08	2641.25±0.04	2687.08±0.05	2731.08±0.09
17	2728.51±0.09	2777.24±0.07	2823.81±0.05	2868.40±0.05
18	2864.28±0.09	2913.51±0.04	2960.62±0.04	3005.54±0.04
19	3000.13±0.05	3049.84±0.05	3097.09±0.05	3142.77±0.04
20	3135.96±0.05	3186.26±0.05	3234.17±0.06	3279.76±0.05
21	3271.74±0.06	3322.67±0.10	3370.91±0.11	3417.35±0.07
22	3408.14±0.09	3459.74±0.12	3508.80±0.12	3554.93±0.15
23	3544.67±0.10	3596.35±0.17	3646.05±0.14	-
24	3682.25 ±0.16	3734.72±0.17	-	-

pronounced dip around the peak of the 5-minute oscillations. This dip is a resonance effect with convection which decreases the damping of the modes, and was predicted by Gough (1980; cf Balmforth, 1992). It is interesting that the IPHIR data did not show this dip, which is suggesting that the resonance effect may be decreased by increasing solar activity.

Detailed studies of the time variation of mode amplitudes for solar p

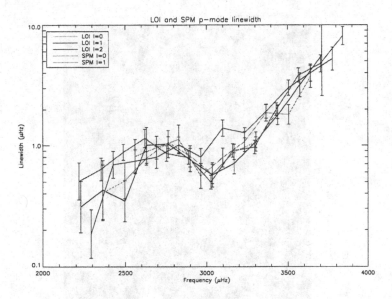

Figure 5.   Linewidth determined from the SPM green channel and LOI for $l = 0 \ldots 2$.

modes are essential for understanding both the excitation mechanisms and
the response of the solar surface to the perturbations caused by the modes.
In this context it is crucial to understand that the properties of the solar
surface acts as a filter between the oscillations and the observer. These
studies require continuous data sets.

Wavelet analysis is an appropriate tool for studying time-dependent
phenomena. With an optimal selection of parameters this method also provides a better combined frequency and time resolution than is provided
by conventional Fourier analysis. These methods have previously been successfully used on IPHIR data (Baudin *et al.*1994, Leifsen *et al.*1995). The
amplitudes of the p modes from the SPM blue channel and the LOI have
been studied with this method. The results from the LOI data on the $l = 1$
indicate no correlation between the different rotationally split components
of each radial order. This is to be expected from our view that the oscillations are excited randomly by the turbulence in the upper boundary layer of
the convection zone, despite the fact that each individual excitation event
applies equal impulses to eastward and westward propagating modes with
like values of l and $|m|$. The results with high frequency resolution for the
$l = 0, n = 16 - 26$ modes from the SPM blue channel are shown in Fig. 6.
From the data we see no clear and consistent correlation of the excitation of
the different modes. This too is as we expect (Chang, 1995). A prominent
phenomenon is the clear variation of the central frequency position of the
modes. Complicated structures with shifts and splittings of the modes are

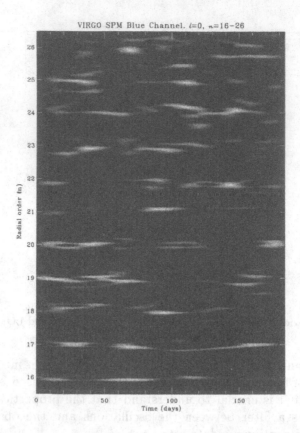

Figure 6. Wavelet analysis of $l = 0, n = 16 \ldots 26$ from the SPM blue channel. The time axis starts on the January 29 1996. The frequency coverage for each radial order is 4 μHz.

clearly seen. This is a result of phase wandering due to the excitation, and is manifest as a power variation in different regions of the mean Lorentzian profile of the modes. It shows the danger in fitting Lorentzians to the power spectra of modes obtained from short time strings of data. Such a fitting will give spurious frequency shifts. The required length of the time series to achieve a reasonable accuracy of the Lorentzian fitting may be so long that without appropriate care solar cycle variations could be masked.

The time dependence of the amplitude variation of the different modes has been studied by calculating the power spectra. For this, wavelet parameters giving a time resolution of about 10 hours were used. The results for $l = 0$ show a dominant peak at approximately the period of the solar rotation and some minor peaks at 16.9 and 11.6 days, whereas the spectrum of the irradiance signal shows also a strong 27 day period and two other peaks at around 9 and 7 days. In the case of the irradiance they are related to the active region modulation as seen from time series (e.g. Fig.1 & 3). The periodicities are also seen in the power spectra of the individual modes, but there the amplitudes of the different peaks vary strongly between the modes. The explanation of the modulation stems from the variation of the solar surface structure, i.e. with active regions, which modulate what we observe. Further observations are needed to see whether the periodicities in the amplitudes correspond to the mean solar rotation or to the rotation of the active regions.

Another interesting property of the oscillations concerns the relative amplitudes of the components of the multiplets. Although the datasets are perhaps not yet long enough to secure reliable statistics, there does appear to be a systematic variation in the power, as is illustrated in Fig. 7. Such a distribution is not inconsistent with preferential excitation of the oscillations in regions of enhanced activity. In particular, if we adopt a latitudinal variation of the excitation rate that has maxima at midlatitudes, we can account for the general trends in the observed amplitudes if the maximum excess rate of energy input into oscillations is about 50% of the uniform background. The best fit to the data of a smooth symmetric function with single maxima in each hemisphere has the maxima at about ±40° latitude, which is similar to though somewhat greater than the mean latitudes of the sunspot belts early in the cycle. Nevertheless, the broad agreement does suggest that magnetic activity plays a role in exciting the oscillations, in addition to modifying their frequencies (cf Kuhn and Libbrecht 1991). The difference between the locations of the maxima in the excitation function that we have deduced and the locations of the sunspots could perhaps be a result of inadequate statistics, or it could perhaps be an artifact of having misjudged the instrumental sensitivity to different modes by ignoring the magnetic perturbation to the eigenfunctions which causes the apparent decrease in power which Bogdan et al.(1993) has observed to occur in active regions. Alternatively, it could simply be that the regions of enhanced excitation extend beyond the observed regions of activity, as we were led to suspect from the variations in the radiance measurements discussed in section 3.2.

8. Conclusions

All instruments within the VIRGO package work very well, and interesting new results have been gathered from our preliminary investigation of the first 4–6 months of observations. The very quiet environment of SOHO, together with the continuity of the observations, enables us to obtain time series of a quality never previously achieved. Our first results confirm broadly what has been found from previous observations: firstly, that the solar irra-

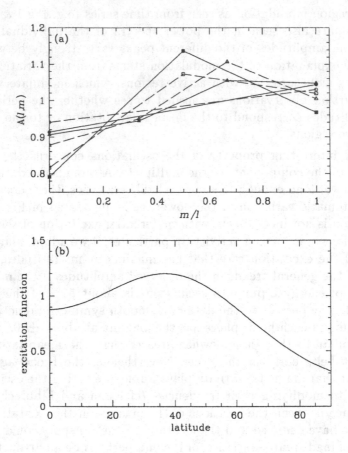

Figure 7. (a) Mean relative amplitudes of the components of dipole (circles), quadrupole (squares) and octupole (triangles) multiplets, plotted against $|m|/l$. The amplitudes of each multiplet are normalized such that the mean power per mode is unity. Filled symbols joined by continuous lines are derived from observations by LOI; the corresponding open symbols joined by dashed lines are theoretical values computed assuming the latitudinal variation of energy input illustrated in (b).

diance is modulated by the passage of sunspots across the disk, but that not all of the modulation is straightforwardly explained in terms of sunspot flux blocking and facular enhancement, and secondly, that the p-mode frequencies are, as indicated by helioseismic inversions, at present more-or-less in agreement with the latest standard solar models. However, we have in addition some new observations that provide hints that profound solar activity is more extensive than its superficial manifestation. This has been obtained partly from a comparison of coarsely resolved measures of radiance with the total irradiance. We also have evidence that magnetic activity plays a significant role in the dynamics of the oscillations beyond its modulation of the resonant frequencies. Pallé *et al* (1990) and Elsworth *et al* (1993) have already shown that the mean linewidths vary with the solar cycle. Comparison of our VIRGO data with IPHIR data reveals that the variation depends on frequency, in a manner that is suggestive of a modulation in the mechanism by which the oscillations interact with the turbulence in the convection zone. Moreover, by comparing the amplitudes of different components of p-mode multiplets, each of which are influenced differently by spatial inhomogeneity, we have found evidence that activity enhances excitation. This appears to be contrary to the temporal variation reported by Pallé *et al* (1990) and Elsworth *et al* (1993), who found that on average mode amplitudes decrease as the global level of activity rises. Further analysis is required to ascertain whether this disaccord is an indication that the excitation process is substantially more complicated than one might suspect or whether the discrepancy is simply a result of the modification of the oscillation eigenfunctions in the solar atmosphere by the magnetic field. Whatever is the case, it is evident that our first results have already raised interesting issues concerning the role played by activity in the global dynamics of the sun.

Acknowledgements

VIRGO is a co-operative effort of many individual scientists and engineers at several institutes in Europe and USA. Without the continuous and concerted efforts of the team this experiment would never have reached the success it demonstrates today.

The VIRGO team has been supported by several national and international funding agencies which are gratefully acknowledged: The PMOD/WRC by the Swiss National Science Foundation under grants 2.860-0.88, 20-28779.90, 20-33941.92, 20-40589.94 and PRODEX, the IRMB by the Fonds de la Recherche Fondamentale Collection d'initiative ministerielle and PRODEX, the SSD/ESA by their annual funds from the Science Directorate, the IAC by the CICYT through PNIE under grants ESP88-0354 and ESP90-0969, the OCA by the CNES and CNRS. Individual contributions to the project have been supported by grants from PPARC of the UK to D. O. Gough, by UCLA under a contract with NASA to J.M.Pap. Thanks are extended to the Mt.Wilson Observatory for the magnetograms retrieved from the SOHO archive.

References

Appourchaux, T., Toutain, T., Telljohan, U., Jiménez, A. and Andersen, B.N. (1995) *Astr. Astrophys.* **294**, L13.

Appourchaux, T., Andersen, B.N., Fröhlich, C., Jiménez, A., Telljohan, U., Wehrli, C. (1997a) *Solar Phys.* **170**, 25

Appourchaux, T., Andersen, B.N., Berthomieu, B., Fröhlich, C., Gough, D.O., Jiménez, A., Provost, J., Sekii, T., Toutain, T., Wehrli, C. (1997b) in Schmider, F.X., Provost, J., ed(s)., *Proc. IAU Symposium 181, Nice, October 1996*, Kluwer Academic Publ., Dordrecht, The Netherlands, this volume

Balmforth, N.J. (1992) *MNRAS* **255**, 603

Baudin, F., Gabriel, A. and Gibert, D. (1994) *Astron.Astrophys.* **285**, L29

Bogdan, T.J., Brown, T.M., Lites, B.W. and Thomas, J.H. (1993) *Astroph.J.* **406**, 723

Chang, H.-Y. (1995) *Ph.D. dissertation*, University of Cambridge

Crommelynck, D., Fichot, A., Domingo, V., Lee III, R. (1996) *Geoph.Res.L.* **23**, 2293

Elsworth, Y., Howe, R., Isaak, G.R., McLeod, C.P., Miller, B.A., New, R., Speake, C.C. and Wheeler, S.J. (1993) *MNRAS* **265**, 888

Fröhlich, C., Romero, J., Roth, H., Wehrli, C., Andersen, B.N., Appourchaux, T., Domingo, V., Telljohann, U., Berthomieu, B., Delache, P., Provost, J., Toutain, T., Crommelynck, D., Chevalier, A., Fichot, A., Däppen, W., Gough, D.O., Hoeksema, T., Jiménez, Gómez, M., Herreros, J., Roca-Cortés, T., Jones, A.R., Pap, J. and Willson, R.C. (1996) *Solar Phys.* **162**, 101

Fröhlich, C., Andersen, B., Appourchaux, T., Berthomieu, G., Crommelynck, D.A., Domingo, V., Fichot, A., Finsterle, W., Gómez, M.F., Gough, D.O., Jiménez, A., Leifsen, T., Lombaerts, M., Pap, J.M., Provost, J., Roca Cortés, T., Romero, J., Roth, H., Sekii, T., Telljohann, U., Toutain, T., Wehrli, C. (1997a) *Solar Phys.* **170**, 1

Fröhlich, C., Crommelynck, D., Chevalier, A., Fichot, A., Finsterle, W., Jiménez, A., Romero, J., Roth, H., Wehrli, C. (1997b) *Solar Phys.* **173**, submitted

Gough, D.O. (1980) in Hill, H.A., Dziembowski, W.A., ed(s)., *Nonradial and Nonlinear Stellar Pulsation*, Lecture Notes in Physics **125**, Springer-Verlag, Berlin, 273

Harvey, J. (1985) *E.Rolfe & B.Battrick* **Future missions in solar, heliospheric & space plasma physics**, SP 235: ESA Publication Division, Noorddwijk 199

Kuhn, J.R. (1996) in T. Roca-Cortes & F. Sánchez, ed(s)., *The structure of the Sun*, Cambridge University Press, 231

Kuhn, J.R. and Libbrecht, K.G. (1991) *ApJ.* **381**, L35

Leifsen, T., Hanssen, A., Andersen, B. N. and Toutain, T. (1995) *A.S.P. Conf. Ser.* **76**, 520

Neckel, H., Labs, D. (1984) *Sol.Phys.* **90**, 205

Solar-Gephysical Data (1996) *Solar-Gephysical Data comprehensive reports* **Number 624, Part II**, 24

Pallé, P.L., Régulo, C. and Roca Cortés, T. (1990) in Y. Osaki & H. Shibahashi, ed(s)., *Progress of Seismology of the Sun and Stars*, Lecture Notes in Physics 367, Springer Verlag, Berlin, 189

Toutain, Th. and Fröhlich, C. (1992) *Astr. Astrophys.* **257**, 287.

Willson, R.C., Gulkis,S.,Janssen,M., Hudson, H.S., Chapman.G.A. (1981) *Science* **211**, 700

NEW TIME-DISTANCE HELIOSEISMOLOGY RESULTS FROM THE SOI/MDI EXPERIMENT

T.L. DUVALL, JR.

Laboratory for Astronomy and Solar Physics, NASA
Goddard Space Flight Center, Greenbelt, MD 20771, USA

AND

A.G. KOSOVICHEV AND P.H. SCHERRER

W.W. Hansen Experimental Physics Laboratory,
Stanford University, Stanford, CA 94305, USA

Abstract. The SOI/MDI experiment on SOHO has some unique capabilities for time-distance helioseismology. The great stability of the images observed without benefit of an intervening atmosphere is quite striking. It has made it possible for us to detect the time of flight for separations of points as small as 2.4 Mm in the high-resolution mode of MDI (0.6 arcsec/pixel). This has enabled us to detect the supergranulation flow beneath the surface. Using a tomographic inversion technique, we can now study the 3-dimensional evolution of the flows near the solar surface.

1. Introduction

In time-distance helioseismology, the time-of-flight of acoustic waves is measured between various points on the solar surface. In the geometrical acoustics approximation, the waves can be considered to follow ray paths that depend only on a mean solar model, with the curvature of the ray paths being caused by the increasing sound speed with depth below the surface. The time-of-flight is affected by various inhomogeneities along the ray path, including flows, temperature inhomogeneities, and magnetic fields. By measuring a large number of times between different locations and using an inversion method, it is possible to construct 3-dimensional maps of the subsurface inhomogeneities.

J. Provost and F.-X. Schmider (eds.), Sounding Solar and Stellar Interiors, 83-89.
© 1997 IAU. Printed in the Netherlands.

2. Technique

We are using a new technique coming to be known as time-distance helioseismology (Duvall *et al.* 1993; Duvall *et al.* 1996*a, b*; Kosovichev 1996; Kosovichev and Duvall, 1996; D'Silva 1996; D'Silva *et al.* 1996), in which the time $\tau(x_1, x_2)$ for acoustic waves to travel between different surface locations (x_1, x_2) is measured from a cross correlation technique. In the first approximation, the time $\tau(x_1, x_2)$ depends on inhomogeneities along a geometric ray path connecting the surface locations. If the temperature is locally high, the waves will traverse the region more quickly leading to a shorter time. If there is a flow with a component in the direction of the ray path, the flow will tend to speed up the wave in the direction of the flow, while slowing it down in the reciprocal direction. In this case, we are

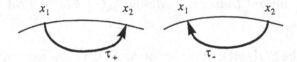

Figure 1. The reciprocal ray paths between two surface points, x_1 and x_2. The difference between the reciprocal travel times, $\tau_+ - \tau_-$, is primarily determined by the flow velocity along the ray path, while the mean, $\frac{1}{2}(\tau_+ + \tau)$ is sensitive to the sound speed and to the second-order effects of the flow.

lead to the very distinctive signature of the travel time being different for the opposite direction of travel between the same points. We have found no other effect which can yield this signature.

A magnetic field in the region of wave travel can also lead to travel time anisotropy (Duvall *et al.* 1996*a*). This is somewhat harder to isolate, as the distinctive signature of the field is that waves travelling along the field direction have a different wave speed than waves travelling perpendicular to the field. For some field geometries, it is possible to set up pairs of rays that intersect at right angles to look for this type of signature, but it is more difficult as the two rays have different paths except at the location where they intersect, and so other inhomogeneities can confuse the issue. The general form of rays below the surface is shown in Fig. 2. In this paper the effect of magnetic anisotropy is not considered.

3. Observations

For this study we have used 8.5 hours of Doppler images from January 27, 1996 taken in the high resolution mode of MDI, which has 0.6 arcsec/pixel. To study convection near the surface, it is necessary to use fairly short intervals in order that the evolution not be too great during the observing

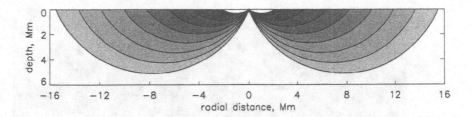

Figure 2.   The rays going from the center to the different annuli. The ray paths are curved because of the increasing sound speed with depth. The sound speed increases from 7 km/s near the surface to 23 km/s at 5 Mm depth. The horizontal separation between surface points is approximately π times the depth. The measured travel times are sensitive to the component of velocity along the particular ray path.

interval. On the other hand, it is necessary to observe for some length of time in order to get a statistical sample of waves. The 8.5 hours is a compromise between these two competing requirements, with the supergranule lifetime of about 1 day putting an upper limit on how long to observe. Another point is that we need to observe for longer than the travel time as the correlation that we are detecting arises from the same wave travelling from one location to the other through the subsurface layers. The travel times used in the present study are for surface point separations in the range 6-30 Mm, and are in the interval 17-34 minutes.

4. Analysis

To get the travel times between different locations, we calculate the temporal cross correlation function between the data at one location and the data within an annulus at some great-circle distance from the point. We look at both positive and negative lags of the correlation function, as this can tell us in which direction the waves are travelling. For example, a signal at location 1 first and at location 2 later will lead to a signal at one sign of lag while a signal at location 2 first will lead to a correlation at the opposite sign of lag.

The cross correlation function, averaged over a number of origins, is shown in Figure 3, along with a curve showing the time versus distance for a simple theory. If we imagine generating a pulse of acoustic energy at the surface that propagates along the ray paths to the distant location, we might expect to see a pulse at the distant location. But in fact we see something that is much broader and really looks like a wave packet with a period of five minutes, and a width in time of about 1/bandwidth of our oscillation power spectrum, or $(1 \text{ mHz})^{-1}=17$ min. This makes sense, because

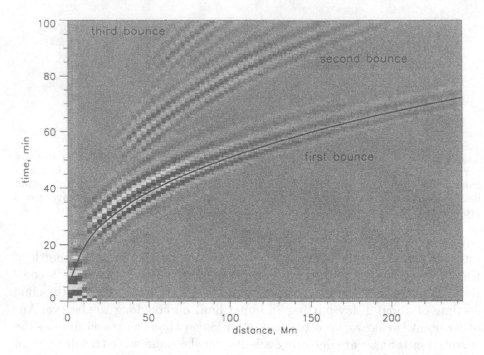

Figure 3.   A mean cross covariance function for the data. The solid line is a theoretical plot of the time for waves to travel along a ray path for the specified great-circle distance. The grayscale picture is the cross covariance function. The first, second and third bounces are visible near 30 min., 60 min., and 90 min. The fine structure in each of the ridges is caused by the finite bandpass of the oscillations.

the pulse that we generated at the surface had an infinite bandwidth and so could make a zero-width feature in the correlation function. As shown earlier by Duvall *et al.* (1993), the correlation function that we observe is the Fourier transform of the power spectrum. The remaining features in Figure 3 correspond to the multiple bounces. If there is a correlation observed at time t and at distance d, there will be a similar signal at $n \times t$ and $n \times d$, where n is integral.

In some earlier work (Duvall *et al.* 1993), the correlation function was rectified by using the analytic signal formalism, and times were measured from this signal which is the envelope of the correlation function. Later it was found that the higher frequency structure in the correlation function actually is a more sensitive measure of subsurface inhomogeneities (Duvall *et al.* 1996a). In that work, the location of one of the fine structure peaks was determined by measuring the zero crossing of the instantaneous phase of the correlation function. For the present work, this has been further refined (Kosovichev and Duvall 1996). A 20-minute interval in the neighborhood of the peak in the correlation function is fit to a Gaussian wave packet, with independent parameters for the location of the envelope, the location of one of the fine structure peaks, the amplitude and width of the Gaussian, and the frequency. All the results shown are from the location of the fine structure peak.

Following Duvall *et al.*(1996a), we compute the mean cross correlation for waves travelling both out from the center to the annuli and the reverse. The difference of the times measured from these cross correlations is a measure of divergence of the flow. It could be due to either a downflow near the central location or to an average horizontal outflow (or inflow) near the annulus. In addition to this divergence signal, a mean time is computed for the inward and outward waves. To get more directional information about flows, we break up the annulus surrounding a point into quadrants, centered on the four directions north, south, east, west (Duvall *et al.* 1996b). We then average the cross correlations for eastward and westward waves separately. The travel times are measured separately for the eastward and westward waves by the above procedure. To get a signal proportional to a flow, we then take the difference of these two times. The same procedure is applied to north-south propagation.

5. Inversions

The images are used in a ray-theory inversion (Kosovichev 1996, Kosovichev and Duvall 1996). The region of the Sun is separated into a 3-D grid of inhomogeneities with the same horizontal sampling interval as the input images (4.3 Mm) and an equal increment in depth covering the same depth range as the input data. The number of grid points in depth is taken to be the same as the number of input images, and, in this case, is 8. Each grid point has four independent variables, the three components of flow velocity and a sound speed inhomogeneity. No attempt is made to satisfy any conservation equations, like the continuity equation.

In Figure 4, we show the inversion results for the subsurface layer. The flow field mainly consists of isolated outflows of the characteristic size of 30 Mm, corresponding to supergranules observed on the surface. One advantage of studying near-surface regions is that we do have extra information at the surface that we would not have for the deep interior. In Figure 5 we show a vertical cut showing the subsurface flows. It would appear that the pattern of horizontal motions at the surface only persists to a few Mega-meters in depth. But it is probably premature to draw strong conclusions yet because of the newness of the method.

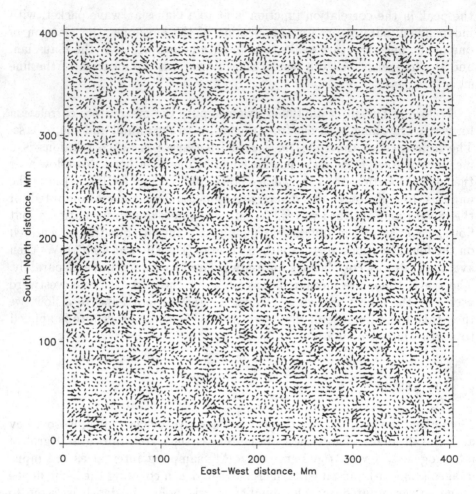

Figure 4. The horizontal flows near the surface from the inversion procedure are shown as vectors.

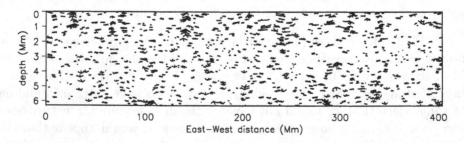

Figure 5. A vertical cut showing the component of flow in the plane. The flow is again shown as vectors.

Conclusions

We have shown that there is detailed information about the subsurface inhomogeneities contained in the helioseismology data. We have also shown it should be possible to extract this information and have shown some first-order attempts to do so. We would appear to have an exciting new window into the solar interior.

Acknowledgments

The authors acknowledge many years of effort by the engineering and support staff of the MDI development team at the Lockheed Palo Alto Research Laboratory (now Lockheed-Martin Advanced Technology Center) and the SOI development team at Stanford University. SOHO is a project of international cooperation between ESA and NASA. This research is supported by the SOI-MDI NASA contract NAG5-3077 at Stanford University.

References

D'Silva, S. (1996) *Astrophys. J.*, **469**, 964.
D'Silva, S., Duvall, T.L.Jr., Jefferies, S.M. & Harvey, J.W. (1996) *Astrophys. J.*, **471**, 1030.
Duvall, T.L.Jr., Jefferies, S.M., Harvey, J.W., & Pomerantz, M.A. (1993) *Nature*, **362**, 430.
Duvall, T.L.Jr., D'Silva, S., Jefferies, S.M., Harvey, J.W., & Schou, J. (1996a) *Nature*, **379**, 235.
Duvall, T.L.Jr., Kosovichev, A.G., Scherrer, P.H., & Milford, P.N. (1996b) *BAAS*, **188**, 49.08
Kosovichev, A.G. (1996) *Astrophys. J.*, **461**, L55.
Kosovichev, A.G. & Duvall, T.L.Jr. (1997) in *Solar Convection and Oscillations and their Relationship*, eds J. Christensen-Dalsgaard, F. Pijpers & C.S. Rosenthal, Proc. of SCORe'96 Workshop, Aarhus (Denmark), May 27-31, 1996, Kluwer Acad. Publ, in press.

Conclusions

We have shown that there is detailed information about the subsurface inhomogeneities contained in [illegible] data. We have also shown it should be possible to extract this information and have [illegible] some first-order attempts to do so. We would appear to [illegible] a new window into the solar interior.

Acknowledgements

The authors acknowledge [illegible] efforts of the engineering and support staff of the MDI development team at the [illegible] Laboratory [illegible] Advanced Technology Center and the SOI development team at Stanford University. SOHO is a project of international cooperation between ESA and NASA. This research is supported by the SOI/MDI NASA contract NAG5-3077 at Stanford University.

References

D[illegible] (19[illegible]) [illegible] *Astrophys. J.* [illegible]
Duvall, T. L. Jr. [illegible] (19[illegible]) *Nature* [illegible]
[illegible] (19[illegible]) *Nature* [illegible]
Duvall, T. L. Jr. [illegible] (19[illegible]) *Nature* [illegible]
[illegible] (19[illegible]) [illegible]
[illegible]
Kosovichev, A. G. [illegible] *Proc.* [illegible]

GOLF RESULTS: TODAY'S VIEW ON THE SOLAR MODES

G. GREC [2], S. TURCK-CHIÈZE [4], M. LAZREK [7] [2]
T. ROCA CORTÉS [3], L. BERTELLO [6], F. BAUDIN [3]
P. BOUMIER [1], J. CHARRA [1], D. FIERRY-FRAILLON [5]
E. FOSSAT [5], A. H. GABRIEL [1], R. A. GARCIA [3] [4]
B. GELLY [5], C. GOUIFFES [4], C. RÉGULO [3]
C. RENAUD [2], J.M. ROBILLOT [8], R. K. ULRICH [6]

[1] *Institut d'Astrophysique Spatiale, Unité Mixte CNRS Université Paris XI, 91405 Orsay, France*
[2] *Département Cassini, URA 1362 du CNRS, Observatoire de la Côte d'Azur 06304 Nice, France*
[3] *Instituto de Astrofísica de Canarias, 38205 La Laguna, Tenerife, España*
[4] *Service d'Astrophysique, DSM/DAPNIA, CE Saclay, 91191 Gif-sur-Yvette, France*
[5] *Département d'Astrophysique, URA709 du CNRS, Université de Nice 06034 Nice, France*
[6] *Astronomy Department, University of California, Los Angeles, U.S.A.*
[7] *CNCPRST, 10102 Rabat, Maroc*
[8] *Observatoire de l'Université Bordeaux 1, BP 89, 33270 Floirac, France*

1. Introduction

The SOHO probe was successfully launched on December 2nd, 1995. The performances of the Atlas II flight, the trajectory and the final injection in the Halo orbit around the L1 Lagrangian point left on board a large amount of hydrazine, allowing the possibility for a mission extension later than the 2 planned years. The operations of the GOLF experiment started on January 16th for a period devoted to the initial tests and to the adjustments of the thermal settings. The effective solar observations started on February 18th and are still running. For the studies presented here below, the data set ends

J. Provost and F.-X. Schmider (eds.), Sounding Solar and Stellar Interiors, 91-102.
© 1997 *IAU. Printed in the Netherlands.*

in mid-September. All tables and figures come from the compilation of the data analysis made in several institutes with different methods, and some complementary or additional results are displayed in the poster booklet published from this symposium.

2. The solar velocity signal

We did use GOLF in nominal mode up to April 1st (A. Gabriel *et al.*, 1995). This mode has been shown to be free from any first order variations and the test conducted for the depointing sensitivity indicates that no perturbing signal is expected from the attitude variations of SOHO. Unfortunately, at the beginning of April, we had to face with mechanical troubles, and due to that, GOLF is no longer used in the nominal differential mode, but is monitoring only the blue resonance windows in the wing of the DI and DII photospheric lines. In these conditions, GOLF data depend at the first order on the seasonal changes of the solar distance, the spacecraft to Sun radial velocity, the solar magnetic activity and also on the possible drifts of the temperature of several instrumental components. Detrending these effects has now become more crucial for the low frequency investigation. In practice, the characteristic time of most drifts is longer than 1 day, so that little real trouble can be confidently expected. On the other hand, the conversion of the photometric signal into velocity is a complex question, because the p-mode oscillations produce other effects than radial velocity changes. This question will be dealt in a future paper. For the study of the p-mode frequencies spectrum, it has been considered of almost no importance and has been addressed as simply as possible.

The data obtained during the time of nominal operation has made possible to compare the solar spectrum obtained for several analyses:

• The "differential photometer mode" analysis, makes use of the successive measurements I_R and I_B in the red or blue wings. The velocity is computed as $V_S = V_0(I_B - I_R)/(I_B + I_R)$, where I_R and I_B are the intensity mesurements for the red and blue channel. V_0 depends on the orbital velocity, but is nearly constant during the February-March run. An approximate value $V_0 = 4000$ m/s comes from laboratory measurements, and a further calibration will come from an analysis of the data versus the orbital velocity.

• The 2 "photometer modes" analyses make use of the fluctuations of either I_R or I_B, converted into velocity signal [1]. Those fluctuations are simply corrected from the photometric sensitivity changes due to the variable distance of the Sun and to the long term thermal drifts, using the low frequency com-

---

[1] The band-passes are $\simeq 35$ mÅ, tuned on the wings of the D photospheric lines, and shifted every 5 sec. from $\simeq 5$ mÅ to monitor the slopes of the lines wings. At this preliminary level of analysis, the information coming from this modulation is not used.

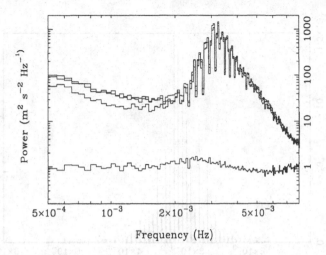

Figure 1. Power spectrum for: velocity in differential mode (the lower of 3 curves), or computed respectively from the blue wing or red wing signals (the 2 upper curves). The noise level in the spectrum below the p-mode frequencies is higher for the 2 "photometer" modes than for the "differential photometer" mode. The curve below displays the ratio V_R/V_B.

ponents of the signal (periods longer than 1 day). This conversion assumes that the measured intensity fluctuations are only sensitive to the weighted sum of the velocity field on the solar disk.

Fig. 1 shows the power spectra obtained after a simulation of the 3 modes with the nominal data set. The 2 "photometer modes" are simply normalized by means of the "differential mode", forcing the integral of the power between 7 and 8 mHz (solar signal, but not p modes) to be the same. This range being close to the Nyquist frequency, the power spectrum folding is avoided by means of a 40 sec sampling of the signal integrated over 80 sec. The 3 spectra contain roughly the same photon noise contribution, they have been smoothed for a better visibility. The increase of power at low frequency (around 1 mHz), where most of the power is expected to be the solar velocity background, can tentatively be interpreted as an increased visibility of the solar granulation. The "photometer mode" using a single wing does not integrate all the solar disk as uniformly as the "differential photometer mode" resulting in a less efficient statistical decrease of the signal due to the granular field. This is true as well for the magnetic activity and supergranules, but remains a small effect.

In the p-mode frequencies range, the ratio V_R/V_B depends on the difference of the photospheric filtering of the modes amplitude, related to the different $\tau = 1$ altitudes for red and blue channel, but also to the contribution of intensity fluctuations in the velocity measurements, which is again an altitude-dependent function.

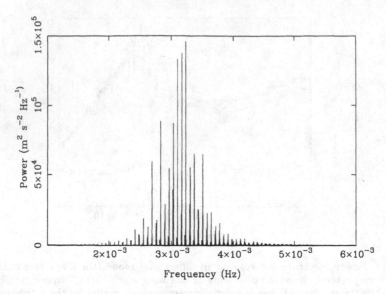

Figure 2. The *p*-mode power spectrum, showing the low level of noise. The time coverage of the data is practically 100 %.

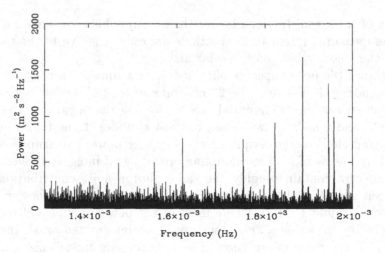

Figure 3. Low frequency part of the *p*-mode solar spectrum. The amplitude of the lower frequency detected *p* modes is below 1cm/s.

3. The *p*-mode analysis

The high resolution power spectrum of the I_B signal measured from April 5th to September 15th, 1996, has been computed with several calibrations techniques. The results are summarized in fig. 2, fig. 3 and fig. 4.

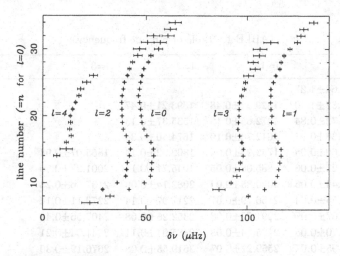

Figure 4. "Echelle" diagram of the p-mode solar spectrum, modulo 135.2 μHz. The frequency ν of a given mode is $\nu = 135.2\,n + \delta\nu + 12$ in μHz. This fig. displays several low amplitude modes not listed in tab. 1.

3.1. P-MODE FREQUENCIES

The p-mode parameters are determined by different methods of numerical fit on the velocity power spectrum, including mostly the maximum likelihood fit of Lorentz profiles, singlet or multiplet, according to the degree. In the fit of multiplet, the splitting is assumed to be known (its measurement is another question, explored independently). The estimations of the uncertainty are deduced from a Monte Carlo simulation. Tab. 1 shows the p-mode frequencies and uncertainties (Lazrek *et al.*, 1996-b). The uncertainties are larger than what can be found in former publications. Taking into account only the statistical uncertainty shown by different methods, they are related to the stochastic character of the p modes: less than 6 months of data have been analysed here, and the p-mode temporal behaviour has proved to be a not quite stationary process on the long time scale. A conservative view has been preferred, and in most cases the real error is probably smaller than the listed uncertainty.

3.2. ROTATIONAL SPLITTING

The line profile of a given p mode depends on the excitation and damping processes, but it also shows a multiplet structure due to the solar rotation, called rotational splitting. Note, however, that other physical processes can be invoked as source of rotational splitting, but are not addressed here. Excepted for the lower frequency part of the p-mode spectrum, the separation of the splitted components is of the same order of magnitude as the

TABLE 1. Table of p-mode frequencies

n	l=0	l=1	l=2	l=3	l=4
07	1116.64±1.37				
08	1260.52±1.40	1329.22±0.48	1393.21±0.47		
09	1406.72±0.84	1472.65±0.16	1535.37±0.15		
10	1548.38±0.05	1612.78±0.19	1674.50±0.34		
11	1686.60±0.05	1749.23±0.04	1809.92±0.14	1865.07±0.94	
12	1822.21±0.06	1885.08±0.05	1945.71±0.11	2001.23±0.50	
13	1957.44±0.06	2020.85±0.04	2082.14±0.09	2137.96±0.23	
14	2093.55±0.04	2156.79±0.07	2217.97±0.14	2273.11±0.14	2321.28±1.00
15	2228.67±0.07	2291.77±0.08	2352.26±0.08	2407.56±0.42	
16	2362.70±0.08	2425.54±0.08	2485.94±0.11	2541.94±0.21	
17	2495.95±0.07	2559.22±0.07	2619.55±0.08	2676.19±0.30	
18	2629.84±0.07	2693.26±0.06	2754.51±0.06	2811.36±0.10	2864.33±0.45
19	2764.03±0.08	2828.03±0.08	2889.64±0.08	2946.86±0.22	3001.37±0.56
20	2898.80±0.06	2963.50±0.05	3024.75±0.12	3082.27±0.14	3137.78±0.47
21	3033.76±0.04	3098.10±0.08	3159.82±0.09	3217.74±0.15	3273.52±0.54
22	3168.61±0.06	3233.28±0.08	3295.35±0.14	3353.36±0.65	
23	3303.12±0.08	3368.59±0.09	3430.75±0.15	3490.28±0.44	
24	3438.88±0.12	3504.08±0.14	3566.71±0.30	3626.16±0.42	
25	3574.61±0.14	3639.93±0.24	3703.74±0.58	3763.10±2.10	
26	3710.64±0.98	3776.29±0.20	3839.16±1.17	3900.47±1.50	
27	3846.28±0.33	3913.53±0.28	3977.39±0.98	4036.00±2.00	
28	3983.56±0.73	4048.98±0.40	4114.14±1.49	4173.10±3.10	
29	4120.40±0.82	4185.73±0.44	4250.75±1.32	4309.82±3.10	
30	4258.08±1.88	4324.77±0.61	4389.03±1.27		
31	4395.44±1.13	4462.27±0.58			
32	4536.01±2.25	4600.83±1.04			
33	4675.00±3.00				

linewidth, and so its measurement is difficult to achieve with the very high level of accuracy required to invert the rotation deep in the solar sphere. It is mostly a matter of statistics: at the preliminary level of the present analysis, it is not attempted to estimate the splitting for individual modes and global methods have been used to improve the statistics. Those methods have been studied earlier for the analysis of the ground based experiments (Lazrek *et al.*, 1996). Tab. 2 shows the results, averaged over the broadest possible range of radial order for each degree, from $l=1$ to $l=3$. Although very preliminary, in term of statistical significance, it can be noticed that, when compared with the ground-based data, the GOLF data show a higher quality level.

TABLE 2. Rotational splitting of p modes

l=1	n=9-21	452 ±14 nHz
l=2	n=10-21	432 ±20 nHz
l=3	n=10-21	449 ±20 nHz

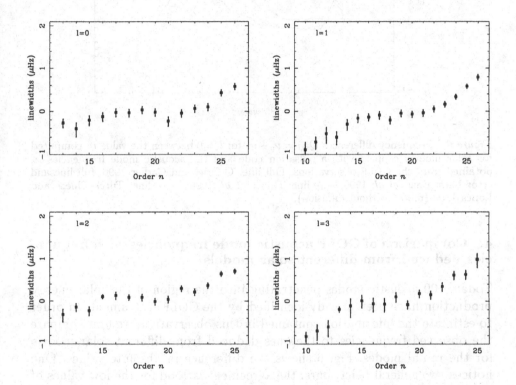

Figure 5. Width of p modes, for degrees $l = 0$ to $l = 3$.

3.3. P-MODE LINEWIDTHS

The maximum likelihood fits provide the linewidths, which have been assumed equal for all components of any multiplet. Fig. 5 shows these linewidths for degrees from $l = 0$ to $l = 3$. The general diagonal increase with frequency, as well as the flat or slightly decreasing part in the middle of the p-mode domain, are consistent with the well-known results obtained for intermediate degrees modes (Libbrecht, 1988).

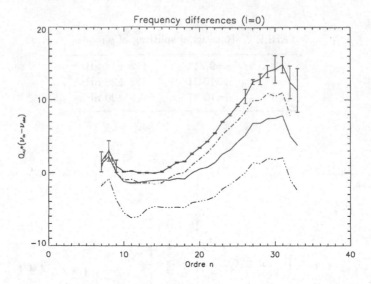

Figure 6. Frequency differences $Q_{n,l} = \nu_c - \nu_o$ for $l = 0$ between the value ν_c computed for solar models coupled with a pulsation code and the acoustic mode frequencies ν_o obtained from the GOLF observations. Full line: Gabriel and Carlier 1996, full line and error bars: Basu *et al.* 1996, $-.-$ line: Guzik *et al.* 1996, $-..-$ line: Turck-Chièze and Lopes 1993 (model without diffusion).

4. Comparison of GOLF acoustic mode frequencies with frequencies deduced from different solar models

Today, 100 acoustic modes penetrating into the region of the solar energy production have been already identified by the GOLF instrument. In order to estimate the information contained in this observation, we first compare the observed frequencies to the ones deduced from different solar models for the radial modes. Fig. 6 shows the difference in absolute values. One notices two general behaviours: the agreement is good for the low values of the radial order for the models including gravitational settling of elements along the solar life, at high frequencies the predictions slightly deviate from the observed values. One may notice that the models for which the slope is reduced, correspond to models which solve the structure equation above the effective radius without introducing a specific atmosphere. This point is illustrated in (Brun *et al.*, 1996) and (Turck-Chièze *et al.*, 1996) which shows that the origin of the discrepancy comes from the description of the very external layers located at ± 200 km around the effective radius. In this thin region of the Sun, different processes are in competition, the radiative and convective transport, the turbulent flows and the non adiabatic effects (Kosovishev, 1995), (Guzik *et al.*, 1996). The observed great difference, mainly sensitive to the size of the cavity and the partial ionization of light

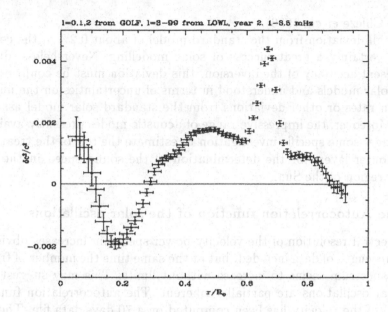

Figure 7. Sound speed difference between the function computed from the acoustic mode frequencies measured with GOLF and LOWL and the recent solar model of Basu *et al.* 1996. Private communication from S. Basu and J. Christensen-Dalsgaard.

elements, together with the observed small difference are well reproduced by the acoustic frequencies deduced from the standard model.

5. Determination of the solar internal sound speed from the GOLF instrument

As far as the nuclear core is concerned, the direct comparison of the frequencies observed and predicted from pulsation models is not very informative and one needs to perform an inversion of the sound speed from the acoustic mode frequencies to better estimate the information obtained by GOLF. The first inversion of the GOLF data (Basu and Christensen-Dalsgaard, 1996) is obtained by coupling these data to the LOWL data set of the second year of measurement from February 95 to February 96 (Tomczyk *et al.*, 1996), which is the most precise and complete list of acoustic mode frequencies presently available. For this inversion, the GOLF list has been limited to the interval from 1.5 mHz to 3.5 mHz, due to the limited available data series for $l > 3$. Fig. 7 shows a precise determination of the sound speed in the solar core. One may notice that the GOLF instrument is able to describe the solar interior with an accuracy in radius better than 0.02 $R\odot$ and a precision on the sound speed better than 0.05 %. We note a good continuity between the GOLF data and the LOWL data in frequency

(Turck-Chièze *et al.*, 1996).

The small deviation from the standard model of about 0.2% in the central core is certainly a great success of solar modelling. Nevertheless, due to the present accuracy of the inversion, this deviation must be confirmed by other solar models and understood in terms of uncertainties on the nuclear reaction rates or other deviations from the standard solar model assumptions. Moreover, the impressive range of acoustic modes presently available encourages some specific investigation to estimate the role of the treatment of the outer layers for the determination of the sound speed in the very central region of the Sun.

6. The Autocorrelation function of the solar oscillations

The spectral resolution of the velocity power spectrum increases obviously with the length of data included, but at the same time the number of fringes in the spectrum seems to increase without limits. This may suggest that the solar oscillations are partially coherent. The autocorrelation function (Fig. 8) of the velocity has been computed on a 70 days data file. The successive maxima of the AC, in the interval $0 < t < 100$ hours are due to the interferences of the p modes, roughly spaced from 68 μHz. For longer delays, the AC reaches a stable value, and this result is consistent with the hypothesis that about 20 % of energy in the p-mode spectrum is due to periodic components (Grec *et al.*, 1996).

To study the statistical life-time of the p modes, we use the more general convolution $B_T(\theta) = V(t) * (V(t)\Pi_T(t))$, where $V(t)$ is the velocity signal and the window $\Pi_T(x) = 1$ for $0 < x < T$ and $\Pi_T(x) = 0$ elsewhere.

Like the AC, the functions $B_T(\theta)$ contain oscillatory components. To compare the amplitudes for several values of T, we extract the absolute value of the envelope of each function, and then filter the result to remove the height frequencies ($\nu_{max} = 1/24h$), this final stage being used to get a readable plot. Fig. 8-b shows the envelope of $B_T(\theta)$ for several values of T. After the initial decrease for $0 < \theta < 100$ h the mean level for each value of T is $\simeq$ proportional to $1/\sqrt{T/15}$, where T is in days. We can then conclude that the p-mode lifetime is close to 15 days, each 15 days time series being independent from each other, value a bit longer than expected from the low frequency p-mode linewidths.

7. The search for g modes

The search for g modes is a major objective of the GOLF experiment. Fig. 9 shows the low frequency part of the velocity spectrum, where a wide number of sharp lines are visible. We need an "Ariane line" to identify the g modes. We made several tests using an asymptotic approximation to

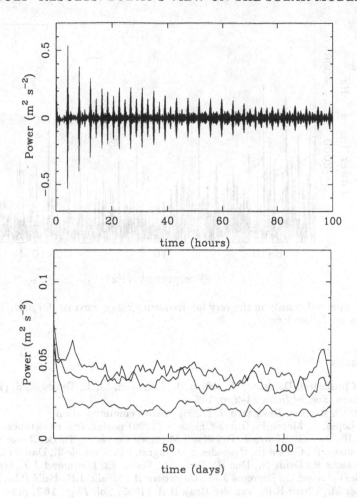

Figure 8. Autocorrelation function of the solar oscillations. Top: the first 100 hours of the AC. Down: the filtered envelope of the amplitude of the B function (see text), from top to bottom for $T = 10$ days, $T = 20$ days, $T = 80$ days.

compute a frequency table (Provost, 1986): changing the model parameters, we can tune the model up to a good correlation level, which may suggest the existence of g modes. Moreover, with the present data, this method seems unable to define a unique set of parameters. A second analysis focuses on the search for modes splitting. No g-mode identification has been achieved. Are the g modes detectable, or even present in the spectrum? The future may give the answers. The analysis of the variations of the instrumental parameters shows no significant signal source at the frequencies observed in the velocity spectrum, in which many low frequency peaks show a long coherence time.

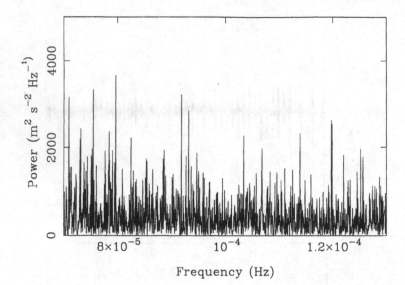

Figure 9. Spectral density in the very low frequency range, next to 100 μHz. The highest amplitudes are below 1 cm/s.

References

Basu, S., Christensen-Dalsgaard J., Schou, J., Thomson, M. J., Tomczyk, S. (1996) *Bull. of Astron. Soc. of India* **24-2**, p. 147.
Basu, S. & Christensen-Dalsgaard J. (1996) private communication.
Brun, S., Lopes, I., Morel, P., Turck-Chièze S. (1996) poster, this conference.
Gabriel A.H., Grec G., Charra J., Robillot J.M., Roca Cortés T., Turck-Chièze S., Bocchia R., Boumier P., Cantin E., Cespédes E., Cougrand B., Crétolle J., Damé L., Decaudin M., Delache P., Denis N., Duc R., Dziko H., Fossat E., Fourmond J.J., Garcia R.A., Gough D., Grivel C., Herreros J.M., Lagardère H., Moalic J.P., Pallé P.L., Pétrou N., Sanchez M., Ulrich R.K., van der Raay H.B. (1995) *Sol. Phys.* **162**, p. 61.
Gabriel, M. & Carlier, F. (1996) *Astron. Astrophys.* to appear.
Grec G., Gabriel M., Renaud C. & the GOLF team (1996) poster, this conference.
Guzik, J., Cox, A.N., Swenson, F.J. (1996) *Bull. of Astron. Soc. of India* **24-2**, p. 161.
Kosovishev, S. (1995) *Fourth SOHO workshop* ESA SP-376 vol. 1, p. 165.
Lazrek, M., Pantel, A., Fossat, E., Gelly, B., Schmider, F.X., Fierry-Fraillon, D., Grec, G., Loudagh, S., Ehgamberdiev, S., Khamitov, I., Hoeksema, T., Pallé, P.L., Régulo, C. (1996) *Sol. Phys.* **166-1**, p. 1.
Lazrek M., Régulo C. R., Baudin F., Bertello L., Garcia R. A., Gouiffes C., Grec G., Roca Cortès T., Turck-Chièze S., Ulrich R. K., Robillot J. M., Gabriel A. H., Boumier P., Charra J. & the GOLF team (1996) poster, this conference.
Libbrecht, K.G. (1988) ESA SP 286, p. 3.
Provost, J. & Berthomieu, G. (1986) *Astron. Astrophys.* **165**, p. 218.
Tomczyk, S., Streander, K., Card, G., Elmore, D., Hull, H., Cacciani, A. (1995) *Sol. Phys.* **159-1**, p. 1.
Turck-Chièze, S. & Lopes, I. (1993) *Astrophys. J.* **408-1** , p. 347.
Turck-Chièze S., Basu, S. Brun S., Christensen-Dalsgaard J., Eff-Darwich A., Gabriel M., Henney C. J., Kosovichev A., Lopes I., Paternò L., Provost J., Ulrich R. K. & the GOLF team (1996) poster, this conference.

PRECISION SOLAR ASTROMETRY FROM SOHO/MDI

J.R. KUHN
National Solar Observatory/Sac Peak
and Michigan State University
P.O. Box 5, Sunspot NM 88349
and Dept. Physics-Astronomy, MSU, E. Lansing, MI 48824,
USA

R. BOGART, R. BUSH, L. SÁ AND P. SCHERRER
Stanford University, CSSA-HEPL, Stanford CA 94305-4055,
USA

AND

X. SCHEICK
Michigan State University, E. Lansing MI, 48824, USA

Abstract. The SoHO/MDI experiment generates a continous record of the solar limb brightness using 1.96" pixels. Because there is no atmospheric blurring, these data allow measurements of solar limb brightness and shape changes with a precision that has not been achieved from the ground. The first results of a 1 month astrometric timeseries from the MDI solar structure program will be described here.

1. Introduction

The solar limb is, potentially, a sharp spatial fiducial reference with which we can hope to detect the effects of: solar oscillations (both p- and g-modes), the gravitational quadrupole moment (or the solar oblateness), and changes in the solar radius (perhaps as a diagnostic for the solar luminosity cycle). Ground-based attempts to measure a changing radius or limb shape have a long history but, unfortunately, corresponding observations from space are virtually nonexistent.

Hill's group (cf. Brown *et al.* 1978) pioneered modern astrometric attempts to measure solar oscillations, although their conclusions were never reproduced. Later astrometric measurements from limb photometry failed

103

J. Provost and F.-X. Schmider (eds.), Sounding Solar and Stellar Interiors, 103-110.
© 1997 IAU. Printed in the Netherlands.

to detect solar g-modes, but did produce interesting upper limits to the possible amplitudes of such modes (cf. Kuhn *et al.* 1986).

A recent review of the solar oblateness problem (Sofia *et al.* 1994) reveals that the ground and balloon-based measurements are not consistent. While Dicke *et al.* (1987) hinted that the sun may be changing its shape with the solar cycle, the recent balloon data do not confirm this. Since atmospheric blurring is orders of magnitude larger than the ultimate accuracy required, these measurements are quite difficult and hours or days of observations (typically) are needed to achieve milliarcsecond accuracy in oblateness measurements from the ground. It seems likely that the apparent differences between many of these experiments are dominated by poorly understood systematic errors.

The solar radius measurements are also in disarray. For example, Ribes *et al.* (1991) and Parkinson *et al.* (1980) describe the inconsistency between the many ground-based astrometric measurements of the sun's radius. While there are hints of periodic solar radius variations over timescales of 1000 days to 80 years, the measurements are generally neither consistent, nor conclusive. The evidence for long term secular variations is similarly inconclusive.

The solar limb position varies at physically interesing angular scales of milliarcseconds (the solar shape) and microarcseconds (due to oscillations). Since atmospheric "angle of arrival" fluctuations due to turbulence occur at a scale of an arcsecond, the advantage of space-based astrometric observations is obvious. Until SoHO, and except for a relative few continuum solar images from the Yohkoh experiment, there have been no space experiments that could provide interesting full-disk solar astrometric data.

2. Data Analysis

The MDI experiment (Scherrer *et al.* 1995) was primarily designed to obtain full-disk and higher resolution doppler "images" of the sun for helioseismic analysis. In full-disk mode, images are formed on a 1024×1024 pixel CCD at a scale of about 1.96 arcseconds/pixel. Doppler and magnetic information is computed from five 94mA passbands, spaced 75mA apart near the 6768 A Ni I line. A continuum image is computed from an algebraic combination of the five filtergrams in an expression that minimizes the "leakage" of the doppler signal (Scherrer *et al.* 1995). The ray path through the instrument traverses literally dozens of optical elements so that flat-field calibration is essential. Spatially displaced images were obtained early in the mission to allow flat-fielding by using the Kuhn *et al.* (1991) technique. Slow drifts in the optical elements produce a temporally increasing, but low spatial frequency flat-fielding error. The differential pixel-to-pixel intensity calibration is estimated to be accurate to about 1% .

Without atmospheric blurring the dominant error in the limb position measurements result from optical aberrations in the instrument and un-calibrated gain variations in the instrument-detector system. The latter effect is minimized by applying a flat-fielding technique that derives the calibration from the data to be calibrated. The former error is minimized because the spacecraft is in a relatively stable environment and most of the optical aberrations are stable, at least over the timescales in which we ob-serve a temporally variable solar limb position. We'll discuss the magnitude of the fixed instrumental aberration problem below in the context of the oblateness measurements. From a least-squares analysis of the solar limb profile, it can be shown then that the dominant time-variable astromet-ric error, corresponding to the 1% flat-fielding uncertainty, is of order 0.01 pixel (about 20 milliarcsec) in the derived limb position. Since there are many independent limb pixels in a single measurement, and a possibility of many independent measurements, the statistical error in the solar shape or the limb position spectrum can be quite small.

Image data from MDI are obtained with a 1 min cadence. Except dur-ing prearranged "campaigns" there is insufficient downlink bandwidth to return all of these data to the ground station. An annulus of approximately 6 pixels around the solar limb is temporally averaged with a 24 min Gaus-sian weighting, and sampled and downloaded every 12 minutes. The data described here were taken after May 1, 1996 during a 34 day period. An 8 hour full-disk intensity dataset was also obtained during a campaign in March. These data were sampled with a 1 min cadence and were reduced to limb observations in the same manner as the 12 min continuous limb observations.

There are approximately 20,000 pixels from each image near the limb which are assigned to radial and angular bins. The image center in pixel units is computed using a robust iterative least-squares technique (Mc-Williams and Kuhn 1992) which repeatably finds the solar image center to about 0.01 pix. Each limb intensity measurement is assigned to one of 40 radial, r_i, and 512 angular, θ_j, bins as $I(r_i, \theta_j)$. An angular average limb profile $I_0(r)$ is also computed for each dataset. A least-squares fit of the data in each angular bin of the form $I(r_i, \theta_j) \rightarrow (1 + \alpha_j)I_0(r_i - \beta_j)$ is performed. From this calculation two parameter sets, α_j and β_j, describe the brightness deviation and spatial limb shift in each of 512 angular bins around the solar image. From a timeseries of limb data we thus derive two-dimensional datasets, $\alpha_j(t_k)$ and $\beta_j(t_k)$, that describe the limb brightness changes and angular limb shift at times, t_k. Because of limited space in this report, the following discussion focuses on the angular limb position timeseries.

3. Observing Limb p-modes and Searching for g-modes

We have used 443 observations with a 1 min cadence to "calibrate" our technique by looking for p-modes in the $\beta_j(t_k)$ timeseries. For this purpose the 2-dimensional limb displacement power spectrum is computed as $F(K_j, \omega_k) = \| \sum_{rs} \beta_r(t_s) \exp it_s\omega_k \exp iK_j\theta_r \|^2$.

The photospheric radial velocity of solar acoustic p-modes are most naturally spatially decomposed into spherical harmonics, $Y_{lm}(\theta, \phi)$, where θ and ϕ correspond to heliocentric colatitude and longitude coordinates. It can be shown that for spherical harmonics l greater than about 40 that a given acoustic mode specified by lm projects, primarily, onto "limb harmonic" $K_j = l$ and $K_j = l+2$. Thus, the ridge structure of the 2-dimensional $k - \omega$ acoustic doppler spectrum should be visible in the $F(K, \omega)$ limb spectrum – which was obtained from fundamentally *one-dimensional* spatial data. Figure 1 shows a greyscale image of $log_{10}[F(K_j, \omega_k)]$ where the limb shape power is plotted in units of pixel^2. For 5-min period acoustic modes with a velocity amplitude of 15 cm/s, the shape amplitude will be about 10^{-5} arcsec or 5 micropixels (corresponding to a power of about 3×10^{-11} in the units of Fig. 1. The excess power and p-mode ridge structure at ω =3mHz is quite clear and the amplitude is consistent with known 5-min period radial velocity amplitudes.

The linear properties of gravity modes have been well studied (cf. Berthomieu and Provost 1990) although they have never been conclusively detected. Observational limits on g-mode velocity amplitudes with frequencies less than 100 microhertz are a few cm/s (Garcia *et al.* 1988). Limits on $l = 2$ modes from limb astrometry (oblateness) observations are at a similar level (Kuhn *et al.* 1986). Some speculative calculations (cf. Kumar *et al.* 1996) suggest that photospheric g-mode velocity amplitudes might only be at the level of 1 mm/s or smaller.

Because solar g-modes have small surface velocity amplitudes and low frequencies, there is an advantage to using astrometric techniques, rather than doppler measurements, to search for them. A fixed velocity amplitude results in an increasing shape amplitude for modes of lower frequency. Similar to the p-mode analysis above, we have analysed 34 days of 12-min cadence limb data. The spectral analysis is sensitive to shape oscillations with frequencies between a few and a few hundred microhertz. Figure 2 shows the mean noise power averaged over $l = 2 - 509$. The strong peaks near periods of 1 hour are harmonics of the 96min periodic interruption of the data when MDI obtains magnetograms. During this time the temperature of the MDI optics changes slightly in a manner which affects the image at the level of 10's of micropixels. A "shoulder" in the power spectrum near periods of 10 hours is significant and appears to be caused by the sun.

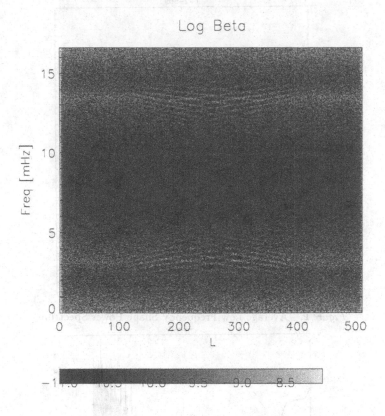

Figure 1. Limb Shape Power Spectrum

Figure 3 shows the averaged shape power for the low l modes expressed in terms of the effective velocity amplitude (in units of mm/s). The shape noise power approaches a few mm/s in these data with no outstanding evidence of g-mode excitation.

The low frequency limb position data has also been analysed by limb angle to determine the power spectrum of the limb displacement as a function of solar latitude. This analysis sheds some light on the origin of the shoulder in the power spectrum in Fig. 2. We find maxima in the limb position power spectra which vary with latitude. At high latitutes the peak occurs near zero frequency and at the equator the peak occurs at a frequency of about 25 microhertz. The amplitude of the peak corresponds to a limb displacement of about 0.8km, and the shape of the spectrum (with latitude) suggests a rotating surface distortion pattern (similar to the supergranulation) with a transverse scale at all latitudes of about 6×10^4km. This unusual signal clearly requires additional study.

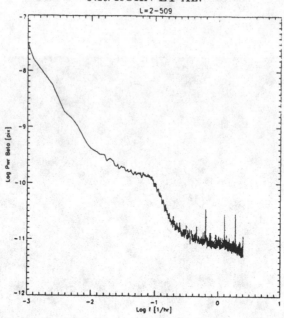

Figure 2.   Average Low Frequency Shape Power Spectrum

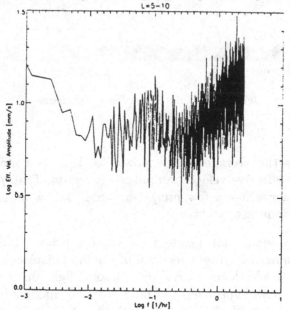

Figure 3.   Low l Effective Shape Velocity Noise

4. The Solar Oblateness and Radius

By rolling the spacecraft the static solar shape can be extracted from the much larger limb distortion caused by the complex MDI optics. If the limb

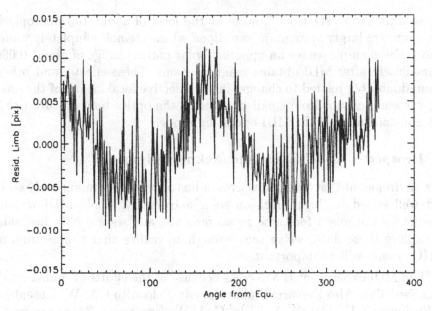

Figure 4. Residual Solar Limb Shape

position as a function of camera-fixed angle θ (and for a given roll angle θ_i) is $d_i(\theta) = n(\theta) + s(\theta + \theta_i)$ then $d_i(\theta)$ is the observed limb shape, $n(\theta)$ is the limb shape due to optical or other instrumental effects, and $s(\theta)$ is the actual solar limb shape. We have analysed only a few minutes of MDI data from roll angles of 0, 30, 60, 90, 137, 180, and 270 degrees. From the data taken at angles spaced by 90 degrees the solar and instrumental contributions can be separated. Thus, it can be shown that, $n(\theta) = 0.25(d_0 + d_{90} + d_{180} + d_{270})$ is accurate for spatial harmonics through the quadrupole distortion. The intrinsic solar limb shape can then be computed from $d_i(\theta - \theta_i) - n(\theta - \theta_i) = s(\theta)$ and averaged over data obtained at all roll angles. Figure 4 shows the solar limb shape, $s(\theta)$, where the solar equator corresponds to the angular origin and the limb position is plotted in pixel units. For comparison, the instrumental contribution to the limb position, $n(\theta)$, has an amplitude of about 0.2 pixels. The function $s(\theta)$ corresponds to a solar oblateness of $(r_{eq} - r_{pole})/r_{eq} = 9.6 \pm 0.8 \times 10^{-6}$. An optimum roll sequence and a few more minutes of data should easily decrease the measurement uncertainties here. Nevertheless this measurement is several times more accurate than the ground-based observations, and suggests that the sun's shape does not change significantly with the solar cycle.

The solar radius has also been determined from the mean radial limb profiles from each limb dataset. A cross-correlation analysis of the radial derivative in the limb profiles has proven to be a robust radius estimator.

These data show "statistical" noise at the level of about 100 micropixels, but there are larger systematic variations which are not completely understood. For example we see an apparent solar radius change of about 0.006" immediately after MDI obtains magnetograms. These effects, and others, are undoubtably related to changes in the effective focal length of the imaging system, perhaps from small changes in the optics temperature due to the non-constancy of the MDI operating mode.

5. Prospects, Conclusions and Acknowledgments

The environment that SoHO now lives in has proven to be remarkably stable and well suited for high precision solar astrometry. While MDI was not specifically optimised for these measurements, and we are only beginning to explore these data, we've seen enough to realize that the astrometric SoHO results will be important.

The MDI experiment is a success because of the efforts of many. At the Lockheed Palo Alto Research Laboratory in Palo Alto CA, W. Rosenberg, L. Springer, T.D. Tarbell, A. Title, C. J. Wolfson and I. Zayer are particularly responsible. Several others (largely responsible for the development of the SOI Science Support Center) have been critical to making this research possible: J. Aloise, L. Bacon, K. Leibrand, V. Johnson, K. Scott, and J. Suryanarayanan. The MDI engineering team consisted of D. Akin, B. Carvalho, R. Chevalier, D. Duncan, C. Edwards, N. Katz, M. Levay, R. Lindgren, D. Mathur, S. Morrison, T. Pope, R. Rehse, and D. Torgerson. Finally, Peter Milford helped with early attempts to measure the solar limb position from Yohkoh data while we were developing the software now used.

References

Berthomieu, G., Provost, J. (1990) *Astron. & Astrophys.*, **227**, 563.
Brown, T.M., Stebbins, R.T., and Hill, H.A. (1978) *Ap. J.*, **223**, 324.
Dicke, R. H. (1981) *Proc. Natl. Acad. Sci*, **78**, 1309.
Dicke, R. H., Kuhn, J.R., and Libbrecht, K.G. (1987) *Ap. J.*, **318**, 451.
Garcia, C., Pallé, P.L., and Cortes, T. (1988) *Seismology of the Sun and Sun-like Stars (ESA SP-286)*, ed. E. J. Rolfe (Paris: ESA), 353.
Kuhn, J.R., Lin, H., and Loranz, D. (1991) *Publ. Astron. Soc. Pac.*, **103**, 1097.
Kuhn, J.R., Dicke, R.H., Libbrecht, K.G. (1986) *Nature*, **319**, 128.
Kumar, P., Quataert, E., Bahcall, J.N. (1996) *Ap.J.*,**458**,83.
McWilliams, T, and Kuhn, J.R. (1992) *SOI Technical Note 83* Stanford, CA.
Parkinson, J.H., Morrison, L.V., Stephenson, F.R. (1980) *Nature*, **288**, 548.
Ribes, E., Beardsley, B., Brown, T.M., Delache, Ph., Laclare, F., Kuhn, J.R., Leister, N.V. (1991) *The Sun in Time* (Sonett, C., Giampapa, M., Matthews, M., eds.) Univ. Ariz. Press, Tucson, 59.
Scherrer, P.H. *et al.* (1995) *Solar Phys.*, **162**, 129.
Sofia, S., Heaps, W., Twigg, L.W. (1994) *Ap. J. Suppl.*, **427**, 1048.

PLANE-WAVE ANALYSIS OF SOI DATA

R. S. BOGART AND L. A. DISCHER DE SÁ
Stanford University

I. GONZÁLEZ HERNÁNDEZ AND J. PATRÓN RECIO
Instituto de Astrofísica de Canarias

D. A. HABER AND J. TOOMRE
JILA

F. HILL
National Solar Observatory

E. J. RHODES, JR. AND Y. XUE
University of Southern California

AND

THE SOI RING DIAGRAMS TEAM

Abstract.
The unprecedented combination of spatial resolution and stability achieved by the Solar Oscillations Investigation/Michelson Doppler Imager on SOHO has opened up new opportunities for the analysis of solar surface oscillations of high spatial frequencies. In this regime the oscillations are essentially plane waves, amenable to the techniques of ring-diagram analysis of their three-dimensional power spectra. This approach holds the promise of measuring fluid motions and possibly magnetic fields in spatially-resolved structures within the uppermost levels of the convective envelope, a region unresolved by the global modes. Atmospheric g-modes trapped above the photosphere may also be detectable. We review the first results of plane-wave analysis of various types of SOI data and comparisons with the analyses of comparable ground-based datasets.

J. Provost and F.-X. Schmider (eds.), Sounding Solar and Stellar Interiors, 111-118.
© *1997 IAU. Printed in the Netherlands.*

1. Introduction

A fundamental goal of local-area helioseismology is to subject regions smaller than the entire visible part of the Sun to analysis for any peculiar velocity fields, radial structure variations, and strong magnetic fields that may exist at certain times and differentiate them from other such regions. The most straight-forward technique is to apply what was historically the earliest helioseismic method, plane-wave analysis, to suitably remapped and tracked observations. The approach has been described and elaborated elsewhere: see for example Hill(1988) and Patrón(1994). The principle justification for plane-wave analysis of eigenmodes trapped in spherically-symmetric cavities is that the regions considered are so small that the curvature of the cavity is negligible and the waves observed are of such short lifetimes that a spherical closure condition scarcely applies. The numerical simplicity of a three-dimensional Fourier transform strongly recommends itself over spherical harmonic decomposition when the effective spherical harmonic degree of the eigenmodes may be well over 1000 and the region sampled less than 1% of the surface of the sphere.

It has been demonstrated that three-dimensional plane-wave analysis can be applied to observations of solar oscillations obtained with high-resolution instruments on the ground (Hill, 1988; Patrón, 1994; González et al., 1995). Inversion of these results for the sub-photospheric velocity fields yields at least plausible pictures of flows in the upper convective envelope, a region inadequately resolved by global-mode helioseismology. Long-period ground-based observations are severely affected by atmospheric seeing on the scale of the wavelengths of the modes of interest, however. In order to verify that ground-based high-resolution oscillation measurements are not systematically biased or distorted by the observing conditions, it is essential to observe the same regions at the same time from several instruments and ideally from space. In the absence of significant pointing jitter and drift, space-based observations will also allow us to begin to construct synoptic views of the Sun, by obtaining data for long periods under virtually unchanging conditions, and to extend the observations to even higher spatial frequencies. Since the commencement of its scientific observing programme in April 1996 the Solar Oscillations Investigation/Michelson Doppler Imager (SOI/MDI) on Soho has been providing a wealth of data ideally suited to these tasks. The development and application of plane-wave analysis techniques has already proceeded from various instrument commissioning observations (Bogart et al., 1995b). Here we summarize the results of the first analyses of calibrated data sequences from SOI/MDI and the comparisons with concurrent ground data.

2. Analysis Techniques

The principle diagnostic of plane-wave analysis thus far is the horizontal flow field, manifested by dipole terms in the expansions of the eigenfrequencies $\omega_n(\theta)$ for a given k. (For a uniform transverse motion of the observer, advection of the wave fronts leads to a shift in the measured mode frequencies of $\Delta\omega_n(k) = v \cdot k$) To remove the 2 km/s signature of solar rotation and to follow identified photospheric features, we select small regions centered at selected Carrington coordinates at a given time. The regions are tracked at the rate of photospheric differential rotation appropriate for the latitudes of their central points (Snodgrass, 1984).

The selected regions are mapped from the plate coordinates via Postel's projection, a standard cartographic projection in which path lengths along lines through the center of the map are proportional to lengths along the corresponding geodesics. Since the regions selected for analysis are small, typically about 15° in diameter, the choice of projection is not critical. The map scale is selected to preserve the instrumental resolution at the center of the disc: for MDI full-disc images this is 1.978 arc-sec per pixel, corresponding to a map scale of 0°.12 heliographic per pixel. The resulting maps of the line-of-sight Doppler velocities (or other observed parameters) are apodized in a circular region using an exponential taper with a characteristic thickness of 3.5 pixels to equalize the spatial frequency resolution in all directions.

In order to improve the convergence of some of the fitting procedures, the tracked and mapped data are generally high-pass filtered in the time domain using a gaussian window of full-width-half-maximum 21 minutes (0.8 mHz), after which the three-dimensional power spectrum is computed. A sample power spectrum of a 2-day interval of SOI/MDI Doppler data is shown in Figure 1. The eigenmodes form the characteristic set of nested trumpet surfaces, the outermost being the lowest order modes.

There are two basic approaches to analyzing the power spectrum. In the standard technique, described by Hill(1988) and substantially improved by Haber et al.(1995) and Patrón et al.(1996), the variations are analyzed along cuts in temporal frequency, where the eigenfrequencies appear as sets of nearly circular concentric rings (hence the name "ring-diagram analysis"). Directional dependencies of the sound speed and/or frequency shift show up as asymmetries in the rings; these asymmetries may be parametrized and measured for particular eigenmodes as functions of frequency. An alternative approach, suggested by Bogart et al.(1995a), is to cut the power spectrum along sets of concentric cylinders in the two-dimensional spatial frequency space. Along such cylinders the frequency shifts of particular eigenmodes depend on the azimuthal angle; they can be Fourier-decomposed

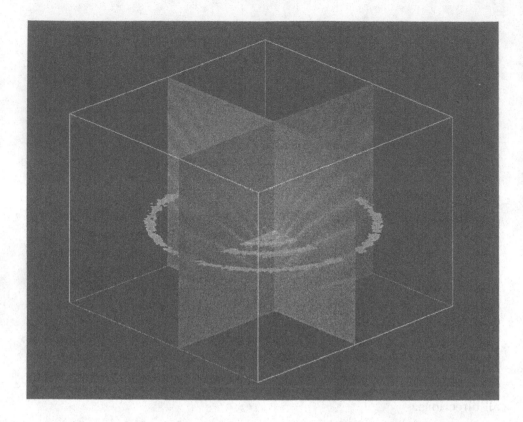

Figure 1.   Three-dimensional Doppler power spectrum from SOI/MDI. The region observed is concentric and coterminous with region F, but of twice the diameter (region F'). Power is shown projected along the horizontal axes, corresponding to orthogonal components of the horizontal wavenumber, and on a single cut in the vertical (temporal frequency) axis.

to isolate the first-degree advection terms from the second-degree sound speed variation terms. The two approaches are obviously complementary, the first having its greatest resolving power for lower spatial frequencies (higher-order modes) and the second for higher spatial frequencies (lower-order modes). Considerable effort has been made to refine and compare these approaches with the aim of systematizing plane-wave analysis in future. The results presented at this meeting rely predominantly on the standard ring-fitting approach, but have been corroborated in a few cases with the Fourier approach.

To infer the depth dependence of the horizontal velocity field from the first-order terms in the frequency shifts, we perform a least-squares inversion with second derivative smoothing, as described by Patrón(1994). This inversion procedure allows us to tune the trade-off parameter between resolution and error magnification, an essential step because of the current uncertainties in our error estimation procedures.

3. Data Analyzed

The various regions analyzed are summarized in Table 1. All are quiet sun regions except for regions B and C, which cover a predominantly unipolar area of a decaying active region and a sunspot, respectively. The latitude and longitude of disc center (b_0, l_0) are nominal, referred to the center of the tracking interval. In all cases the regions analyzed were circles on the photosphere of diameter $15°.36$ heliographic or 186.6 Mm.

TABLE 1. Regions Selected for Analysis

	Start Time	End Time	b_C	l_C	b_0	l_0
A	96.06.06 14:17:30	96.06.07 07:20:30	3.04	291.0	0.04	287.8
B	96.06.06 14:17:30	96.06.07 07:20:30	-4.96	268.2	0.04	287.8
C	96.06.06 14:17:30	96.06.07 07:20:30	5.04	305.8	0.04	287.8
D	96.06.02 12:59:30	96.06.03 06:02:30	-0.45	341.4	-0.45	341.4
E	96.06.02 12:59:30	96.06.03 06:02:30	-0.45	26.4	-0.45	341.4
F	96.06.01 06:09:30	96.06.03 06:44:30	-0.53	349.7	-0.53	349.7

Doppler data were extracted for all regions. In addition, line-depth (equivalent-width) measurements exist for the same intervals, but these have not yet been analyzed. Concurrent ground-based observations have been obtained during the selected intervals by various combinations of the Taiwan Oscillations Network (TON), the High-l Helioseismometer at Kitt Peak, and the High-Degree Helioseismology Network (Mt. Wilson). So far, the only detailed comparison made has been with a 512-minute subsequence of region F (González Hernández et al., 1996).

4. Results

The inversions for the first-order terms in the expansion of the local frequency shifts, sensitive to the horizontal velocity profile, are in general agreement with those seen before in ground data, at least for quiet sun regions. The detailed comparison with the results from a set of concurrent ground-based obsevations, reported elsewhere in these proceedings (González Hernández et al., 1996), is quite close. The results for regions of moderate magnetic activity, also reported elsewhere (Haber et al., 1996) are less clear. There is apparent spatial asymmetry in the observed power

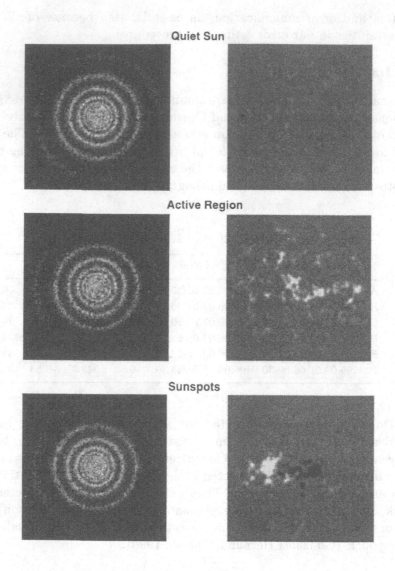

Figure 2.   Ring diagrams and the corresponding magnetograms for regions A–C.

distribution in the ring diagrams for the magnetically active regions in Figure 2, but similar effects are seen when quiet-sun observations at different distances from disc center are compared (regions D & E). Some of this effect is undoubtedly due to foreshortening, and the extent to which it affects measurement of the ring-shift parameters is uncertain.

Attempts to measure the second-order terms, sensitive to directional

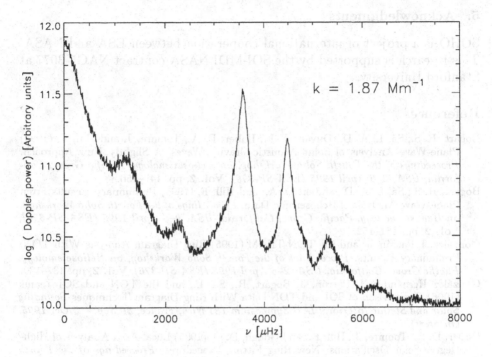

Figure 3. Temporal frequency spectrum of the azimuthally averaged Doppler power for region F′ for a value of k equal to 1.87 Mm^{-1}. No temporal filtering was performed.

sound speed variations and magnetic field effects, are also reported elsewhere here (Haber *et al.*, 1996). There is no conclusive detection of such effects in the samples studied.

At very high spatial frequencies we observe spectral features at temporal frequencies below those of the acoustic waves as the period of the f-mode decreases to below the 5-minute band. In the azimuthally-averaged power spectrum shown in Figure 3, corresponding to an effective degree of $l \approx 1300$ for global modes, two such features are clearly visible. There are hints in the preliminary analysis that the higher-frequency feature, here appearing at about 2.3 mHz, is dispersive; it might be a signature of atmospheric g-modes. The other feature, at a frequency of about 1.2 mHz ($P \approx 14$ min) nearly independent of k, seems to vary in strength significantly with magnetic activity, being enhanced by a factor of roughly 30 in regions B & C over region A. Formal identification of these features is premature, but there is at least a strong hint that we may be beginning to systematically sound not just the sun's interior, but its atmosphere as well.

5. Acknowledgments

SOHO is a project of international cooperation between ESA and NASA. This research is supported by the SOI-MDI NASA contract NAG5-3077 at Stanford University.

References

Bogart, R. S., Sá, L. A. D., Duvall, T. L., Haber, D. A., Toomre, J. and Hill, F. (1995) Plane-Wave Analysis of Solar Acoustic-Gravity Waves: A Slightly New Approach. *Proceedings of the Fourth Soho Workshop, on Helioseismology, Pacific Grove, California, USA, 2–6 April 1995 (ESA SP-376)*, **Vol. 2**, pp. 147–150.

Bogart, R. S., Sá, L. A. D., Haber, D. A., and Hill, F. (1995) Preliminary Results From Plane-Wave Analysis of Helioseismic Data. *Proceedings of the Fourth Soho Workshop, on Helioseismology, Pacific Grove, California, USA, 2–6 April 1995 (ESA SP-376)*, **Vol. 2**, pp. 151–152.

González, I, Patrón, J, and the TON TEAM (1995) Ring Diagram Analysis With TON: Preliminary Results. *Proceedings of the Fourth Soho Workshop, on Helioseismology, Pacific Grove, California, USA, 2–6 April 1995 (ESA SP-376)*, **Vol. 2**, pp. 137–139.

González Hernández, I, Patrón, J, Bogart, R., Sá, L., and the TON and SOI Teams (1996) Comparison of SOI and TON Data With Ring Diagram Techniques *Sounding Solar and Stellar Interiors: IAU Symposium 181 Nice, France, 30 Sep. – 3 Oct 1995*. (*in press*).

Haber, D. A., Toomre, J., Hill, F., and Gough, D. O. (1995) Local-Area Analysis of High-Degree Solar Oscillations: New Ring Fitting Procedures. *Proceedings of the Fourth Soho Workshop, on Helioseismology, Pacific Grove, California, USA, 2–6 April 1995 (ESA SP-376)*, **Vol. 2**, pp. 141–146.

Haber, D., Zweibel, E., Toomre, J., Bogart, R. S., Sá, L., Burnette, A., Xue, Y., Rhodes, E., and Hill, F. (1996) Possible magnetic field effects seen with local-area analysis of MDI and Mount Wilson data. *Sounding Solar and Stellar Interiors: IAU Symposium 181 Nice, France, 30 Sep. – 3 Oct 1995*. (*in press*).

Hill, F. (1988) Rings and Trumpets— Three Dimensional Power Spectra of Solar Oscillations. *Ap.J.*, **333**, pp. 996–1013.

Patrón Recio, J. (1994) *Tridimensional Distribution of Horizontal Velocity Flows Under the Solar Surface.* Doctoral Dissertation, Departamento de Astrofisica, Universidad de La Laguna.

Patrón, J, González Hernandez, I, and the TON TEAM (1995) Comparisons on fitting ring diagrams. *Sounding Solar and Stellar Interiors: IAU Symposium 181 Nice, France, 30 Sep. – 3 Oct 1995*. (*in press*).

Snodgrass, H.B. (1984) Separation of Large-Scale Photospheric Doppler Patterns. *Solar Phys.*, **94**, pp. 13–31.

INTERNAL STRUCTURE AND ROTATION:
SEISMIC INVERSIONS

THEORETICAL SOLAR MODELS

J. PROVOST

Département Cassini, UMR CNRS 6529, OCA
BP 229, 06304 Nice CEDEX 4, France

Abstract. In the last two decades the large amount of accurate helioseismic data has open a new possibility to sound the stellar interiors and to test the stellar evolution theory. This has particularly impulsed a critical reexamination of the basic assumptions and physical inputs in solar modelling. The present status of theoretical solar models is presented and discussed in relation with the observational constraints: helioseismology, lithium depletion and measured neutrino fluxes.

1. Introduction

The Sun is a normal star of one solar mass which is burning its central hydrogen. So theoretical solar models are obtained using the theory of stellar evolution which attempts to describe the evolution of the structure of a star during its life. A success of this theory was to explain the statistical properties of stars like the accumulation of stars along the so called main sequence in the Herzsprung-Russell diagram relating the stellar luminosity and effective temperature. But up to recently this theory remained not much constrained. In the last 30 years two possibilities to test the solar interior appeared: solar neutrino measurements and helioseismic observations.

The results of neutrino experiments have much questioned the validity of solar models hence that of the stellar evolution theory. This theory relies on simplifying basic assumptions. The star, hence the Sun, is assumed spherically symmetric: rotation, magnetic field, macroscopic motions are ignored. The Sun is in hydrostatic equilibrium, i.e. pressure balances gravity. It is in thermal equilibrium: the energy loss at the surface by radiation is equal to the energy produced in the core. The energy source is nuclear reactions, mainly the hydrogen burning in the solar case. The tempera-

121

J. Provost and F.-X. Schmider (eds.), Sounding Solar and Stellar Interiors, 121-136.

ture gradient along the radius results from the energy transport, assumed to be stationary, from the center to the surface. This transport occurs by means of radiation except in the convectively unstable zone, which for the Sun extends on the outer 30% of the radius . The radiative transport is controlled by the atomic absorption coefficients which together with the chemical composition determine the opacity of the stellar material. In convection zone the energy transport is modeled by the so called mixing length theory of the convection. It introduces a free parameter, α that measures the convection efficiency.

The solar evolution results from changes of chemical composition, mainly due to the fusion of hydrogen into helium. The composition changes also in the radiative zones of the Sun, due to the element diffusion which is driven by the gradients of the structure. The evolution is started from a chemically homogeneous model either during the pre main sequence contraction phase or on the zero age main sequence. The initial abundances are assumed those observed at the photosphere. Mixing processes outside the convectively unstable zone, which is assumed perfectly mixed, are generally ignored. In the case of the Sun the knowledge of the luminosity, radius and age allows to obtain with the above assumptions a theoretical solar model.

The large amount of very accurate helioseismic data has open a new possibility to sound the solar and stellar interiors. This has impulsed a critical reexamination of the basic assumptions and of the physical inputs in solar and stellar modelling. A lot of work has been made specially on the sensitivity to the physics of the solar structure, oscillations and neutrino fluxes. This makes impossible to give an exhaustive presentation. Many complete reviews on the stellar and solar modelling have been presented recently (Turck-Chièze et al. 1993, Berthomieu 1996, Christensen-Dalsgaard et al. 1996, Lebreton & Baglin 1996, Roxburgh 1996; see also the references therein). This review is devoted to theoretical solar modelling and direct comparisons with helioseismic data, while the inferred seismic model is discussed in these proceedings by Basu (see also e.g. Gough et al., 1996)

The recent development concerning the basic input physics (Section 2) as well as the present status of theoretical solar models (Section 3) are presented. The comparison between the theoretical oscillation frequencies and the helioseismic data is given in Section 4. Finally the solar neutrino problem is briefly considered (Section 5).

2. Physics of Solar Models

Given the basic assumptions of the stellar evolution theory, the structure of the model is determined by the microscopic properties of the solar material: the equation of state (EOS) which relates the pressure and thermodynamical quantities to the density, temperature and chemical composition, the opacity, the nuclear reactions rates and microscopic diffusion parameters. There have been important improvements in the description of the microscopic physics of the solar interior. On the contrary the macro-physics, needed to describe motions like convection, remains poorly accounted.

Opacity

The opacity of the stellar interior depends both on the radiative atomic parameters and on the composition of the stellar mixture. Significant revisions of the opacities calculations have been obtained by two groups, with results in good agreement. The OPAL project (e.g. Rogers 1994) and the OP project (e.g. Seaton *et al.* 1994) have obtained an increase of opacity by over a factor 3 compared to earlier calculations for temperature from 100 000 K to 10^6K due to improved atomic physics for partially ionized Iron. Additional improvements in the bound bound transition of Iron using intermediate coupling rather than pure LS coupling led to further increase that results in successful modelling of Cepheid stars.

This increase of opacity has much changed the solar structure and oscillation frequencies providing deeper convection zone and theoretical frequencies closer to the observed ones (e.g. Guzik & Cox 1991, Bahcall & Pinsonneault 1992, Guenther *et al.* 1992, Berthomieu *et al.* 1993). It results also in a strong increase of the initial helium abundance estimated from the solar calibration from 0.24 to 0.28. It was suspected by Christensen-Dalsgaard *et al.* (1985) and Korzennik & Ulrich (1989) that such an opacity increase will improve the agreement with the observations. Last updated OPAL opacities with improvements in physics, specially the equation of state, and refinements of the element abundances, including more metals, have shown that the decrease due to a better EOS is almost compensated by the increase due to the new composition except near the base of the convection zone (log T=6.2) (Iglesias & Rogers 1996).

Low temperature opacities, less accurate due to molecular contributions, have been recomputed by several groups (Kurucz 1991, Alexander Ferguson 1994, Neuforge 1993, Sharp & Turck-Chièze 1996). They influence much the structure of the superficial layers, hence the oscillation frequencies, but they have no significant effect on the depth of convection zone as long as the model is calibrated to the solar radius.

Equation of state

The solar plasma at first approximation is a mixture of almost fully ionized perfect gases except close to the surface around $T \sim 10^5$K. There is very small departure from perfect gas due to dynamical interaction between the components of the plasma. But the very accurate helioseismic data allow to study these small effects due to the sensitivity of the oscillations specially to the Γ_1 index which relates the changes in pressure and density of the oscillations. Γ_1 depends much on chemical composition in the partially ionized zones, thus it provides a possibility of diagnostic of the helium content in the convection zone (Gough 1984).

Two detailed description of the EOS MHD (Mihalas *et al.* 1990) and OPAL (Rogers *et al.* 1996) have been developed in parallel with the OP and Livermore opacity calculations. There is a good agreement between the two EOS, except a pressure difference of 1.2 percents around 100 000K (Rogers *et al.* 1996). Its origin has been explained by Baturin *et al.* (1996), as the manifestation of some "τ" correction for the Debye-Hückel theory which is used in the MHD formalism. The fact that such a small difference can be seen from helioseismology (as shown by Guzik *et al.* 1996) open to plasma physicists a possibility to test their theory in a domain not accessible to laboratory experiments.

Microscopic diffusion

In the radiative zones of the Sun, in absence of macroscopic motions, the composition changes due to the element diffusion which is mainly driven by pressure gradient, temperature gradient and composition gradients. The effect of radiative pressure is negligible in the solar case (e.g. Vauclair 1996). The heavier elements tend to sink towards the center and sharp composition gradients are smoothed by the diffusion. In convection zones the motion is assumed rapid enough to make the chemical composition homogeneous and to inhibit the segregation of elements through diffusion process.

Computation of diffusion velocities are mostly based on the work of Burgers (1969) who provided a complete set of equations to describe the evolution of a multicomponent fluid. This diffusion process has been introduced in solar modelling first by Cox *et al.* (1989). Different formulations have been used: Thoul *et al.* 1994 solve exactly the Burgers equations and then represent the results by simple analytical functions, while Michaud & Proffitt (1993) search analytical solutions by approximations. Their descriptions agree within 15% according to Bahcall & Pinsonneault (1995).

Uncertainties of the solar macro-physics

Let us now point out the main uncertainties of the solar structure which results from the rough description of the solar macro-physics.

First, at the surface, the structure of the super adiabatic layer where almost the whole energy is transported by convection depends much of the theory of convection which gives the temperature gradient. The convection flux is estimated from mixing length theory which has no real physical basis. One ignores the turbulent pressure which represents 10% of the total pressure and which lowers the gas pressure and density (e.g. Balmforth 1992, Kosovichev 1995). A first step towards a more realistic description of the convection has been made using Canuto & Mazzitelli (1991) theory of convection (see Basu & Antia 1994) to compute solar models (Paterno *et al.* 1993). This theory tries to account for the whole spectrum of turbulent eddies. Next improvement may come from time dependent hydrodynamical simulations (e.g. Brummell *et al.* 1995) which may lead to improved parameterized formulations of the convection and to better description of the convection in solar conditions.

Concerning the radiative interior, it is likely that convective motions penetrate below the convection zone changing the temperature gradient (Zahn 1991). Different types of mixing can also take place in the radiative interior. Induced gravity waves may lead to significant mixing (Montalban 1994). Additional mixing may be caused by rotationally induced instabilities related to meridional circulation or to the spin down of solar interior from initial rapid rotation (Zahn 1997). These mixing processes can counteract the diffusion processes and result in modification of the abundance profile. Possible mixing of the nuclear core will change the evolution by bringing more hydrogen in the solar core and will result in a lower central temperature as proposed by Schatzman in relation with the solar neutrino problem (see Section 5).

3. Properties of Theoretical Solar Models

The computed model has to satisfy global solar constraints. The mass, radius and luminosity of the Sun are known with a substantial accuracy. The solar age has been estimated from radioactive datation of meteorites and spectroscopic measurements provide detailed chemical abundances of the solar photosphere relatively to the hydrogen. This is usually measured by the ratio of the mass fraction of heavy elements Z to the mass fraction in hydrogen X. The helium abundance Y is unknown.

Hence the solar model is calibrated to solar age, radius and luminosity by adjusting the two unknown parameters, Y initial helium abundance and the convection parameter α. Finally the calibrated model is determined by the input physics and the input parameters.

TABLE 1. Global solar constraints

$M_\odot$	(1.9891± 0.0004)	10^{33} g	planetary motions
$R_\odot$	(6.9599±0.001)	10^{10} cm	measure of visible surface
$L_\odot$	(3.846±0.005)	10^{33} erg s-1	Willson *et al.* (1988)
$t_\odot$	(4.52±0.004)	10^9 y	Wasserburg (1995)
$(Z/X)_\odot$	0.0245±0.0015		Grevesse & Noels (1993)

The properties of some solar models computed with updated physics and with and without microscopic diffusion by P. Morel with CESAM code

TABLE 2. Global characteristics of theoretical solar models (from Morel *et al.* 1997a) compared to observed values: the helium content of the outer layers Y_{surf} and the depth of the convection zone r_{ZC} inferred from helioseismic data (see respectively Basu & Antia 1995 and Christensen-Dalsgaard *et al.* 1991); the lithium photospheric abundance at solar age $Li_{\odot}$ (=log (n_{Li}/n_H) with $n_H = 10^{12}$); the measured neutrino capture rates Φ_{Ga} (Hampel *et al.*, 1996), Φ_{Cl} (Davis, 1993) and Φ_{Ka} (Fukuda *et al.*, 1996) (see Section 5); the low degree frequencies differences $\overline{\delta\nu}_{02}$ and $\overline{\delta\nu}_{13}$ characterizing the solar core properties correspond to a mean of recent observations (see Section 4).

	S	D	$D_{massloss}$	$\odot$
Y_{init}	0.268	0.274	0.273	–
Y_{surf}	0.268	0.245	0.0245	0.246; 0.249
$r_{ZC}/R_{\odot}$	0.723	0.711	0.711	0.713±0.003
T_c (10^6K)	15.38	15.65	15.62	–
ρ_c (g cm^{-3})	146.5	151.2	150.6	–
Y_c	0.617	0.638	0.639	–
Li_{ZAMS}	3.08	2.95	3.13	–
$Li_{\odot}$	3.08	2.88	1.92	1.12
Φ_{Ga} (SNU)	121± 9	130	144	$69^{+7.8}_{+8.1}$
Φ_{Cl} (SNU)	6.2± 2	8.27	8.93	2.55±0.23
Φ_{Ka} (10^6cm^{-2} s^{-1})	4.49± 1.4	6.20	6.67	2.8±0.4
$\overline{\delta\nu}_{02}(\mu$Hz)	9.34	9.09	9.13	9.01±0.05
$\overline{\delta\nu}_{13}(\mu$Hz)	16.40	16.04	15.98	15.90±0.08
P_0(mn)	36.45	35.75	35.76	–

are given in Table 2. These models are computed with OPAL equation of state and opacities, nuclear reaction rates from Caughlan & Fowler (1988) (see Morel *et al.* 1997a for details). The initial helium abundance Y_{init} required to calibrate the Sun at the present luminosity and radius is slightly larger for models with microscopic diffusion. The main effect of settling is a decrease of the surface helium abundance Y_{surf} by about 10%, in agreement with helioseismic measurement $Y_{\odot} \sim 0.25$ (see Basu these proceedings); there is an increase of the depth of the convection zone to a value close to the depth estimated from the helioseismology $r_{ZC}/R_{\odot} \sim 0.713$ (see Basu these proceedings). As the helium sinks toward the center, Y_c and ρ_c increase, hence X_c decreases which requires a higher central temperature to obtain the solar luminosity at solar age. Consequently the predicted neutrino capture rates are larger when helium settling takes place and this process slightly increases the discrepancy with the neutrino experimental measurements (see Section 4).

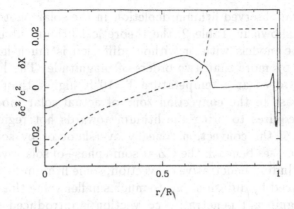

Figure 1. Difference of hydrogen content (dashed line) and relative square sound speed difference (full line) between the models with and without microscopic diffusion D and S, as a function of the solar radius.

Figure 1 illustrates the effect of the microscopic diffusion by the variation along the solar radius of the difference in hydrogen content and the relative sound speed difference between solar models with and without diffusion. Dominant effect is settling of Helium out of the convection zone. It results an increase in hydrogen content in the convection zone leading to an increase of its depth. The square of sound speed difference presents a large bump of about 1% in the layers below the convection zone.

Now many groups have computed models including settling of Helium (Proffitt & Michaud 1991, Bahcall & Pinsonneault 1992, Christensen-Dalsgaard *et al.* 1993, Chaboyer *et al.* 1995a, Gabriel & Carlier 1997) or settling of Helium and of heavier elements (Cox *et al.* 1989, Proffitt 1994, Bahcall & Pinsonneault 1995, Henney & Ulrich 1995, Basu *et al.* 1996, Morel *et al.* 1996 - 1997, Richard *et al.* 1996). The comparison of these models reveals a rather good agreement even if different formulations of the diffusion have been used.

Of course uncertainties still remain on the diffusion coefficients whose consequences on solar modelling have to be discussed (see Gabriel & Carlier in the posters proceedings of this meeting). Nevertheless microscopic diffusion can be now considered as standard assumption in solar modelling. Another uncertainty of all these models results from the fact that the available opacity tables are computed for fixed relative chemical compositions of the heavy elements entering in the considered mixture, while these values vary in time and space due to the nuclear burning and to the different velocity of diffusion of each element. A detailed discussion of that is given in Morel *et al.* (1997a).

These solar models computed according the standard assumptions do not account for the observed lithium depletion in the solar photosphere at the solar age: as shown in Table 2, the theoretical lithium abundance at solar age for the models with or without diffusion is much larger than the observed one by more than two orders of magnitude. The lithium can be nuclearly destroyed at a temperature $T=10^6K$ higher than the temperature at the base of the convection zone of actual solar models. Thus its destruction requires to bring the lithium towards hot regions, either by an extension of the convection zone by overshoot or by some mixing or transport processes beneath the CZ at some phase of solar evolution. In solar models including penetrative convection, some lithium depletion occurs, slightly enhanced by diffusion, but to much smaller value than the observed one, unless significant penetrative convection is introduced (Ahrens *et al.* 1992, Morel *et al.* 1996).

Rotationally induced mixing has been introduced in solar modelling particularly in order to try to explain the observed lithium depletion. The combined effects of microscopic diffusion and rotational mixing treated through some effective diffusion coefficients have been investigated (Chaboyer *et al.* 1995b, Richard *et al.* 1996). The mild mixing occurring below the convection zone inhibits the helium diffusion in the outer layers resulting in a slightly thinner convection zone. The observed lithium abundance at the solar age is reproduced by adjustment of some parameters which characterize the amount of mixing. The corresponding solar models are in agreement with the seismic model within $5 \ 10^{-3}$.

Alternative models with a mass loss during the first stage of solar evolution have also been considered in relation with the lithium depletion. There is no reliable theory of mass loss along the solar evolution, but some attempts have been made to evaluate its effect on lithium abundance and on solar oscillation frequencies (Guzik & Cox 1995, Morel *et al.* 1997). The lithium abundance at solar age is significantly decreased if mass loss occurs, but not in a sufficient way to account for the observations. Table 2 shows the characteristics of such a model. It appears that such models with decreased lithium abundance are rather close to the seismic models if a mass loss of about $0.1 \ M_\odot$ occurs on a time-scale smaller than 0.2 Gy.

4. Helioseismic Constraints

The solar models have to be compared with helioseismic observations. Given the structure of the model it is straightforward to compute the theoretical frequencies of the model, assuming that the oscillations are small perturbations around the equilibrium structure and that they behave adiabatically, i.e. δp is related to $\delta \rho$ through the adiabatic index Γ_1.

There are two approaches to understand the solar interior structure from

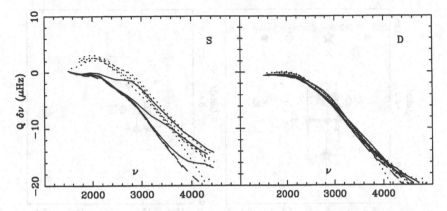

Figure 2. Scaled frequencies differences $Q\delta\nu$, in μHz, between GONG observations and models with and without helium settling as a function of the frequency, for different degrees. The points corresponding to modes of a given degree are linked by continuous curves: $\ell = 2, 3, 4, 5, 10, 20$ dotted line; $\ell = 30, 40, 50$ full line; $\ell = 70, 100, 120, 140$ dashed line. The scaling factor Q is the ratio of the energy of the mode n, ℓ to the energy of the radial mode with the same frequency and the same surface amplitude.

the solar oscillation data: the forward method and the inverse method. This paper is concerned only by the first one. Basu (these proceedings) describes how to obtain a seismic model by inverse method.

The forward method consists in a direct comparison between the observed frequencies and theoretical frequencies of a model evolved with a set of constitutive physics. Since low degree modes are sensitive to the entire solar structure while the highest degree modes probe only the solar surface, the differences between observed and computed frequencies for modes of varying degrees are used as guides to the differences between the model and the Sun, thus as guides to the deficiencies of the input physics. Then new models and frequencies with improved physics are computed, hopefully improving the agreement and informing on the sensitivities of the model structure and frequencies to the input physics. This method has been used to validate improvements in the equation of state and opacities as well as to emphasize the importance of settling of helium and heavy elements in solar modelling.

The actual scaled differences between the observed p modes frequencies, here the initial results of the GONG network (Harvey *et al.*, 1996), and the theoretical frequencies are very small, less than 20 μHz (see Figure 2), i.e. an agreement of a few 10^{-3}. It has been shown that the smooth frequency dependence which appears in the differences is due to problems with the model localized at the surface. An arbitrary increase of the surface opacities may reduce the difference between computed and observed frequencies by decreasing strongly the "slope" in Figure 2 (Christensen-Dalsgaard 1990,

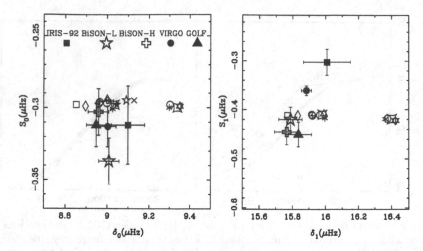

Figure 3. Fit of the frequencies differences $\delta\nu_{n,\ell} = \nu_{n+1,\ell} - \nu_{n,\ell+2}$ for $\ell = 0$ (left) and $\ell = 1$ (right), for recent observations with their error bars and for CESAM solar models (see text). IRIS data are 1992 data. For BiSON L and H refer respectively to low and high solar activity data. The smaller error bars for the VIRGO data are probably due to the fact that the frequencies of modes of degree $\ell = 3$ are determined in an easiest way by the LOI experiment than in full disk observations.

Turck-Chièze & Lopes 1993, Guzik *et al.* 1996). This "slope" is also very sensitive to different treatments of the convection in the solar near surface layers, which is usually described by the mixing length theory. Namely the Canuto-Mazitelli treatment of the convective flux (Basu & Antia 1994) or the inclusion of turbulent pressure (Kosovichev 1995) or a treatment of the convection zone based on results of hydrodynamical simulations (Rosenthal *et al.* 1995) decreases much the differences seen in Figure 2. The contribution of the outer layers to the solar frequencies and their capacity of diagnosing the structure of these layers have been much analysed (e.g. Pérez Hernández & Christensen-Dalsgaard 1994, Brodsky & Vorontsov 1988).

For models with diffusion the dependence of the relative frequencies differences shown in Figure 2 on the degree of the oscillation mode is very small. Nevertheless when subtracting the small frequency dependence it has been shown that the dispersion with degree is of order 1 μHz (Christensen-Dalsgaard *et al.* 1996) with a small separation between the modes which are reflected above or below the base of the convection zone. This is an indication of some problem in the model at these layers that is quantified by seismic inversions (see Basu 1997). The effect of uncertainties on the diffusion coefficients has been discussed, e.g. by Gabriel & Carlier, who have shown that a slight increase of these coefficients cannot be distinguished from a slight increase in opacities, both improving the agreement between observed and theoretical oscillations frequencies.

A strong constraint on the structure of the solar core is provided by low degree p modes frequencies. The small separations between frequencies of modes of consecutive radial order n and of degrees ℓ and $\ell + 2$ are predominantly determined by the conditions in the solar core as shown by the asymptotic analysis

$$\delta\nu_{n\ell} = \nu_{n+1,\ell} - \nu_{n,\ell+2} \approx \frac{2(2\ell + 3)}{n + \ell/2 + \epsilon} \int_0^R (-\frac{dc}{dr})\frac{dr}{r}$$

where c is the sound speed. The variation of this quantity for radial order high enough can be roughly approximated as:

$$\delta\nu_{n\ell} \approx \delta_\ell + S_\ell(n - n_0).$$

Figure 3 shows the values of δ_ℓ and S_l for $\ell = 0$ and $\ell = 1$ corresponding to solar models including or not the microscopic diffusion (models of Table 2 and from Morel *et al.* (1997)) derived by least square analysis for $n_0 = 21$ and for modes such that $15 \leq n \leq 26$. Results obtained from the same modes for the recent observations by the ground based networks BiSON (Elsworth *et al.*, 1994) and IRIS (Gelly *et al.*, 1997) and by the spatial experiments GOLF (Grec *et al.*, 1997) and VIRGO (Fröhlich *et al.* 1997a-b) are reported with their observational uncertainties. Combining all these data we obtain mean values $\delta_0 = (9.01 \pm 0.05)\mu Hz$ and $\delta_1 = (15.90 \pm 0.08)\mu Hz$, which have been reported in Table 2 and are useful to constrain the theoretical models. In Figure 3 the points corresponding to models with (on the left) and without microscopic diffusion (on the right) are well separated and there is a better agreement with the observations for the models including microscopic diffusion. Note that the "slope" S_ℓ which characterizes the variation of the small separation $\delta\nu_{n\ell}$ with the radial order is the same for the two sets of models.

These constraints provided by low degree p mode have to be satisfied by any solar model particularly those proposed to solve the neutrino problem. Up to now it has revealed to rule out most of the proposed models (see Section 5).

5. The Solar Neutrinos

Another way to look at the core of the Sun and to test nuclear physics involved in stellar main sequence phase is to measure the solar neutrinos. The neutrino fluxes, produced by nuclear reactions inside the Sun, are measured by 4 experiments. The Chlorine experiment (Davis, 1993) is mainly sensitive to ^8B and ^7Be high energy neutrinos, which depend much of the temperature of the solar core while the Kamiokande water Cerenkov experiment (Fukuda *et al.*, 1996) measures only the ^8B neutrinos. The two Gallium experiments GALLEX (Hampel *et al.*, 1996) and SAGE (Abdurashitov *et al.*, 1994) are sensitive to all neutrinos, including low energy pp neutrinos which number is strongly related to the solar luminosity.

The comparisons with theoretical predictions soon revealed a strong deficit of neutrinos measured by the Chlorine experiment and confirmed by the other ones (see Table 2). Consequently much work has been done to modify the physical inputs, specially the nuclear reaction parameters and electron screening for nuclear reactions have been fully reexamined with consequences on solar modelling and neutrino predictions (e.g. Bahcall & Pinsonneault 1992 1995, Turck-Chièze *et al.* 1993). Most of the reactions of the p-p chain which are of importance in the generation of energy are known at a level of 5%. Due to the calibration process to the actual solar luminosity, the structure of the model is not much sensitive to the exact specification of nuclear parameters. On the contrary the reaction rates values are important for the predicted neutrino fluxes and the uncertainties of some of them can be as large as up to 30%: this has been used by Dzitko *et al.* (1995) to obtain a solar model with low ^{8}B and ^{7}Be neutrino fluxes, but as expected with a pp neutrino flux larger than observed by Gallium experiments.

In the frame of the standard description of the solar model given in this review and as far as the more realistic microphysics is used to describe the Sun's interior, each different neutrino experiment has revealed to be incompatible with the others. An alternative explanation which accounts for the different neutrino measurements lies in the physics of the neutrino itself, which, if it has a mass, may oscillate between different states (Mikheyev & Smirnov 1986, Wolfenstein 1978).

As the neutrino fluxes depend much on the core temperature (e.g. Bahcall 1989, Bahcall & Ulmer 1996), many efforts have been developed to explain the neutrinos deficit by lowering the Sun's central temperature and hence to add new physics in the solar models. Different processes have been proposed like turbulent mixing in the core (Schatzman *et al.*, 1981) or energy transport through hypothetical weakly interactive particles named WIMPS (Faulkner *et al.*, 1986). It results respectively too high or too low values of δ_ℓ related to the small low degree p modes differences of frequencies (Provost, 1984; Elsworth *et al.*, 1990). A possible diffusion near the solar core due to stochastic gravity waves has been also investigated (Schatzman & Montalban 1995) and schematized (Morel & Schatzman 1996). More recently Cumming and Haxton (1996) proposed a model with a transport of ^{3}He which results in mixing the Sun on time scales characteristic of ^{3}He equilibrium. All these processes succeed to lower the neutrino fluxes but the corresponding solar models do not satisfy the low degree p modes constraints.

In fact the question is what is the real constraint on the central core provided by helioseismology. The oscillations are mainly determined by the mechanical properties of the Sun through the sound speed and density stratification. Thus they constrain the ratio of the temperature T to the mean molecular weight μ but not each separately. On the contrary the neutrinos fluxes depend crucially from the temperature and the chemical composition independently. Models in hydrostatic equilibrium having the same sound speed as the seismic model have been constructed and the possible range of temperature and neutrino fluxes consistent with helioseismic results has been explored (Dziembowski *et al.* 1994, Antia & Chitre 1995 1997, Shibahashi & Takata 1996 1997, Roxburgh 1997). These so-called "secondary" inversions provide the temperature and helium abundance profiles (see Basu 1997). Antia & Chitre have shown that the reduction of the neutrino fluxes to their observed values requires large unrealistic reduction of opacities. Shibahashi & Takata obtain low boron and berylium neutrino fluxes but their models do not have the solar luminosity. Roxburgh explores the predicted values of neutrino fluxes taking arbitrary profiles of hydrogen abundance and varying the nuclear cross sections well beyond the current estimates. He recovers the observed values, but his approach do not attempt to give a coherent picture of the solar structure and evolution and supposes a slow diffusive mixing and changes of the opacities or an extra contribution to the energy transport across the Sun.

In conclusion all these attempts are not able to reproduce simultaneously the neutrinos fluxes measured by the different experiments and there is no theoretical solar models compatible both with helioseismic and neutrino measurements.

6. Conclusion

Theoretical solar models are in good agreement with the helioseismological constraints, except for the surface layers for which the complexity of the real Sun is not included in the models. There are also significant differences between the Sun and the models specially at the base of the convection zone that remain to be removed. At the present moment no consistent solar model satisfies the constraints provided by the helioseismology and by the measured neutrino fluxes.

To obtain solar models closer of the seismic model inferred from helioseismology, one has specially to improve the description of the physics, which determines the variation of the chemical abundances along the radius, like the treatment of the diffusion and the mixing processes induced by the rotation or by internal gravity waves. This requires to take into account the detailed chemical abundances, which largely influence the opacity. For that purpose one needs opacity tables for a large variety of chemical mixtures, which are not yet available. Another important improvement of the solar model relies in a better description of the structure of the outer layers which are dominated by convection.

The accuracy of helioseismic data which increases with the arrival of the measurements by the spatial experiments on board SOHO may lead to a detection of asphericity and of macroscopic motions and one needs to develop consistent theoretical solar models including rotation and taking fully into account transport of material and of angular momentum, in agreement with the rotation observed inside the Sun. The detection of some solar gravity modes would put strong constraint on the deeper solar interior and increase greatly our knowledge of the solar core.

Acknowledgments. The author is grateful to G. Berthomieu for stimulating discussions and to P. Morel for providing solar models. Thanks to all the colleagues who sent work before publication. The scientific teams of the VIRGO and GOLF spatial experiments on board SOHO and the GONG project are acknowledged for permission to use provisional data. SOHO is a mission of international cooperation between ESA and NASA. GONG is managed by the National Solar Observatory, a division of the National Optical Astronomy Observatories, which is operated by AURA Inc. under a cooperative agreement with the National Science foundation.

References

Abdurashitov J.N. *et al.* (1994) *Phys. Lett. B* **328**, 234
Ahrens B., Stix M., Thorn M. (1992) *Astron. Astrophys.* **264**, 673
Alexander D.R., Ferguson J. W. (1994) *Astrophys. J.* **437**, 879
Antia H.M., Chitre S.M. (1995) *Astrophys. J.* **442**, 434
Antia H.M., Chitre S.M. (1997) *MNRAS* submitted
Bahcall J.N. (1989) *Neutrino Astrophysics*, Cambridge University Press
Bahcall J.N., Pinsonneault M.H. (1992) *Rev. Mod. Physics* **64**, No 4, 885
Bahcall J.N., Pinsonneault M.H. (1995) *Rev. Mod. Physics* **67**, 781
Bahcall J.N., Ulmer A. (1996) *Phys Rev. D*, sous presse
Balmforth N.J. (1992) *MNRAS* **255**, 603
Basu, S. (1997) these proceedings
Basu, S., Antia, H.M. (1994) *J. Astrophys. Astr.* **15** 143
Basu, S., Antia H.M. (1995) *MNRAS* **276**, 1402
Basu, S., Christensen-Dalsgaard, J., Schou, J., Thompson, M.J., Tomczyk, S. (1996) *Astrophys. J.*, **460**, 1064
Baturin V.A., Däppen W., Wang Xiaomin, Yang Fan (1996) in *Stellar evolution, what should be done?*, Eds A. Noels, D. Fraipont-Caro, M. Gabriel, N. Grevesse & P. Demarque, p 33
Berthomieu G. (1996) in *Stellar evolution, what should be done?*, Eds A. Noels, D. Fraipont-Caro, M. Gabriel, N. Grevesse & P. Demarque, p 263
Berthomieu G., Provost J., Morel P., Lebreton Y. (1993) *Astron. Astrophys.* **262**, 775
Brodsky M.A., Vorontsov S. (1988) in *Advances on Helio and Asteroseismology*, Ed J. Christensen-Dalsgaard and S. Frandsen, p 137
Brummel N., Cattaneo F., Toomre J. (1995) *Science* **269**, 1370
Burgers J.M. (1969) *Flow equations in composite gases* (Academic Press New York)
Canuto V.M., Mazitelli I. (1991) *Astrophys. J.* **370**, 295

Caughlan G.R., Fowler W.A. (1988) Atomic Data an Nuclear Data Tables, **40**, 284
Chaboyer B, Demarque P., Pinsonneault M.H. (1995a) *Astrophys. J.* **441**, 865
Chaboyer B, Demarque P., Ghenther D.B., Pinsonneault M.H. (1995b) *Astrophys. J.* **446**, 435
Christensen-Dalsgaard J. (1990) in *Inside the Sun*, ed. G.Berthomieu, M. Cribier, p 305
Christensen-Dalsgaard, J., Gough, D.O., Thompson, M.J. (1991) *Astrophys. J.* **378**, 413
Christensen-Dalsgaard J., Proffitt C.R., Thompson M.J. (1993) *Astrophys. J.* **403**, L75
Christensen-Dalsgaard J., Duvall Jr T.L., Gough D.O., Harvey J.W., Rhodes Jr E.J. (1985) *Nature* **315**, 378
Christensen-Dalsgaard J., Däppen W., *et al.* (1996) *Science* **272**, 1286
Cox, A.N., Guzik, J.A., Kidman, R.B. (1989) *Astrophys. J.* **342**, 1187
Cumming A., Haxton W.C. (1996) *Phys. Rev Lett.* **77**, 4286
Däppen W. (1994) in *Equation of state in Astrophysics*, ed G. Chabrier and E. Schatzman, Cambridge University press, p 368
Däppen W. (1996) *Bull. Astron. Soc. India* **24**, 151
Davis, R. Jr. (1993) *Frontiers of Neutrino Astrophysics*, ed. Y. Suzuki and K. Nakamura, Universal Acad. Press Inc., Tokyo, Japan. p 47
Dziembowski W.A., Goode P.R., Pamyatnykh, Sienkiewicz R. (1994) *Astrophys. J.* **432**, 417
Dzitko H., Turck-Chièze S., Delbourgo-Salvador P., Lagrange C. (1995) *Astrophys. J.* **447**, 428
Elsworth Y., Howe R., Isaak G.R., McLeod C.P., New R. (1990) *Nature* **347**, 536
Elsworth Y., Howe R., Isaak G.R., McLeod C.P., Miller B.A., New R., Speake C.C., Wheeler S. J. (1994) *Astrophys. J.* **434**, 801
Faulkner J., Gough D. O., Vahia M.N. (1986) *Nature* **321**, 226
Fröhlich C., *et al.* (1997a) *Solar Phys.* **170**, 1
Fröhlich C., *et al.* (1997b) these proceedings
Fukuda Y. and the Kamiokande Collaboration (1996) *Phys. Rev. Lett.* **77**, 1683
Gabriel M., Carlier F. (1997) *Astron. Astrophys.* **317**, 580
Gelly B., Fierry-Fraillon D., *et al.* (1997) *Astron. Astrophys.* in press
Gough D.O. (1984) *Mem. Soc. Astron. Ital.* **55**, 13
Gough D.O., *et al.* (1996) *Science* **272**, 1296
Grec G., *et al.* (1997) these proceedings
Grevesse N., Noels A. (1993) in *Origin and evolution of the elements* Eds Prantzos N. Vangioni-Flam, Casse M. Cambridge University Press, p 15
Guenther D.B., Demarque P., Kim Y.C., Pinsonneault M.H. (1992) *Astrophys. J.* **387**, 372
Guzik J.A., Cox A.N. (1991) *Astrophys. J.* **381**, 333
Guzik J.A., Cox A.N. (1995) *Astrophys. J.* **448**, 905
Guzik J.A., Cox A.N., Swenson F.J. (1996) *Bull. Astron. Soc. India* **24**, 161
Hampel, W. and the GALLEX collaboration (1996) *Phys. Lett.* B **388**, 384
Harvey J.W., *et al.* (1996) *Science* **272**, 1284
Henney, C.J., Ulrich, R.K. (1995) in *4th SOHO Workshop: Helioseismology*, eds J.T. Hoeksema, V. Domingo, B. Fleck &B. Battrick, ESA SP 376 vol 1, p 3
Iglesias C.A., Rogers F.J. (1996) *Astrophys. J.* **464**, 943
Korzennik S. G., Ulrich R. K. (1989) *Astrophys. J.* **339**, 1144
Kosovichev A.G. (1995) in *4th SOHO Workshop: Helioseismology*, eds J.T. Hoeksema, V. Domingo, B. Fleck &B. Battrick, ESA SP 376 vol 1, p 165
Kurucz R.L. (1991) in *Stellar Atmospheres: Beyond Classical Models*, L. Crivallery, I. Hibeny and D.G. Hammer (eds), NATO ASI Series, Kluwer, Dordrecht
Lebreton Y., Baglin A. (1996) in *Stellar evolution: what should be done?*, Eds A. Noels, D. Fraipont-Caro, M. Gabriel, N. Grevesse & P. Demarque, p 1
Michaud, G., Proffitt, C.R. (1993) *Inside the Stars*, ed. A. Baglin & W.W. Weiss (San Francisco: ASP), p 246
Mihalas D., Hummer D.G., Mihalas B., Däppen W. (1990) *Astrophys. J.* **350**, 300

Mikheyev S.P., Smirnov S. P. (1986) *Sov. J. Nucl. Phys.* **42**, 913
Montalban, J. (1994) *Astron. Astrophys.* **281**, 421
Morel P., Schatzman E. (1996) *Astron. Astrophys.* **310**, 982
Morel P., Provost J., Berthomieu G. (1997a) submitted to *Astron. Astrophys.*
Morel P., Provost J., Berthomieu G., Audard N. (1997b) in *"Sounding solar and stellar interiors* eds J. Provost and F.X. Schmider, in press
Morel P., Provost J., Berthomieu G., Matias J., Zahn J.P. (1996) in *Stellar evolution: what should be done?*, Eds A. Noels, D. Fraipont-Caro, M. Gabriel, N. Grevesse & P. Demarque, p 395
Neuforge C. (1993) *Astron. Astrophys.* **274**, 818
Paterno L., Ventura R., Canuto V.M., Mazitelli I. (1993) *Astrophys. J.* **402** 733
Pérez Hernández F., Christensen-Dalsgaard J. (1994) *MNRAS* **267**, 111
Proffitt. C.R., Michaud G. (1991) *Astrophys. J.* **380**, 238
Proffitt C.R. (1994) *Astrophys. J.* **425**, 849
Provost J. (1984) in *Observational tests of the stellar evolution theory*, eds A. Maeder and A. Renzini, p 47
Richard O., Vauclair S., Charbonnel C., Dziembowski W.A. (1996) *Astron. Astrophys.* **312** 1000
Rogers F.J. (1994) in *Equation of state in Astrophysics*, ed G. Chabrier and E. Schatzman, Cambridge University press, p 16
Rogers F.J., Swenson F., Iglesias C.A. (1996) *Astrophys. J.* **456**, 902
Rosenthal C.S., Christensen-Dalsgaard J., Houdek G., Monteiro M., Nordlund A., Trampedach R. (1995) in *4th SOHO Workshop: Helioseismology*, eds J.T. Hoeksema, V. Domingo, B. Fleck &B. Battrick, ESA SP 376 vol 1, p 459
Roxburgh I.W. (1996) *Bull. Astron. Soc. India* **24**, 89
Schatzman, E., Montalban, J. (1995) *Astron. Astrophys.* **305**, 513
Schatzman E.,Maeder A.,Angrand F., Glowinski R. (1981) *Astron. Astrophys.* **96**, 1
Seaton M.J., Yan Y., Mihalas D., Pradhan A.K. (1994) *MNRAS* **266**, 805
Sharp C., Turck-Chièze S. (1996) submitted to *Astron. Astrophys.*
Shibahashi H., Takata M. (1996) *Pub. Astr. Soc. Jap.* **48**, 377
Shibahashi H., Takata M. (1997) these proceedings
Thoul A.A., Bahcall J.N., Loeb A. (1994) *Astrophys. J.* **421** 828
Turck-Chièze S., Lopes I. (1993) *Astrophys. J.* **408**, 347
Turck-Chièze S., Däppen W., Fossat E., Provost J., Schatzman E., Vignaud D. (1993) *Physics Reports*, **230**, No 2-4, p 57
Vauclair S. (1996) in *Stellar evolution: what should be done?*, Eds A. Noels, D. Fraipont-Caro, M. Gabriel, N. Grevesse & P. Demarque, p 57
Wasserburg G.J. (1995) in Bahcall Pinsonneault 1995 *Rev. Mod. Physics* **67**, 781
Willson R.C., Hudson H.S. (1988) *Nature* **332**, 810
Wolfenstein L (1978) *Phys. Rev. D.* **17**, 2369
Zahn, J.-P. (1991) *Astron. Astrophys.*, **252**, 179.
Zahn, J.-P. (1997) these proceedings

THE SEISMIC SUN

SARBANI BASU

Teoretisk Astrofysik Center, Danmarks Grundforskningsfond
Aarhus Universitet, DK-8000 Aarhus C, Denmark

Abstract. Helioseismic techniques allow us to probe the interior of the Sun
with very high precision and in the process test the physical inputs to stellar
models. The picture of the Sun that has been built in this manner may be
termed "The Seismic Sun". After a brief discussion of some of the inversion
techniques used in the process, our current view of the seismic Sun shall
be reviewed. What we know so far suggests that the internal structure of
the Sun can be represented by a standard model, however, one which has a
smoother sound-speed and abundance variation than the solar models with
the usual treatment of diffusion.

1. Introduction

Helioseismology is a powerful tool to study the internal structure of the
Sun. The observed solar frequencies provide an unprecedented quantity of
data than can be used to deduce the solar structure. This ability to determine the properties of solar interior is providing more stringent tests of
stellar structure and evolution theory than those provided by the knowledge of just the global properties of the stars like luminosity, mass, radius
etc. The value of solar oscillation frequencies as diagnostics of the interior
of the Sun lies in the fact that they can be determined to very high accuracy, with the most precise observations having a relative error of less than
10^{-5}. Hence comparison of computed and observed frequencies provides a
stringent test of the model. In fact, none of the solar models constructed
so far are able to reproduce the observed frequency spectrum at the level
of accuracy provided by the observations. Thus in order to exploit the full
potential of the available frequency measurements, one has to resort to
inversions. The model of the Sun built by helioseismic inversions may be
termed the "Seismic Sun".

J. Provost and F.-X. Schmider (eds.), Sounding Solar and Stellar Interiors, 137-150.
© 1997 IAU. Printed in the Netherlands.

In this review, I restrict myself to the spherically symmetric structure of the Sun. Solar rotation is therefore ignored. The presence of magnetic fields is also ignored. A discussion about solar rotation can be found in Sekii (*this volume*). Techniques for determining asphericities arising from other sources are discussed by Duvall et al. in this volume. The forward problem of building a solar model has been discussed by Provost (*this volume*). A review of how observed frequencies are related to solar structure can be found in Christensen-Dalsgaard (1996) and Christensen-Dalsgaard et al. (1996). The results quoted are based on different data sets – BBSO (Libbrecht et al. 1990), LOWL (Schou & Tomzcyk, 1996), BiSON (Elsworth et al. 1994), IRIS (cf., Gelly et al. 1996), GONG (cf., Harvey et al. 1996), etc.

2. Inversion Techniques

Solar oscillations eigenfunctions can be expressed in terms of spherical harmonics and described by three "quantum numbers" — the degree ℓ, the radial order n, and the azimuthal order m. In the absence of asphericities, all modes with the same value of ℓ and n have the same frequency. The main assumption involved while performing an inversion is that the mean frequency of an (ℓ, n) multiplet depends only on the spherically symmetric structure of the Sun. The modes that have been observed so far are acoustic modes and hence depend primarily on the sound speed c. They depend to a much lesser extent on the density ρ. The modes have small amplitudes and periods involved are much shorter than the thermal time scales in the Sun (except at the outermost layers) and hence the oscillations are linear and mostly adiabatic. The oscillations can be described as superposition of acoustic waves, each traveling in a resonant cavity. The upper boundary of the cavity is near the surface. The lower boundary of the cavity is determined by the frequency and degree of the wave — the higher the degree, the less deeply the waves penetrate (for details see Unno et al. 1989, Christensen-Dalsgaard & Berthomieu 1993). The fact that different waves travel to different depths enables us to determine the structure of that region of the Sun over which the waves travel (see Gough et al. 1996).

The first attempts at inversion were using the asymptotic dispersion relation of solar frequencies, the Duvall law (cf. Duvall 1982):

$$F(w) = \int_{r_t}^{R} \left(1 - \frac{c^2}{w^2 r^2}\right)^{-1/2} \frac{dr}{c} = \frac{(n + \alpha(\omega))\pi}{\omega}, \qquad (1)$$

where $w = \omega/L$ and $L^2 = \ell(\ell + 1)$. In this approximation the mode frequencies depend only on the sound speed. Once $F(w)$ is determined by e.g., a least squares fit to the data, it can be inverted to determine the sound speed implicitly:

$$r = R_\odot \exp\left[-\frac{2}{\pi} \int_{a_s}^{a} (w^{-2} - a^{-2})^{-1/2} \frac{dF}{dw} dw\right], \qquad (2)$$

where $a = c/r$.

This inversion method is however not very accurate, and, when applied to artificial data, the inverted sound-speed deviates substantially from the exact sound-speed, particularly near the centre. The accuracy of the inversion can be improved by taking higher order terms in Eq. (1) (cf., Vorontsov & Shibahashi 1991). The next logical step therefore was to assume that the Sun is not very different from a standard solar model (SSM) and linearize Eq. (1) around the model, so that the difference in sound speed between the model and Sun is related to the difference in frequency (cf., Christensen-Dalsgaard et al. 1988). Thus

$$S(w)\frac{\delta\omega}{\omega} = H_1(w) + H_2(\omega), \qquad (3)$$

where the function $S(w)$ is a known function of the reference model alone, while $H_1(w)$ can be determined along with $H_2(\omega)$ by a fit to the data. $H_1(w)$ can be expressed in terms of the sound-speed difference and can be inverted to find the sound-speed difference,

$$\frac{\delta c}{c} = -\frac{2r}{\pi}\frac{da}{dr}\int_{a_s}^{a}(a^2 - w^2)^{-1/2}\frac{dH_1}{dw}w\ dw. \qquad (4)$$

The function $H_2(\omega)$ contains information about the surface layers of the Sun, where our assumption of adiabaticity breaks down. The functions $H_1(w)$ and $H_2(\omega)$ have also been used on their own to study various aspects of solar structure (e.g., Pérez Hernández & Christensen-Dalsgaard 1994, Basu & Antia 1995).

Although the differential asymptotic relation (Eq. 3) gives much better sound-speed results than those obtained with Eq. (1), there are still substantial inaccuracies in the core. Thus for more detailed work, one often uses a complete numerical inversion. Detailed descriptions of these asymptotic techniques and an in-depth analysis of the errors in the different methods can be found in Gough (1985) and Gough & Thompson (1991). Better results can also be obtained by alternative higher order asymptotic approximations (cf., Roxburgh & Vorontsov 1996, Marchenkov et al. 1996).

For numerical inversion for solar structure (e.g., Gough & Kosovichev 1990, Dziembowski et al. 1990; Däppen et al. 1991, Antia & Basu 1994, Basu et al. 1996a) the variational principle for the frequencies of adiabatic oscillations (cf., Chandrasekhar 1964) is used to express the frequency differences between the Sun and a model in terms of corresponding differences in structure. To this are added the effects of near surface errors. Thus, we can write

$$\frac{\delta\omega_i}{\omega_i} = \int K^i_{c^2,\rho}\frac{\delta c^2}{c^2}dr + \int K^i_{\rho,c^2}\frac{\delta\rho}{\rho}dr + \frac{F_s(\omega_i)}{E_i} \tag{5}$$

(cf. Dziembowski et al. 1990). Here $\delta\omega_i$ is the difference in the frequency ω_i of the ith mode between the solar data and a reference model. The functions c and ρ are the sound speed and density respectively. The kernels $K^i_{c^2,\rho}$ and K^i_{ρ,c^2} are known functions of the reference model which relate the changes in frequency to the changes in c^2 and ρ respectively; and E_i is the inertia of the mode, normalized by the photospheric amplitude of the displacement. The term F_s results from the near-surface differences, not taken into account by the adiabatic oscillation equations. The kernels for the (c^2, ρ) combination can be easily converted to kernels for others pairs of variables like (Γ_1, ρ), (u, Γ_1) with no extra assumptions (cf., Gough 1993), where $u \equiv p/\rho$.

There are two complementary methods of using Eq. (5) to determine $\delta c^2/c^2$ or $\delta\rho/\rho$: the regularized least squares (RLS) and the optimally localized averages (OLA). In the former, one tries to fit the given data under the constraint that the solution is smooth. The latter involves finding a linear combination of the kernels localized in spatial coordinates. The complementary nature of the two techniques is discussed by Sekii (*this volume*). Details of the RLS method can be found in Dziembowski et al. (1990), Antia & Basu (1994), Antia (1996). The details on how different versions of OLA are implemented can be found in Kosovichev et al. (1992), Christensen-Dalsgaard & Thompson (1995), Basu et al. (1996b) etc. A combination of RLS and OLA has also been used (cf., Dziembowski et al. 1994).

3. Inversion results — sound speed and density

Fig. 1 shows the relative squared sound-speed and density differences between the Sun and a SSM — model S of Christensen-Dalsgaard et al. (1996). The model was constructed with OPAL equation of state (EOS) (Rogers, Swenson & Iglesias, 1996), OPAL opacities (Iglesias, Rogers & Wilson 1992), observed surface Z/X (from Grevesse & Noels 1993) and incorporates diffusion of helium and heavy elements. The inversion was performed using a combination of modes from the BiSON and LOWL groups, and was done using a subtractive OLA technique (cf., Pijpers & Thompson 1992,1994; Basu et al. 1996a,b). Note that the difference between the Sun and the model is extremely small – fractions of a percent in the case of sound speed. However, the differences are still significant. The most no-

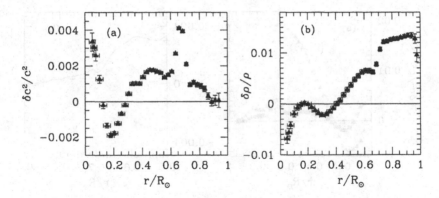

Figure 1. The relative squared sound-speed [panel (a)] and density [panel (b)] differences between the Sun and solar model S. The vertical error bars are 1σ propagated errors, while the horizontal bars are the distance between the quartile points of the averaging kernel and are a measure of the resolution.

ticeable difference is the larger sound-speed in the Sun just below the base of the convection zone. This could be due to the accumulation of excess helium below the convection zone and is a signature of mixing below the convection zone (cf., Gough et al. 1996). Model S does not incorporate any mixing below the convection zone (CZ). If there were mixing, the helium abundance locally would be reduced, decreasing the mean molecular weight, and hence increasing the sound-speed, thereby reducing the difference between the Sun and the model. The difference at the CZ base can also be removed by selective changes in the opacity (Tripathy et al. 1996).

The other region of large sound-speed difference is the core. The structure of the core is, however, still quite uncertain. Gough & Kosovichev (1993) and Gough et al. (1995), showed that the different sets of low-degree modes that are available give different results for the core, while Basu et al. (1996a) demonstrated that an indiscriminate combination of modes from different data sets could result in an erroneous interpretation of the core structure. The reason that the solar core is so uncertain is that only a few modes — those with very low degrees — penetrate to the core, and even those sample the core for a comparatively short time because of the large sound speeds there. It is hoped that more precise data from the SOHO instruments and the GONG network can improve the situation.

Model S is accepted as a SSM. But it does include diffusion of helium and heavy elements. Not so long ago, the term SSM was restricted to models without diffusion. There is evidence that diffusion of helium and heavy elements is important in the context of solar structure (cf., Cox et al. 1989, Christensen-Dalsgaard et al. 1993, etc.) For illustration we have shown the squared sound-speed difference between the Sun and a model without diffusion in Fig. 2(a). Also shown for comparison is the sound-speed difference

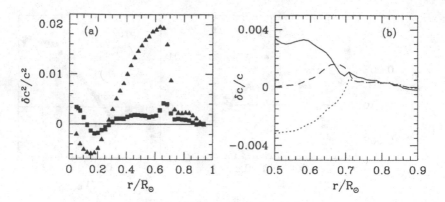

Figure 2. (a) The relative squared sound-speed difference between the Sun and a model without diffusion (triangles) and model S (squares). (b) The relative sound-speed difference between the Sun and solar envelope models without diffusion constructed with the Cox-Tabor opacities (solid line), OPAL opacities (dotted line), and a model with OPAL opacities which incorporates diffusion (dashed line). These models have a CZ depth of $0.287R_\odot$

with model S. The improvement on addition of diffusion is quite obvious.

Since the main cause of improvement in models with diffusion is a change in the convection zone depth – the model without diffusion has a shallower convection zone than the Sun, it can be argued that a change in the depth of CZ by any means will improve the results. This is not quite correct, as shown by Basu & Antia (1994a). As an illustration, in Fig. 2(b) the sound-speed difference between three solar-envelope models is shown. All models have identical CZ depths of $0.287R_\odot$. One model without diffusion has been constructed with the Cox-Tabor opacities (Cox & Tabor, 1976), the other two models were constructed with OPAL opacities, one with diffusion and one without. Thus we see that unless solar opacities are substantially lower than OPAL opacities, models need diffusion to achieve a better match to the solar sound-speed profile. Of course, it could still be argued that opacity can be changed and changing opacity gradient sufficiently could still be a way to match the solar sound-speed profile. Baturin & Ajukov (1996a) have found that substantial opacity changes are needed to construct a no-diffusion model with a sound-speed profile which agrees with the seismically determined profile. The indirect evidence that still goes against models without diffusion is the helium abundance in the solar envelope.

4. The solar helium abundance

Spectroscopic measurements of the abundance of helium in the Sun are very uncertain, hence, helioseismology plays a major role in determining the helium abundance in the solar envelope. The abundance is obtained

TABLE 1. Recent Estimations of the Solar Helium Abundance

Authors	MHD EOS	OPAL EOS	Method
Pérez Hernández, Christensen-Dalsgaard (1994)	0.242	-	Asymptotic
Basu, Antia (1995)	0.246	0.249	Asymptotic
Richard et al. (1996)	0.244	0.2505	Variational
Kosovichev (1996)	0.232	0.248	Variational
Baturin, Ajukov (1996b)	0.25	0.23	Asymptotic

from the variation of the adiabatic index of the solar material in the second helium ionization zone. This variation causes a change in the sound speed of that region. The changes in sound speed leave their signature on the function $H_1(w)$ and since the helium and hydrogen ionization zones are close to the surface, the function $H_2(\omega)$ is affected too.

Helioseismic techniques to determine the solar helium abundance can be roughly divided into two categories: the asymptotic and the variational. In the asymptotic case either the function $H_1(w)$ or $H_2(\omega)$, or the sound-speed as derived from $H_1(w)$ are used to determine the helium abundance, (cf., Gough 1984, Pérez Hernández & Christensen-Dalsgaard 1994, Vorontsov et al. 1992, Basu & Antia 1995). The variational method involved converting the kernels for c^2 and ρ (cf., Eq. (5)) to kernels for u and Y using the derivatives of the equation of state, and then using either RLS or OLA techniques to determine the helium abundance in the solar CZ (cf., Däppen et al. 1991, Kosovichev et al. 1992).

However, the results obtained by different workers have not been fully consistent with each other. It is found that the helioseismic measurement of helium abundance is sensitive to the equation of state of stellar material, which is used in translating the variation of Γ_1 to the difference in Y, and most of the difference in the older determinations can be attributed to this fact. However, some discrepancy still remains. Table 1 shows some of the more recent determination of the solar helium abundance. Results for the MHD (Hummer & Mihalas 1988; Mihalas et al. 1988, Däppen et al. 1988) and OPAL EOS (Rogers 1994, Rogers, Swenson & Iglesias, 1996) are shown, these being the two detailed EOS available for solar applications. The errors in the determination due to errors in the observed frequencies is much smaller than the differences due to EOS effects, and hence have not been included. The main point to note is that the helium abundance is quite low, between 0.24–0.25. This is compatible with solar evolution theories only if helium settles out of the envelope into the radiative zone. In absence of settling, the present day helium abundance in the solar envelope has to be about 0.27–0.28 to satisfy solar luminosity constraints.

TABLE 2. Position of the base of the solar convection zone

Authors	Position of CZ base ($R_\odot$)
Vorontsov (1988)	0.70 ± 0.01
Christensen-Dalsgaard, Gough, Thompson (1991)	0.713 ± 0.003
Kosovichev, Fedorova (1991)	0.713 ± 0.003
Guzik, Cox (1993)	0.712 ± 0.001
Basu, Antia (1996)	
Calibration models without diffusion	0.7141 ± 0.0002
Models with usual treatment of diffusion	0.7105 ± 0.0002
Models with X-profile from inversions	0.7133 ± 0.0002
Final value (including systematic errors)	0.713 ± 0.001

5. Depth of the solar convection zone

The transition of the temperature gradient from the adiabatic to radiative values at the base of the solar convection zone (CZ) leaves its signature on the sound speed. Thus if there are two otherwise similar solar models with different depths of the convection zone, then the model with a deeper convection zone will have an excess of sound speed over the other just below the base of the convection zone. Thus helioseismic measurement of the sound speed enables a determination of the position of the base of the convection zone. This has been used by a number of authors to estimate the position of the base of the convection zone. The results of various determinations of the CZ depth are shown in Table 2. All results roughly agree with each other within the error bars. All authors, except Guzik & Cox (1993), use the inverted sound-speed to determine the CZ depth. Guzik & Cox (1993) use a direct comparison of frequencies to estimate the CZ depth.

Unfortunately, the change in temperature gradient is not the only factor which leaves its imprint on the sound speed near the base of the convection zone. The abundance profiles also affect the sound-speed profile, and can confuse the signal due to change in temperature gradients. The excess helium just below the CZ due to settling causes an increase in the mean molecular weight below the base of the convection zone, and this reduces the sound speed. Thus a model with helium diffusion will appear to have a shallower convection zone, i.e., in regions just below the base of CZ, it will have sound speed similar to a no settling model with shallower CZ. Most authors have not taken this into account, since most CZ-depth determinations were made before the importance of settling was realized.

TABLE 3. Extent of overshoot below the solar convective zone

Authors	Overshoot
Gough, Sekii (1993)	" No convincing evidence"
Roxburgh, Vorontsov (1994)	$< 0.25\, H_p$
Monteiro et al. (1994)	$< 0.07\, H_p$
Basu et al. (1994)	$< 0.10\, H_p$
Basu, Antia (1994b)	$0.05^{+0.08}_{-0.05}\, H_p$
Christensen-Dalsgaard et al. (1995)	$< 0.10 H_p$
Basu (1996)	$< 0.05 H_p$

6. Overshoot below the solar convection zone

Theoretical estimates of the extent of overshoot are uncertain since that requires a non-local theory of turbulence. Recently, such theories have become available (e.g. Canuto & Dubovikov 1996) which naturally lead to overshoot. Work currently in progress (Canuto et al. 1996) seems to indicate that the extent of penetration is small.

The discontinuity in the derivatives of the sound speed at the base of the overshoot layer below the solar convection zone (CZ) introduces a characteristic oscillatory component in the frequencies of solar p-modes as a function of the radial order n (Gough 1990). The amplitude of these oscillations depends on the 'severity' of the discontinuity, which in turn depends on the extent of overshoot, while the period of the signal gives an estimate of the position of the discontinuity. This signal can be extracted and calibrated to find the depth of overshoot, as has been done by a number of groups and the results are shown in Table 3. The consensus seems to be that any overshoot below the convection zone is small. These studies however, assume that the overshoot layer is adiabatically stratified. This is probably true in regions where the convective velocity is large enough to transmit significant convective flux, but if convective velocity becomes too small then the temperature gradient is likely to approach the radiative value and the resulting structure will not be different from the radiative layers and it may not be possible to detect such layers helioseismically.

Like the case of the position of the CZ base, the abundance gradients caused by diffusion at the CZ base confuse the signal from overshoot. In this case any gradient in the helium abundance increases the signal due to the discontinuity since the helium abundance gradient causes a sharp change in the sound-speed. Thus models with a sharp abundance profile have a larger signal than models with smooth or no abundance gradients. This fact was used by Basu & Antia (1994b), Basu (1996) to determine the validity of the abundance profiles produced by different formalisms of diffusion. Only models which have a smooth abundance profile at the CZ base are consistent with observations. These include models constructed with the abundance profiles obtained from models which have mixing below the CZ (e.g., Richard et al. 1996), or from secondary inversions. Model S does not satisfy the observational constraint and this is consistent with the sharp feature found in the sound-speed difference with the Sun (cf. Fig. 1). Models which appear to satisfy the sound-speed constraint, but have steep abundance gradients (e.g., the gradual mass-losing model 3b of Guzik & Cox 1995) do not satisfy the solar constraint either. Thus this appears to be a fairly sensitive test of the abundance gradient.

7. The solar equation of state

Tests for the equation of state so far have only been indirect – either through a comparison of the frequency differences between the Sun and models constructed with different EOS, or through a comparison of the sound-speed differences. Reviews of the solar EOS can be found in Christensen-Dalsgaard & Däppen (1992), and Däppen (1996).

For reasonably simple EOS's, like the EFF (cf., Eggleton et al. 1973), a simple comparison of frequencies is enough to know that the equation of state is not good enough to satisfy solar constraints (Christensen-Dalsgaard et al. 1988). With more sophisticated equations of state, like MHD and OPAL, the frequency differences are dominated by the signature of the improperly modeled solar surface and hence one needs to look at the inversion results.

Sound-speed inversion results first showed that the MHD equation of state was deficient in the CZ, just below the HeII ionisation zone (cf., Dziembowski et al. 1992, Antia & Basu 1994). The OPAL EOS does not show this deficiency (Basu & Antia 1995). In fact, there is now evidence that models constructed with the OPAL EOS give a better fit to the solar data in the lower convection zone and below the CZ too (Basu et al. 1996c). Recently Basu & Christensen-Dalsgaard (1996) have shown how one can invert for the intrinsic difference in Γ_1 (i.e., difference at fixed pressure, density and composition) between the Sun and a model. Unfortunately, the propagated errors are still quite large, so although the EFF EOS can be ruled out, one cannot make a significant distinction between the MHD and OPAL equations of state. The expected increase in data precision should enable us to use this inversion in the future.

8. Secondary Inversions

Helioseismology gives direct constraints on only the mechanical properties of the Sun. If the equation of state is assumed, or where solar plasma is fully ionized (e.g. the core), the sound-speed constraint gives a constraint on T/μ. To be able to extract information of the thermal structure or the composition, additional input, such as equation of state, nuclear reaction rates, and opacities are required. Thus inversions for T or μ are termed "secondary inversions". Secondary inversions are particularly important in the context of the solar neutrino problem, since neutrino flux predictions require a knowledge of the temperature and chemical composition profiles in the solar core. Three techniques have been used for secondary inversions so far: (1) Use the additional input to convert sound-speed and density kernels to kernels for X and Z (cf., Gough & Kosovichev 1990, Kosovichev 1996). This method assumes that the additional inputs are exact and have no errors. (2) Assume that inputs such as opacity may have uncertainties and determine the temperature and composition which satisfies the sound-speed (i.e., T/μ) and ρ constraints, but which requires a minimum change in opacities to satisfy the thermal balance equations (cf., Antia & Chitre 1995, 1996). (3) Solve the stellar structure equations, but instead of specifying composition profiles, as is usual, specify the sound-speed profile. The sound-speed profile is obtained by primary inversions. (cf., Shibahashi & Takata 1996).

Inversions for the helium abundance profile (Shibahashi et al. 1995, Kosovichev 1996, Antia & Chitre 1996) show clear evidence of helium settling, even though the results do not completely agree with one another. Kosovichev's results show the change in the helium abundance earlier than the accepted position of the solar CZ base. That is probably a reflection of the fact that the reference model used was one without diffusion and hence had a shallow convection zone depth. Antia & Chitre (1996) find that the profile is very close to that of a SSM like model S, however, the profile is much smoother below the CZ base, and the change in the helium abundance is not as sharp as in the model, which supports the scenario of turbulent diffusion. The inverted helium profile is similar to that of model 5 of Richard et al. (1996).

One of the aims behind investigating the thermal structure and abundance gradients in the Sun is to be able to predict neutrino fluxes. The earliest results on neutrino fluxes were quite uncertain. Whereas Gough & Kosovichev (1990) said that seismic constraints led to a lower neutrino flux than SSMs (which invariably have higher neutrino fluxes than observed values, see Bahcall & Pinsonneault 1992, 1995), Dziembowski et al. (1991) claimed that seismic constraints increased the neutrino fluxes. The later inversions still show that with the current input physics the neutrino fluxes are larger than the observed values. Shibahashi & Takata (1996) claim that ^{8}B and ^{7}Be neutrino fluxes can be reduced, but that model fails to satisfy the solar luminosity constraint. Antia & Chitre (1995, 1996) show that assuming the nuclear reaction rates are not uncertain, large changes in opacity are required to lower the neutrino fluxes, however, they say that even arbitrary changes in just opacity are not sufficient to satisfy any two of the three solar neutrino constraints (i.e., Chlorine, Gallium and Boron) simultaneously. Roxburgh (1996) is of the view that neutrino fluxes may be lowered in models that have a structure consistent with helioseismic results if there is slow diffusive mixing, the opacities are changed, and there are some other contributions to energy transport within the Sun. But it is fair to say that no solar model exists thus far which satisfies both helioseismic and neutrino constraints.

9. Conclusions

So what is the current seismic model of the Sun? Helioseismic results show that the structure of the Sun is remarkably close to that of a standard solar model. The structure of the solar core is still somewhat uncertain and it is hoped that the new data will help towards reducing this uncertainty. The sound-speed profile of the Sun is, however, smoother than that of a standard solar model. The most visible difference lies just below the solar convection zone and is most probably a consequence of mixing below the solar convection zone base.

Helioseismic estimations of the abundance of helium in the solar convection zone yields a value of between 0.24 and 0.25 in most cases. There is some uncertainty caused by the uncertainty in the equation of state. The depth of the solar convection zone is $0.287 \pm 0.001 R_\odot$, and it appears that there is very little overshoot ($< 0.05 H_p$, i.e., < 2800 Km) below the solar convection zone. Although direct inversions for equation of state are not very precise yet, indirect evidence shows that of the equations of state available today, OPAL gives the best results, though discrepancies still remain.

Since helioseismology puts constraints only on T/μ, additional inputs are required to estimate the solar temperature and neutrino fluxes. From the results available so far, it does appear that solar neutrino constraints cannot be satisfied without changes in the opacity, nuclear reaction rates, or unless other contributions to the energy transport are present.

Acknowledgments. This work was supported by the Danish National Research Foundation through its establishment of the Theoretical Astrophysics Center. This work utilizes data obtained by the Global Oscillation Network Group (GONG) project, managed by the National Solar Observatory, a Division of the National Optical Astronomy Observatories, which is operated by AURA, Inc. under a cooperative agreement with the National Science Foundation.

References

Antia, H.M. (1996) A&A, 307, 609
Antia, H.M., Basu, S. (1994) A&AS, **107**, 421
Antia, H.M., Chitre, S.M. (1995) ApJ, **442**, 434
Antia, H.M., Chitre, S.M. (1996) ApJ, submitted
Bahcall, J.N., Pinsonneault, M.H. (1992) Rev. Mod. Phys., **60**, 297
Bahcall, J.N., Pinsonneault, M.H. (1995) Rev. Mod. Phys., **67**, 781
Basu, S. (1996) in preparation
Basu, S., Antia, H.M. (1994a) JAA, **15**, 143
Basu, S., Antia, H.M. (1994b) MNRAS, **269**, 1137
Basu, S., Antia, H.M. (1995) MNRAS, **276**, 1402
Basu, S., Antia, H.M. (1996) MNRAS, submitted
Basu, S., Christensen-Dalsgaard, J. (1996) in proc: IAU Symp. 181, Posters Volume, eds.
    J. Provost, F.-X. Schmider (Nice Observatory), in press
Basu, S., Antia, H.M., Narasimha, D. (1994) MNRAS, **267**, 209
Basu, S., Christensen-Dalsgaard, J., Schou, J., Thompson, M.J., Tomczyk, S. (1996a)
    ApJ, **460**, 1064
Basu, S., Christensen-Dalsgaard, J., Pérez Hernández, F., Thompson, M.J. (1996b) MN-
    RAS, **280**, 651
Basu, S., Christensen-Dalsgaard, J., Schou, J., Thompson, M.J., Tomczyk, S. (1996c)
    Bull. Astron. Soc. India, **24**, 147
Baturin, V.A., Ajukov, S.V (1996a) Astron. Reports, **40**, 259
Baturin, V.A., Ajukov, S.V (1996b) in Solar Convection and Oscillations and their Re-
    lationship, eds., F.P. Pijpers, J. Christensen-Dalsgaard, C. Rosenthal, (Dordrecht:
    Kluwer), in press
Canuto, V.M., Dubovikov. M.S. (1996) Phys. Fluids, **8**, 12
Canuto, V. M. et al (1996) (in preparation)
Chandrasekhar, S. (1964) ApJ, **139**, 664
Christensen-Dalsgaard, J. (1996) Bull. Astron Soc. India, **24**, 103
Christensen-Dalsgaard, J., Berthomieu, G. (1991) in Solar Interior and Atmosphere, eds.
    A.N. Cox et al. Space Science Series, (Tucson: University of Arizona Press) p401
Christensen-Dalsgaard, J., Däppen, W. (1992) A&AR, **4**, 267
Christensen-Dalsgaard J., Thompson M.J. (1995), in Proc. GONG'94, eds., R.K. Ulrich,
    E.J. Rhodes Jr., W. Däppen, PASPC, **76**, p 144
Christensen-Dalsgaard, J., Däppen, W., Lebreton, Y. (1988) Nature, **336**, 634
Christensen-Dalsgaard, J., Gough, D.O., Pérez Hernández, F. (1988) MNRAS, **235**, 875
Christensen-Dalsgaard, J., Gough, D.O., Thompson, M.J. (1991) ApJ, **378**, 413
Christensen-Dalsgaard, J., Proffitt, C.R., Thompson, M.J. (1993) ApJ, **403**, L75
Christensen-Dalsgaard, J., Monteiro, M.J.P.F.G., Thompson, M.J., (1995) MNRAS, **276**,
    283
Christensen-Dalsgaard, J., Däppen, W., Ajukov, S.V., et al. (1996) Science, **272**, 1286
Cox, A.N., Tabor, J.E. (1976) ApJS, **31**, 271
Cox, A.N., Guzik, J.A., Kidman, R.B. (1989) ApJ, **342**, 1187
Däppen, W. (1996) Bull. Astron. Soc. India, **24**, 151
Däppen, W., Mihalas, D., Hummer, D.G., Mihalas, B.W. (1988), ApJ, **332**, 261
Däppen, W., Gough, D.O., Kosovichev, A.G., Thompson, M.J. (1991) in Lecture Notes
    in Physics, eds., D.O. Gough, J. Toomre, **388** (Heidelberg: Springer), p 111
Duvall, T. L. (1982) Nature, **300**, 242
Dziembowski, W.A., Pamyatnykh, A.A., Sienkiewicz, R. (1990) MNRAS, **244**, 542
Dziembowski, W.A., Pamyatnykh, A.A., Sienkiewicz, R. (1991) MNRAS, **249**, 602
Dziembowski, W.A., Pamyatnykh, A.A., Sienkiewicz, R. (1992) Acta Astron., **42**, 5
Dziembowski, W.A., Goode, P.R., Pamyatnykh, A.A., Sienkiewich, R., (1994) ApJ, **432**,
    417
Eggleton, P.P., Faulkner, J., Flannery, B.P. (1973) A&A, **23**, 325

Elsworth, Y., Howe, R., Isaak, G.R., et al. (1994) *ApJ*, **434**, 801
Gelly, B., Fierry-Fraillon, D., Fossat, E., et al. (1996) *A&A*, submitted
Gough, D.O. (1984) *Mem. Soc. Astron. Ital.*, **55**, 13
Gough, D.O. (1985) *Sol. Phys.*, **100**, 65
Gough, D.O. (1990) in *Lecture Notes in Physics*, **367**, eds., Y. Osaki, H. Shibahashi, (Berlin: Springer) p 283
Gough D.O. (1993) in *Astrophysical Fluid Dynamics: Les Houches, Session XLVII*, eds. J.-P. Zahn, J. Zinn-Justin (Elsevier)
Gough, D.O., Kosovichev, A.G. (1990) in *Proc. IAU Colloq. 121: Inside the Sun*, eds., G. Berthomieu & M. Cribier, (Dordrecht: Kluwer), p 327
Gough, D.O., Kosovichev, A.G. (1993), *MNRAS*, **264**, 522
Gough, D.O., Sekii, T., in *GONG 1992:* , ed., T.M. Brown, ASPCS **42**, p 177
Gough., D. O., Thompson, M. J. (1991) in *Solar Interior and Atmosphere*, eds. A.N. Cox et al. Space Science Series, (Tucson: University of Arizona Press) 519
Gough, D.O., Kosovichev, A.G., Toutain, T. (1995) *Solar Phys.*, **157**, 1
Gough, D.O., Kosovichev, A.G., Toomre, J., et al. (1996) *Science*, **272**, 1296
Grevesse, N., Noels, A. (1993) in *Origin and evolution of the Elements*, eds., N. Prantzos, E. Vangioni-Flam, M. Cassé, (Cambridge: Cambridge Univ. Press), p 15
Guzik, J. A., Cox, A. N. (1993) *ApJ*, **411**, 394
Guzik, J. A., Cox, A. N. (1995) *ApJ*, **448**, 905
Hummer, D.G., Mihalas, D. (1988) *ApJ*, **331**, 794
Harvey, J.W., Hill, F., Hubbard, R.P., et al. (1996) *Science*, **272**, 1284
Iglesias, C.A., Rogers, F.J., Wilson, B.G. (1992) *ApJ*, **397**, 717
Kosovichev, A.G. (1996) *Bull. Astron. Soc. India*, **24**, 355
Kosovichev, A.G., Fedorova, A.V. (1991), *Sov. Astron.*, **35**, 507
Kosovichev, A.G., Christensen-Dalsgaard, J., Däppen, W., Dziembowski, W.A., Gough, D.O., Thompson, M.J. (1992) *MNRAS*, **259**, 536
Libbrecht, K.G., Woodard, M.F., Kaufman, J. M. (1990) *ApJS*, **74**, 1129
Marchenkov., K, Roxburgh, I.W., Vorontsov, S.V., (1996) in *proc: IAU Symp. 181, Posters Volume*, eds. J. Provost, F.-X. Schmider, (Nice Observatory), in press
Mihalas, D., Däppen, W., Hummer, D.G. (1988) *ApJ*, **331**, 815
Monteiro, M.J.P.F.G., Christensen-Dalsgaard, J., Thompson, M. J. (1994), *A&A*, **283**, 247
Pérez Hernández, F., Christensen-Dalsgaard, J. (1994) *MNRAS*, **267**, 111
Pijpers, F.P., Thompson, M.J., (1992) *A&A*, **262**, L33
Pijpers, F.P., Thompson, M.J., (1994) *A&A*, **281**, 231
Richard, O., Vauclair, S., Charbonnel, C., Dziembowski, W.A. (1996) *A&A*, **312**, 1000
Rogers, F.J. (1994) in *The Equation of State in Astrophysics, IAU Colloq. 147*, eds., G. Chabrier & E. Schatzman (Cambridge: Cambridge University Press), p 16
Rogers, F.J., Swenson, F.J., Iglesias, C.A. (1996) *ApJ*, **456**, 902
Roxburgh, I.W. (1996) *Bull. Astron. Soc. India*, **24**, 89
Roxburgh, I.W., Vorontsov, S.V. (1994) *MNRAS*, **268**, 880
Roxburgh, I.W., Vorontsov, S.V. (1996) *MNRAS*, **278**, 940
Schou, J., Tomczyk, S. (1996) in preparation
Shibahashi, H., Takata, M. (1996) *Bull. Astron. Soc. India*, **24**, 301
Shibahashi, H., Takata, M., Tanuma, S. (1995) in *Proc. Fourth SOHO Workshop, Vol. 2*, eds. J.T. Hoeksema, V. Domingo, B. Fleck, B. Battrick, (ESA SP-376), p 9
Tripathy, S.C., Basu, S., Christensen-Dalsgaard, J. (1996) in *proc: IAU Symp. 181, Posters Volume*, eds. J. Provost, F.-X. Schmider, (Nice Observatory), in press
Unno, W., Osaki, Y., Shibahashi, H. (1989) *Non-radial Oscillations of Stars, 2nd ed.*, (Tokyo: Univ. of Tokyo Press)
Vorontsov, S.V. (1988) in *Seismology of the Sun & Sun-like Stars*, eds., V. Domingo, E.J. Rolfe, ESA SP-286, p 475
Vorontsov, S.V., Shibahashi, H. (1991) *PASJ*, **43**, 739
Vorontsov, S.V., Baturin, V.A., Pamyatnykh, A.A. (1992) *MNRAS*, **257**, 32

THE SEISMIC STRUCTURE OF THE SUN FROM GONG

E. ANDERSON[1], H.M. ANTIA[2], S. BASU[3], B. CHABOYER[4],
S.M. CHITRE[2], J. CHRISTENSEN-DALSGAARD[3,5],
A. EFF-DARWICH[6], J.R. ELLIOTT[7], P.M. GILES[8],
D.O. GOUGH[9,10], J.A. GUZIK[11], J.W. HARVEY[1], F. HILL[1],
J.W. LEIBACHER[1], A.G. KOSOVICHEV[8],
M.J.P.F.G. MONTEIRO[12], O. RICHARD[13], T. SEKII[9],
H. SHIBAHASHI[14], M. TAKATA[14], M.J. THOMPSON[15],
J. TOOMRE[7], S. VAUCLAIR[13] AND S.V. VORONTSOV[15]

[1] *National Solar Observatory, Tucson, Arizona, USA*
[2] *Tata Institute of Fundamental Research, Bombay, India*
[3] *Theoretical Astrophysics Center, Danish National Research Foundation*
[4] *Canadian Institute for Theoretical Astrophysics, Toronto, Canada*
[5] *Institute of Physics and Astronomy, Aarhus University, Denmark*
[6] *Center for Astrophysics, Cambridge, Massachusetts, USA*
[7] *JILA, Boulder, Colorado, USA*
[8] *HEPL, Stanford University, California, USA*
[9] *Institute of Astronomy, University of Cambridge, UK*
[10] *DAMTP, University of Cambridge, UK*
[11] *Los Alamos National Laboratory, New Mexico, USA*
[12] *University of Porto, Portugal*
[13] *Observatoire Midi-Pyrénées, Toulouse, France*
[14] *University of Tokyo, Japan*
[15] *Queen Mary & Westfield College, University of London, UK*

1. Introduction

This paper is an interim report of our inferences about the hydrostatic structure of the Sun, following the first report of the GONG team in *Science* (Gough *et al.*, 1996). That work confirms that the spherically averaged structure of the Sun is more or less in agreement with current standard solar models. However, there remain some significant deviations which we

J. Provost and F.-X. Schmider (eds.), Sounding Solar and Stellar Interiors, 151-158.
© 1997 IAU. Printed in the Netherlands.

regard as important clues to the existence of dynamical phenomena which are not taken into account in standard solar modelling.

2. Frequency data

The acoustic spectrum, as obtained from the GONG data, is described by Hill *et al.* (1996). The data used for the inversions reported here for the spherically averaged structure of the Sun were the frequencies of modes with degrees l between 0 and 150. Only frequencies between 1.5 and 3.0 mHz were used for the inversions. Data from GONG months 1,2,4 and 5 were used. (GONG months are contiguous intervals of 36 days, the first being centred about 24 May, 1995.) For the study of asphericity, the even component of degeneracy splitting was determined by fitting the formula $\nu_{nlm} = \nu_{nl} + L\Sigma_i\, a_i\, P_i\, (m/L)$ to the $m - \nu$ power spectra from GONG months 1 and 2.

3. The spherically averaged structure

The spherically averaged structure of the Sun is inferred by comparing the multiplet oscillation frequencies (uniformly weighted averages of the components with like order n and degree l, but with different azimuthal degrees m) of the Sun with degenerate eigenfrequencies of a (spherically symmetrical) standard solar model. A variety of inversion procedures have been carried out. Most commonly the frequency differences between Sun and model are expressed as averages of the differences of two state variables in the Sun, such as $c^2 = \gamma p/\rho$ and ρ, or ρ and γ, where p, ρ and γ are respectively pressure, density and the first adiabatic exponent $(\partial \ln p/\partial \ln \rho)_s$. Those averages are then combined linearly to produce more localized averages of a desired variable, using either one of the optimally localized averaging procedures or a regularized least-squares data-fitting algorithm.

The reliability of the inversion procedures has been tested using artificial data computed from a theoretical model of the Sun whose structure was unknown to the inverters. Some of the results of those tests are summarized in the contributed proceedings of this symposium (Antia *et al.*, 1997). On the whole the results of the inversions agreed with the proxy Sun to within about a quoted standard error, except very near the surface and very near the centre of the Sun.

4. The solar sound-speed

We express the inferences as differences between averages of the sound speed in the Sun and corresponding sound-speed averages of a reference standard theoretical solar model. In all cases the reference model is that of Christensen-Dalsgaard *et al.* (1996) in which gravitational settling of helium and heavy elements is taken into account, but in which no other mechanism for redistribution of helium in the radiative zone is included. The model has an initial uniform heavy-element to hydrogen abundance ratio $Z/X = 0.02768$, an initial helium abundance $Y = 0.2713$, and an age of $4.60 \times 10^9 y$.

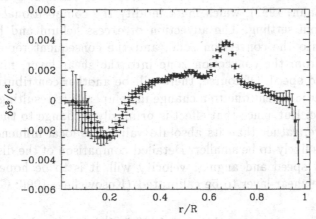

Figure 1. Relative difference $\delta c^2/c^2$ between the square of the sound speed c in the Sun and that in the standard reference theoretical model. The horizontal bars represent the widths of the averaging kernels; the vertical bars represent standard errors.

A sample inversion is illustrated in Figure 1. Because there is reason to be wary of the accuracy of the higher solar frequencies, which have greater line widths, and also the lower frequencies, for which the signal-to-noise ratio is relatively low, modes in somewhat different frequency ranges have been used.

There are three prominent features of the inversions illustrated in Figure 1 which are common to all the inversions and which merit attention:

1. there is a sharp peak in the sound-speed difference immediately beneath the base of the convection zone;
2. the sound speed in the radiative envelope, outside the core, is somewhat greater in the Sun than it is in the model;
3. except possibly in the very central region, the sound speed in the energy generating core of the Sun is lower than in the model.

We suspect that the differences in the outer layers of the convection zone may be a product of inaccurate modelling of the surface layers of the Sun, and we refrain from discussing them.

The peak in $\delta c^2/c^2$ immediately beneath the convection zone is hardly a surprising feature of the inversions given that the only mechanism in the reference model for transporting material, aside from convection, is gravitational settling. In reality, motion beneath the convection zone, such as might be induced by convective penetration, is likely to cause significant mixing. In particular, helioseismic inversion has shown the presence of a thin region of rotational shear immediately beneath the convection zone (e.g. Thompson *et al.*, 1996; Korzennik *et al.*, these proceedings; Sekii, these proceedings), which has been dubbed the tachocline; here there must be a circulatory flow driven by the pressure imbalance on level surfaces (e.g. Spiegel and Zahn, 1992), which must modify the compositional gradients that result from settling. The advection of excess helium and heavy elements back into the convection zone, and the consequent replenishment of hydrogen from the convection zone into the shear layer, must surely raise the sound speed. Of course, there will be another contribution to the sound-speed adjustment due to a change in temperature resulting from the opacity change. But, since that effect is primarily a change to the gradient of temperature, rather than its absolute value, the local influence on the sound speed is likely to be smaller. Detailed comparison of the distribution of both sound speed and angular velocity will, it is to be hoped, enable theories of the shear layer to be calibrated (Kosovichev, 1996; Gough and Sekii, 1997).

A second contributor to the peak in $\delta c^2/c^2$ immediately beneath the convection zone could have been substantial mass loss (about $0.1 M_\odot$) by the Sun. If mass loss were to have taken place, material that is now immediately beneath the convection zone would previously have been relatively deeper and more compressed. Under those conditions the augmentation of the abundances of helium and heavy elements would have been less than it is in the standard model, thereby contributing to an augmentation in c^2. The solar sound speed is compared with a model suffering mass loss in Figure 2. Notice that the peak in $\delta c^2/c^2$ is hardly present. However, the oscillation frequencies appear not to be reproduced perfectly.

There are many potential causes for the remaining sound-speed excess in the bulk of the radiative zone: $0.3 \lesssim r/R \lesssim 0.65$, and it is not unlikely that the actual cause is a combination of several of them. Some could act locally, such as errors in the value of γ which influence c^2 directly without having a direct influence on the structure of the radiative zone in the model. Others act more globally, such as errors in opacity, and errors in the equation of state that influence quantities other than γ. (Such errors must surely be present if there is an error in γ.) Their influence is spread partly by the global nature of the differential equations that determine the overall stratification, and partly by the changes in initial conditions demanded by the calibration of the radius and luminosity of the present Sun. Christensen-Dalsgaard (1996) has discussed the influence on model structure of a wide variety of changes to many of the structure variables.

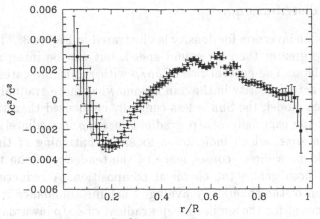

Figure 2. Relative difference $\delta c^2/c^2$ between the square of the sound speed c in the Sun and that in a model computed by Guzik and Cox (1995) that has suffered mass loss.

The excess sound speed in the outer part of the core ($0.1 \lesssim r/R \lesssim 0.3$) of the reference model is particularly interesting because it is a symptom of erroneous modelling of the region in which the nuclear reactions are taking place. The relevance of that to the solar neutrino problem is obvious. Knowledge of this discrepancy is not new, though the details have been in question, and perhaps should be regarded as still being in some doubt. Although the discrepancy is substantially greater than the formal standard errors that reflect the random errors in the data, it must be borne in mind that it arises from relatively small contributions to the oscillation frequencies. We must be aware that it might therefore be the product of some small systematic error in the data analysis, either in the procedure for converting raw data to oscillation frequencies, or in the assumptions upon which the frequency inversion schemes rely. How the discrepancy arises has not yet been determined, but we remind the reader that one possibility is that, as in the case of the rotational shear layer at the base of the convection zone, it could be a consequence of macroscopic material motion redistributing the helium that is produced by the nuclear reactions. If that were to be so, then one would expect the decline in the mean sound-speed difference in the outer layers of the core to be associated with a corresponding increase at the centre, caused by the hydrogen-rich material that has been brought from greater radii to displace the central core. Such a possibility is hinted at by the GONG data, but it is not demonstrated. We note also that an error in the assumed age of the Sun can also produce a deviation in the core and the radiative envelope not wholly unlike that implied by the inversions (Gough and Novotny, 1990), but it does not account for the entire discrepancy.

5. Density distribution

An example of an inversion for density is illustrated in Figure 3. The relative differences are greater than for sound speed, but can be interpreted in a consistent fashion. The general rise of $\delta\rho/\rho$ with radius indicates that the decline in the actual density in the Sun is somewhat more gradual than it is in the reference model; the Sun is less centrally condensed than the model. In particular, the especially sharp gradient in $\delta\rho/\rho$ immediately beneath the convection zone, which indicates a localized flattening of the density gradient, could be a direct consequence of the tendency of the tachocline circulation to homogenize the chemical composition. A reduction in the radial variation of the spherically averaged helium abundance in the core might also account for the similar steep gradient of $\delta\rho/\rho$ inwards of $r/R \simeq 0.15$.

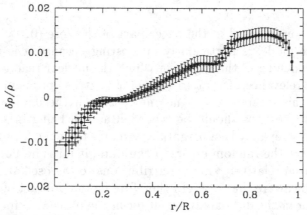

Figure 3.   Relative difference $\delta\rho/\rho$ between the density ρ in the Sun and that in the standard reference theoretical model.

6. Asphericity

One cannot distinguish from frequency data alone between the possible aspherical mechanisms that might be responsible for the even component of degeneracy splitting (Zweibel and Gough, 1995). We therefore represent the asphericity as a, possibly fictitious, variation of the wave propagation speed.

The even splitting coefficients a_i vary with frequency in approximately the same manner as does the inverse of the modal inertia. This is consistent with the findings of Woodard and Libbrecht (1993). Therefore most of the contribution to the even component of degeneracy splitting must be

confined to the outer layers of the star, near and above the upper turning
points of the modes.

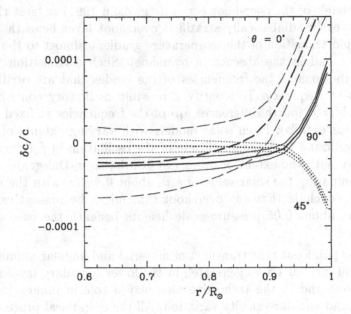

Figure 4. Relative wave-propagation-speed deviation from the spherical average, plotted
against radius, at the equator ($\theta = 90°$), at latitude 45° and at the poles: thick lines
represent expectations of the averages and the thin lines ± one standard deviation from
them.

Figure 4 illustrates the wave-speed deviation from the spherical average
at three values of latitude. These deviations are positive at high and low
latitudes, and negative at midlatitudes. That is in qualitative agreement
with the findings of Woodard and Libbrecht for 1986, which was at a similar
epoch in the solar cycle. We expect, as the sunspot cycle progresses, that
this behaviour will reverse once there is substantial midlatitude magnetic
activity.

7. Thermal stratification at the base of the convection zone

The meridional velocities induced by baroclinicity in the tachocline (Spiegel
and Zahn, 1992; Gough and Sekii, 1997) are too small to perturb significantly the radiative heat flux. Consequently, the issue of the extent of the
convective overshoot – by which we mean motion beneath the level at which
the stratification is neutrally stable to convection and which is driven by
pressure gradients induced by buoyancy forces within the convection zone
– may be studied separately. Our interest is in the extent of the motion

that is sufficiently vigorous to modify the thermal stratification, bringing it closer to being adiabatic.

Simple models of the overshoot region (e.g. Zahn 1991) suggest the existence of a nearly adiabatically stratified overshoot layer beneath which there is a rapid transition of the temperature gradient almost to the value it would have had in the absence of overshoot. Such a transition would induce contributions to the frequencies of the modes that are oscillatory with respect to frequency. To identify a possible oscillatory contribution we have considered fourth differences $\delta_4\nu$ of the frequencies at fixed l (see Basu, 1997, for details). From solar models with varying extents of overshoot we have found that the amplitude of the oscillation in $\delta_4\nu$ increases monotonically with the extent of overshoot (Christensen-Dalsgaard et al., 1995). By comparing the solar value of $\delta_4\nu$, about 0.8μHz, with the model sequence, we conclude that any overshoot that might be present extends no more than about 0.05 pressure scale heights beneath the base of the convection zone.

Finally we point out that transport of material and angular momentum by gravity and inertial waves generated in the lower boundary layer of the convection zone and in the tachocline also play a role in influencing the sound-speed and angular-velocity variation. All the dynamical processes in this region control not only the H-He abundance ratio but also the abundances of the light elements Li and Be in the convection zone.

References

Antia, H.M. et al. (1997) Sounding solar and stellar interiors, (ed. J. Provost and F-X. Schmider, Observatoire de la Côte d'Azur) in press
Basu, S. (1997) Mon. Not. R. Astron. Soc., in press
Christensen-Dalsgaard, J. (1996) in Proc. VI IAC Winter School: The structure of the Sun, (eds T. Roca Cortés & F. Sánchez, Cambridge University Press), p. 47
Christensen-Dalsgaard, J., Monteiro, M.J.P.F.G. and Thompson, M.J. (1995) Mon. Not. R. astron. Soc., 276, 283
Christensen-Dalsgaard, J. et al. (1996) Science, 272, 1286
Gough, D.O. and Novotny, E. (1990) Solar Phys., 128, 143
Gough, D.O. et al. (1996) Science, 272, 1296
Gough, D.O. and Sekii, T. (1997) Sounding solar and stellar interiors, (ed. J. Provost and F-X. Schmider, Observatoire de la Côte d'Azur) in press
Guzik, J. and Cox, A.N. (1995) Astrophys. J., 448, 905
Hill, F. et al. (1996) Science, 272, 1292
Kosovichev, A.G. (1996) Astrophys. J., 469, L61
Spiegel, E.A. and Zahn, J-P. (1992) Astron. Astrophys., 265, 106
Thompson, M.J et al. (1996) Science, 272, 1300
Woodard, M.F. and Libbrecht, K.G. (1993) Astrophys. J., 402, L77
Zahn, J.-P. (1991) Astron. Astrophys., 252, 179
Zweibel, E.G. and Gough, D.O. (1995) Proc. Fourth SOHO Workshop: Helioseismology, (ed. J.T. Hoeksema, V. Domingo, B. Fleck and B. Battrick, European Space Agency SP-376, Noordwijk), vol 2, 73

STRUCTURE INVERSIONS WITH THE VIRGO DATA

THIERRY APPOURCHAUX
Space Science Department
ESTEC, NL-2200 AG Noordwijk, The Netherlands

TAKASHI SEKII
Institute of Astronomy, University of Cambridge
Cambridge CB3 0HA, UK

DOUGLAS GOUGH
Institute of Astronomy, University of Cambridge
Cambridge CB3 0HA, UK
and
Department of Applied Mathematics and Theoretical Physics
University of Cambridge, Cambridge CB3 9EW, UK

UMIN LEE
Astronomical Institute, Tohoku University
Sendai, Miyagi 980, Japan

CHRISTOPH WEHRLI
Physikalisch-Meteorologisches Observatorium Davos
World Radiation Center, CH-7260 Davos Dorf, Switzerland

AND

THE VIRGO TEAM

Abstract. The p-mode frequencies obtained by the VIRGO instrument have been inverted to derive the solar core structure. Two sets of frequencies have been used in the inversions. The sets are derived from different time series (the second containing the first), using different procedures for fitting the spectra. The influence of the correlations between the p-mode frequencies has been implemented in the inversion procedure. The two data sets are in good agreement with each other, and show no evidence that the sun is significantly different from the latest available standard solar model.

J. Provost and F.-X. Schmider (eds.), Sounding Solar and Stellar Interiors, 159-166.

1. Introduction

The VIRGO (Variability of solar IRradiance and Gravity Oscillations) investigation on board SOHO (SOlar and Heliospheric Observatory) aims at determining the characteristics of pressure and internal gravity oscillations by observing irradiance and radiance variations.

VIRGO contains two different active-cavity radiometers for monitoring the solar 'constant', two three-channel sunphotometers and a low-resolution imager (Luminosity Oscillation Imager, LOI) with 12 'scientific' and 4 guiding pixels, for measuring the radiance distribution over the solar disk at 500 nm. The instrumentation has been described in detail by Fröhlich *et al.*, 1996, and the observed in-flight performance by Fröhlich *et al.*, 1997 (also in these proceedings). Discussion of the performance of the LOI is provided by Appourchaux *et al.*, 1997. Here we report on the result of inversions carried out using the data from the LOI.

2. P-mode analysis

After the production of the LOI level-1 data, the 12 scientific pixels are ready to be used for extracting the p modes. The times series from each pixel is detrended using a triangular smoothing with a full width of 1 day, and then the residuals are converted to relative values. For extracting a given degree, the 12 pixels are combined using optimal filters derived by Appourchaux & Andersen, 1990. Since these filters are complex they allow each m to be separated in an l, n multiplet. These filters have been successfully used by Appourchaux *et al.*, 1995b,c and by Rabello-Soares *et al.*, 1996. The optimal filters were changed weekly to account for the change in the real size of the solar image (calibrated in flight) and the orientation of the Sun (only the B angle as P is maintained zero by the spacecraft orientation).

The spectra obtained have been utilized to derive the p-mode parameters. We use two sets of frequencies for the inversion. Each set was derived from different time series using different fitting procedures. We now describe how the two sets were obtained.

2.1. SET A

This is the same set as the set that was presented by Fröhlich *et al.*, 1997. The times series starts on 27 March 1996 and ends on 10 August 1996. The resolution is 84 nHz.

A given (m, ν) diagram is fitted using the maximum-likelihood method as described by Appourchaux *et al.*, 1995b. As outlined in that paper, the different $2l+1$ components spectrum of a given l, n multiplet are correlated

with each other because of the imperfect isolation of the modes (see also
Rabello-Soares, 1996). This additional complication is taken into account
in the fitting of the spectra, higher-degree modes leaking into the lower-
degree modes, and *vice versa*. For example, in the $l = 1$ signal we can
detect $l = 6$ and *vice versa*; the $l = 4$ signal is contaminated by $l = 7$
(Fig. 1) and *vice versa*, the $l = 5$ by $l = 8$ and *vice versa*. The full leakage
between the $2l + 1$ modes of a multiplet and between mode of the higher
(or lower) degree has been computed using the computed optimal filters in
a manner similar to that of Rabello-Soares, 1996. This is an improvement
of the procedure adopted by Appourchaux *et al.*, 1995b, who regarded the
mode leakage matrix not to be known a priori and fitted it to the data. The
amplitudes of all the modes in the spectrum, together with the frequencies
of the target and leaked modes, were fitted simultaneously. The frequencies
of the unwanted leaked modes were sometimes fixed using the frequencies
obtained from BBSO. This simplification did not influence the frequencies
of the target modes substantially, yet it increased the rate of convergence of
the iterations in the fitting. In this way the bias on the frequency estimates
due to spatial mode leakage was reduced. A total of 92 modes with $0 \leq l \leq 7$
were obtained in the frequency range 2.1 mHz$\leq \nu \leq$ 3.8 mHz.

2.2. SET B

This is a data set from a longer time series using a different fitting strategy.
The times series starts on 27 March 1996 and ends on 5 October 1996. The
resolution is 60 nHz.

For $l \leq 3$ the correlation matrix between the $2l + 1$ components of a
multiplet is used not to fit the power spectra, but the Fourier spectra, as
described by Schou, 1992. We did not implement this fitting into the Fourier
spectra for $l \geq 4$, because of the many more correlations between the various
l, n and l', n'. Instead we fitted simultaneously both power spectra of the
$l = 1$ and $l = 6$, of the $l = 4$ and $l = 7$, and of the $l = 5$ and $l = 8$, taking
into account the leakage of the modes. This improved the consistency of the
fitted frequencies. The estimate of the $l = 1$ frequencies given by this fit was
not used; instead we used the value obtained from the simultaneous fittinh
of all the modes with $l \leq 3$. The correlation between the uncertainties in
the frequencies was retained for use in the inversion. A total of 100 modes
with $0 \leq l \leq 8$ were obtained in the frequency range 2.1 mHz$\leq \nu \leq$ 3.8
mHz

3. Inversions of the VIRGO frequencies

P-mode frequencies from the two sets have been inverted to investigate the
structure of the solar core. The mode with the shallowest inner turning

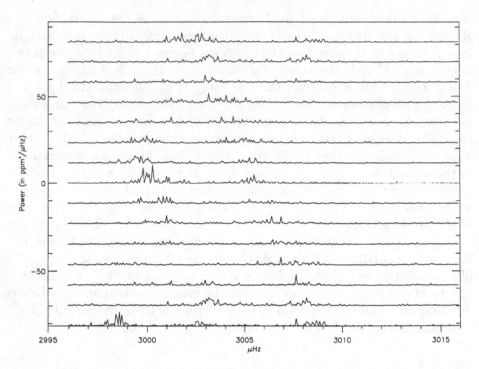

Figure 1. The m, ν diagram for the $l = 7$, $n = 18$ multiplet from the times series of set B; $m = -7$ is at the top. It shows an example of the effect of the leakage of the $l=4$, $n=19$ into the $l=7$. We see clearly the splitting of the $l=7$ mode and also the effect of the leakage of unwanted modes with different values of m. The splitting of the $l = 4$ is not visible because the leakage is more complex: the $l = 4, m = +4$ mode is partially transmitted by the filter for $l = 7, m = -7$, and the $l = 4, m = -4$ is transmitted by the $l = 7, m = +7$.

point penetrates to $r/R_\odot \simeq 0.3$, and our investigation concentrates on the layers below these depths, since outside these regions the datasets do not provide much information about the solar structure.

The standard strategy of linearization was adopted (e.g., Gough, 1996): we use a theoretical model of the sun as a reference, and then consider the difference between the sun and the model. We attempt to determine how the sun differs from the reference model by considering the difference between the observed frequencies and those of the reference model. The relative frequency difference is equated to a linear functional of the difference in structures:

$$\frac{\delta \nu_i}{\nu_i} \equiv \frac{\nu_{obs,i} - \nu_i}{\nu_i} = \int \left[K_{u,\gamma_1}^i (r) \frac{\delta u}{u} + K_{\gamma_1,u}^i (r) \frac{\delta \gamma_1}{\gamma_1} \right] dr + \frac{F(\nu_i)}{I_i} , \qquad (1)$$

where ν_i and $\nu_{obs,i}$ are respectively the frequency of the model and the observed frequency of the mode i; also K_{u,γ_1}^i and $K_{\gamma_1,u}^i$ are the kernels for

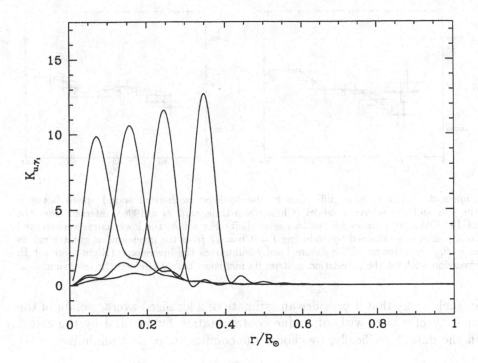

Figure 2. Averaging kernels for the squared isothermal sound speed u obtained from 1105 combined LOI+LOWL frequencies from set A.

the squared isothermal sound speed u and the adiabatic exponent γ_1, which are our choice of inversion variables. The two functions δu and $\delta \gamma_1$ are the differences between the values of u and the values of γ_1 in the sun and in the model. We present below inversions for δu. The term $F(\nu_i)/I_i$, where $F(\nu_i)$ is an arbitrary function of frequency (which we represent as an expansion in terms of Legendre polynomials) and I_i is the mode inertia (normalization is such that the rms vertical component of the displacement eigenfunction is unity at the surface), is added to represent all the uncertainties that arise from our lack of knowledge of the subsurface layers (for detailed discussion see Gough 1995).

To carry out the inversions we used optimally localized averaging (e.g., Gough, 1996; also Däppen *et al.*, 1991); to estimate δu at radius r_0, we construct a linear combination of the relative differences,

$$\sum c_i(r_0) \frac{\delta \nu_i}{\nu_i} = \int \left(\sum c_i K^i_{u,\gamma_1} \right) \frac{\delta u}{u} dr + \int \left(\sum c_i K^i_{\gamma_1,u} \right) \frac{\delta \gamma_1}{\gamma_1} dr +$$

$$\sum c_i \frac{F(\nu_i)}{I_i} \qquad (2)$$

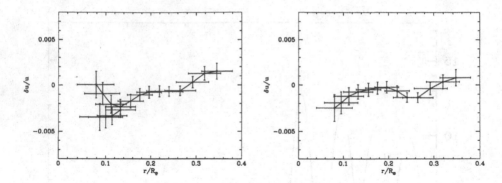

Figure 3. The relative difference in the squared isothermal sound speed between the sun and the reference model (Christensen-Dalsgaard *et al.*1996), inferred from the LOI+LOWL frequencies for the two sets. (Left) For set A: the lower curve represented by 4 points was obtained by excluding $l = 0, n = 15$ from the inversion; at greater values of $r/R_{\odot}$ the removal of the datum hardly influences the inversions. (Right) For set B: inversion without the correlation matrix; its inclusion barely affected the inversion.

in such a way that it provides an estimate of a localized average of δu in the vicinity of $r = r_0$ without undue contamination by $\delta\gamma_1$ and by the errors in the data. Specifically, we choose the coefficients c_i that minimize

$$S \equiv \int K_{u,\gamma_1}(r,r_0)^2(r-r_0)^2 dr + \alpha \int K_{\gamma_1,u}(r,r_0)^2 dr + \beta \sum_{i,j} c_i c_j E_{ij} \quad (3)$$

where E_{ij} is the covariance matrix between the uncartainties in the frequencies. Expression (3) is minimized subject to

$$\int K_{u,\gamma_1} dr = 1 \quad (4)$$

and

$$\sum_i c_i \frac{P_\lambda(\nu_i)}{I_i} = 0 \quad , \quad \lambda = 0, \ldots, \lambda_{\max} , \quad (5)$$

where

$$K_{u,\gamma_1}(r,r_0) \equiv \sum_i c_i K_{u,\gamma_1}^i(r) , \quad (6)$$

$$K_{\gamma_1,u}(r,r_0) \equiv \sum_i c_i K_{\gamma_1,u}^i(r) \quad (7)$$

are the averaging kernel for u and the kernel determining the contamination by $\delta\gamma_1$ of the inference, and $P_\lambda(\nu_i)$ is a Legendre polynomial.

The first term in S encourages localization of the averaging kernel, which has unit modulus imposed by the first constraint; the second suppresses the amplitude of the contaminating kernel, and the third suppresses error magnification and the correlation between the frequencies. Thus the parameters α and β are used to balance the weights between resolution, contamination from $\delta\gamma_1$, and error. The purpose of the second set of constraints is to eliminate the contribution from the unknown surface layers, with a resolution determined by the value of λ_{max} in the Legendre decomposition of $F(\nu_i)$.

Owing to the small number of LOI modes, the degree of localization is not very good. To obtain a better localization, it is essential to supplement the dataset with higher-degree modes. We used the LOWL dataset (Tomczyk et al.1996) for this purpose. From the LOWL 2-year average dataset, 1013 modes with $l \geq 8$ in the frequency range $1.5\text{mHz} \leq \nu \leq 3.5\text{mHz}$ were chosen and added to the LOI frequencies. The improved averaging kernels obtained from the combined dataset are shown for set A in Fig. 2, which illustrates the importance of having higher-degree modes. Fig. 3 (Left) shows the result of the inversion for set A, together with a result of a second inversion, where we excluded a single mode ($l = 0, n = 15$) because a problem with that mode was recognized during the analysis. We note how sensitive our inference of the deepest part of the sun can be to just a single mode. Fig. 3 (Right) shows the result of the inversion of set B.

There seem to be systematic differences between the LOWL and the LOI frequencies, which have a noticeable influence on the results of the inversion. We have confirmed that this does not result principally from the correlation between the uncertainties in the low-degree frequencies by carrying out an inversion with mode set B in which the off-diagonal components of E_{ij} were ignored: the results were superficially similar. However, we suspect that the fact that the LOWL and the LOI data were accumulated in different time intervals is significant; in addition the instruments observe in velocity and in intensity, respectively. The influence of the surface uncertainties is known to vary with time over the solar cycle, and to assume that the function $F(\nu_i)$ in equation (1) is the same for both the LOWL and the LOI data is therefore inconsistent. A discrepancy of this kind caused by temporal variations of the surface layers can be manifest in the inversions as an error in the structure of the core.

4. Conclusion

We have carried out a more careful analysis of the LOI p-mode frequencies than that which we reported previously (Fröhlich et al., 1997) with no significant effect on the results of inversion. If there are any systematic errors in the analysis of the spectra, a yet more careful analysis will be required to remove them. We must point out that the improved treatment of the spectra is yet to be performed for $l \geq 4$. As a result of fitting the power spectra, the frequency correlation matrix might have been underestimated. We acknowledge that systematic errors could play a very important role in the inferences of the internal solar structure. Nevertheless, we can conclude that the fine details of the spherically averaged structure of the solar core is unlikely to be very different from the latest standard solar model.

Acknowledgements

The VIRGO team is led by Claus Fröhlich (PMOD/WRC, CH) and includes the following co-investigators (in alphabetical order): Bo Andersen, Thierry Appourchaux, Gabrielle Berthomieu, Dominique Crommelynck, Werner Däppen, Vicente Domingo, Alain Fichot, Douglas Gough, Todd Hoeksema, Antonio Jiménez, Judit Pap, Janine Provost, Teodoro Roca Cortés, José Romero, Thierry Toutain, Christoph Wehrli, Richard Willson.

We are grateful to Laurent Gizon on commenting an earlier version of the paper. Many thanks to S.Tomczyk and J.Schou for providing the LOWL data through the web.

References

Appourchaux, T., Andersen, B.N., Fröhlich, C., Jiménez, A., Telljohan, U., Wehrli, C. (1997) Submitted to *Solar Phys.*

Appourchaux, T., Telljohann, U., Martin, D., Lévêque, S. and Fleur, J. (1995a) in *Fourth SOHO workshop on helioseismology*, V.Domingo and T.Hoeksema eds., ESA SP-376, p. 359

Appourchaux, T., Toutain, T., Telljohann, U., Jiménez, A., Rabello-Soares, M.C., Andersen, B.N. and Jones, A.R. (1995b) *Astron. Astrophys.*, **294**, L13

Appourchaux, T., Toutain, T., Telljohann, U., Jiménez, A., Rabello-Soares, M.C., Andersen, B.N. and Jones, A.R. (1995c) in *Fourth SOHO workshop on helioseismology*, V.Domingo and T.Hoeksema eds., ESA SP-376, p. 265

Christensen-Dalsgaard *et al.*(1996) *Science*, *272*, 1286

Däppen, W., Gough, D.O., Kosovichev, A.G. and Thompson, M.J. (1991) in *Challenges to Theories of the Structure of Moderate-Mass Stars*, Gough, D.O. & Toomre, J. eds., Springer Verlag, Berlin, p. 111

Fröhlich, C. and the VIRGO team (1997) Submitted to *Solar Phys*

Gough, D.O. (1995) in *Fourth SOHO workshop on helioseismology*, V. Domingo and T.Hoeksema eds., ESA SP-376, p. 249

Gough, D.O. (1996) in *The structure of the Sun*, T. Roca Cortés & F. Sánchez eds., Cambridge University Press, p 141

Rabello-Soares, M.C. (1996) Helioseismological study of the solar interior, PhD thesis, Universidad de La Laguna: La Laguna, Tenerife, Spain

Rabello-Soares, M.C., Roca-Cortes, T., Jiménez, A., Appourchaux, T. and Eff-Darwich. (1996) Submitted to *Astrophys. J*

Schou, J. (1992) On the analysis of helioseismic data, PhD Thesis, Aarhus University: Aarhus, Denmark

Tomczyk, S., Schou, J. and Thompson, M.J. (1996) *Astrophys. J.*, **448**, L57

MAKING A SEISMIC SOLAR MODEL
AND AN ESTIMATE OF THE NEUTRINO FLUXES

H. SHIBAHASHI AND M. TAKATA

Department of Astronomy, University of Tokyo,
Bunkyo-ku, Tokyo 113, Japan

Abstract. We present a method of making a solar model based on the helioseismic data. We first invert the observed eigenfrequencies to determine the sound speed profile, and then solve the basic equations governing the stellar structure with the imposition of the determined sound-speed profile. This approach is different from that of the standard solar model in the sense that the 'seismic' solar model is a snapshot model of the sun constructed without any assumption about the history of the sun. We invert the data obtained at the South Pole by the Bartol/NSO/NASA group along with BISON, HLH, and LOWL data. Finally we estimate the neutrino fluxes of the seismic model.

1. Introduction

One of the scientific goals of helioseismology is to discriminate between the possible solutions of the solar neutrino problem; a defect in the modeling of the sun or one in particle physics. It will be helpful for this purpose to examine quantitatively whether the helioseismic data and the neutrino flux measurements are consistent each other. If the predicted neutrino fluxes of a model based on the helioseismic data are consistent with the neutrino flux measurements, evolutionary solar models are the likely source of the neutrino problem. It is thus desirable to determine the solar-interior structure from the helioseismic data, and to compare the expected neutrino fluxes based on such a model with the detected neutrino fluxes. In our previous work we have deduced the solar-interior structure subject to the constraint that the sound-speed profile is that determined by helioseismic data of Libbrecht et al. (1990) (Shibahashi and Takata 1996). In that work, however, we adopted the sound-speed profile from a previous inversion of the data by

167

J. Provost and F.-X. Schmider (eds.), Sounding Solar and Stellar Interiors, 167–174.
© 1997 *IAU. Printed in the Netherlands.*

Vorontsov and Shibahashi (1991). In order to estimate the error level more precisely, we should have performed an inversion to obtain the sound-speed profile itself as a part of the work to reconstruct a seismic solar model. In this paper, we perform inversion of the observed eigenfrequencies of the sun to determine the sound speed profile, and, then we deduce the density, pressure, temperature, and hydrogen profiles in the solar interior by solving the basic equations governing the stellar structure constrained by the sound-speed profile. The error levels are estimated by a Monte-Carlo simulation using Gaussian noise added to the frequency data. We invert the data obtained by the Bartol/NSO/NASA group along with data from BISON, HLH, and LOWL, and estimate the neutrino fluxes.

2. Methodology of Making a Seismic Solar Model

The standard solar models are based on assumptions about the evolutionary history of the sun (cf. Provost, in these proceedings). Although the standard theory of stellar evolution has succeeded in explaining many observational properties of stars, its success has been in treating stars as a statistical group. There is no guarantee that a specific star, the sun in this case, follows this theory precisely. In the following we depart from the standard construction of a solar model, and try to reconstruct a solar model by using only experimentally measured quantities. These quantities are the mass $M_\odot$, radius $R_\odot$, photon luminosity $L_\odot$, and the sound-speed distribution $c(r)$ obtained from helioseismology (Shibahashi 1993, Shibahashi and Takata 1996). We assume that the sun is in hydrostatic equilibrium. Whether the sun is in thermal balance is uncertain. In this paper, however, we assume that the sun is in thermal balance. The model is spherically symmetric and we ignore the effects of rotation and the magnetic field. We want to emphasize that we do not need any assumptions concerning the history of the sun, and that the seismic solar model constructed in this way is a snapshot model of the present day sun.

The basic equations for constructing a model with the above assumptions are the same as those used in theory of stellar structure:

$$dM_r/dr = 4\pi r^2 \rho, \tag{1}$$

$$dP/dr = -GM_r\rho/r^2, \tag{2}$$

$$dL_r/dr = 4\pi r^2 \rho\varepsilon, \tag{3}$$

$$dT/dr = \begin{cases} -\frac{3}{4ac}\frac{\kappa\rho}{T^3}\frac{L_r}{4\pi r^2} & \text{if radiative} \\ (dT/dr)_{conv} & \text{if convective} \end{cases}. \tag{4}$$

The above expressions differ slightly from the usual ones in that the gravitational energy release is ignored in equation (3): $\varepsilon_g = 0$. A more important difference is the treatment of the auxiliary equations. Since the sound speed, which we regard as a known function of r, is a thermodynamically determined quantity, it is a function of two other thermodynamical quantities such as ρ and P along with the chemical composition X_i. We can express the pressure as a function of c, ρ and X_i;

$$P = P[\rho, X_i, c(r)]. \tag{5}$$

Similarly, $T = T[\rho, X_i, c(r)]$, $\kappa = \kappa(\rho, T, X_i) = \kappa[\rho, X_i, c(r)]$, and $\varepsilon = \varepsilon(\rho, T, X_i) = \varepsilon[\rho, X_i, c(r)]$. Then the basic four equations (1)-(4) are a set of equations for M_r, ρ, L_r, and X_i for the given $c(r)$.

We assume that the abundance ratios of the various heavy elements in the solar interior are the same as those observed spectroscopically near to the solar surface. We adopt $Z/X = 0.0277$ (Grevesse 1984). To fix X and Z, we fix the helium abundance $Y = 0.23$, which is consistent with the helioseimologically determined value of Y (e.g., Basu and Antia 1995). The convection zone is assumed to be chemically homogeneous, and then, X_i is fixed in the convection zone. The extent of the convection zone is helioseismologically determined from the kink of $c(r)$.

Equations (1)–(4) form a boundary value problem with the following boundary conditions: $M_r = 0$ and $L_r = 0$ at $r = 0$, $M_r = M_\odot$ at $r = R_\odot$ and

$$X_i = \textit{the given value} \quad \text{at } r = R_\odot. \tag{6}$$

One of the boundary conditions (6) is used instead of $L = L_\odot$ at $r = R_\odot$, which is obviously required as a solar model. It should be remembered that the luminosity and radius should be determined as eigenvalues in solving the equations for the stellar structure. Hence, there is not always a solution which satisfies $L = L_\odot$ at $r = R_\odot$. If there is no such solution, this means (i) the inverted sound speed is incorrect due to defect of the inversion method, or (ii) the observed frequencies involve large errors, or (iii) our knowledge about micro-physics (the nuclear reaction rates, the opacity, and the equation of state) is poor, or (iv) the sun is not in thermal balance. We have not considered the possibility (iv), and we take account of errors involved in the inversion method and those in the frequency data and in micro-physics. Our policy at the moment is to accept the reasonable micro-physics as it is, while we estimate the error levels due to inversion or the frequency data by performing a Monte-Carlo simulation with Gaussian noise on the frequency data. Among solutions of (1)-(4) obtained in this way, we admit only the solution satisfying $L = L_\odot$ as a seismic solar model.

3. Method of Inversion of the Frequency Data to the Sound Speed Profile

Let us now turn to the problem of inversion of the eigenfrequencies to the sound speed profile. There are various methods (cf. Basu, in these proceedings), and we adopt here the asymptotic method developed by Vorontsov (1990). According to this theory, the function T defined by $T \equiv (n + 1/2)\pi/\omega$ is decomposed into a combination of a function of $\tilde{w} \equiv (\ell + 1/2)/\omega$ and a function of ω, and the cross term, which is separated into a function of $\tilde{w}$ over ω^2; —that is,

$$T(\tilde{w}, \omega) = F(\tilde{w}) + G(\omega) + \Phi(\tilde{w})/\omega^2, \tag{7}$$

where n is the radial order, ℓ is the degree of the mode, and ω denotes the eigenfrequency. The frequency is considered to be continuous function of the continuous variables n and ℓ, and then the function T is treated as a continuous function of ω and $\tilde{w}$. The function of $\tilde{w}$, $F(\tilde{w})$, is given by

$$F(\tilde{w}) \equiv \int_{r_1}^{R_\odot} \left(c^{-2} - \tilde{w}^2 r^{-2}\right)^{1/2} dr, \tag{8}$$

and, hence, once the function $F(\tilde{w})$ is discriminated, the sound-speed profile $c(r)$ is obtained by solving an Abel-type integral equation, which is led by differentiation of (8) with respect to $\tilde{w}$. Therefore, a key process is decomposition of $(n+1/2)\pi/\omega$. We adopt the following method. The function of ω alone, $G(\omega)$, is first eliminated by taking the partial derivative of T with respect to $\tilde{w}$, and then, the functions $F(\tilde{w})$ and $\Phi(\tilde{w})$ are obtained by minimizing

$$\chi^2 \equiv \int \int \left(T_{\tilde{w}} - F_{\tilde{w}} - \Phi_{\tilde{w}}/\omega^2\right)^2 d\tilde{w} d\omega, \tag{9}$$

where the subscript $\tilde{w}$ means the derivative with respect to $\tilde{w}$. Taking variation of χ^2 associated with a slight change of $F_{\tilde{w}}$ and $\Phi_{\tilde{w}}$ and requiring $\delta\chi^2 = 0$ for any $\delta F_{\tilde{w}}$ and $\delta\Phi_{\tilde{w}}$, we get two equations, by which $F_{\tilde{w}}$ and $\Phi_{\tilde{w}}$ are separated from $T_{\tilde{w}}$. Here the derivative $T_{\tilde{w}}$ is evaluated by $T_{\tilde{w}} = -\pi(\partial\omega/\partial\ell)/(\partial\omega/\partial n)$. Since the asymptotic inversion is a nonlinear inversion, it may produce undesirable, spurious features. To eliminate the possibility of spurious results, we invert the frequencies of a theoretical solar model to get the sound speed and calibrate the inverted result of the observed frequencies by seeing how well the sound speed of the model is reproduced from the theoretical frequencies.

Vorontsov's asymptotic formula (7) has been used by Vorontsov and Shibahashi (1991) and the sound speed thus obtained was used in our previous work (Shibahashi and Takata 1996). However, decomposition of

$\mathcal{T}(\omega, \tilde{w})$ has been carried out by using cubic B-spline, which does not necessarily lead a unique solution. The present method of determining $F(\tilde{w})$ is essentially a nonlinear least square method and it is more objective. The adoption of a new decomposition method is a major difference from our previous work.

4. Actual Inversion Using the Observational Data

The data obtained at the South Pole by Jefferies, Pomerantz, Harvey, and Duvall in 1987, 1988, and 1990 cover a wide range of ℓ ($1 \leq \ell \leq 700$), and are suitable for our purpose (cf. Jefferies et al. 1990, Duvall 1995). (Hereafter we call these data SP87, SP88, and SP90, respectively.) We use only the frequencies $2.2\text{mHz} \leq \nu \leq 4.8\text{mHz}$. The higher degree modes data are supplemented by the HLH data taken in 1993 at Kitt Peak (Bachmann et al. 1995). From the HLH data of $100 \leq \ell \leq 1200$, we selected only the modes which are not present in the South Pole data and $\ell \leq 750$. The low degree modes are important for determining the structure of the nuclear reacting core. We combine the data with the low-degree frequency data obtained by the BISON group (Elsworth et al. 1994), since their error estimates are lower than those of the low-degree mode of the South Pole data. It is known that the p-mode frequencies change with solar activity, and the BISON group presented a frequency data set which was corrected to the minimal level of radio flux. We adopt this corrected frequency data set.

The LOWL provides us another uniform data set of frequencies for $0 \leq \ell \leq 99$ (Tomczyk et al. 1995). The data adopted here is the weighted average frequencies obtained in the period 2/26/94 - 2/25/96 and computed by Schou and Tomczyk (1995). The higher degree mode data are again supplemented by the HLH data. We also constructed data sets by combining the LOWL data with the SP data and the HLH data.

Figure 1 shows the sound-speed profile obtained by using these data sets. The lines show the profiles for the most likely values of the frequencies along with the 1-σ level error bars estimated by a Monte-Carlo simulation. In solving the basic equations (1)-(4) with the imposition of the sound-speed profile, we adopted the OPAL opacity library (Rogers and Iglesias 1992) and a subroutine written by Bahcall et al. (1995) to provide the opacity and the nuclear reaction rates, respectively. We treated ^{3}He distribution as being in equilibrium in the deep interior, and assumed that the distribution in the outside follows the accumulation of ^{3}He due to the $D(p,\gamma)^3$He reaction without destruction. The CNO cycle is ignored. We adopted a simple perfect gas law as the equation of state. It should be remembered that there is not always a solution which satisfies $L = L_\odot$ at $r = R_\odot$ and that our

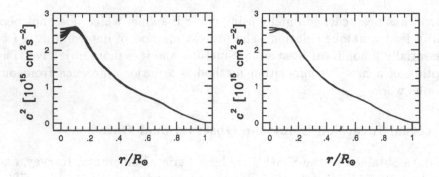

Figure 1.   Figure 1. Squared sound-speed profile inverted from the data of various combinations of SP/HLH, BISON, and LOWL (left) and LOWL/HLH (right).

present policy is to accept the reasonable micro-physics as it is, while we perform a Monte-Carlo simulation with Gaussian noise on the frequency data. Among solutions of (1)-(4) obtained in this way, we admit only the solution satisfying $L = (1 \pm 0.01)L_\odot$ as a seismic solar model. The density profile and and pressure profile obtained as solutions are shown in figure 2. By using these profiles, we confirmed that the core is convectively stable. Figures 3 shows our estimates for the temperature and the hydrogen abundance profiles. The latter is fairly constant outside the nuclear reacting core though there remains a wiggly feature. It should be emphasized that such a constancy is not assumed in making a seismic model as in the case of a standard solar model. This means that roughly speaking the OPAL opacity is correct. The slight decrease of X with depth from the base of the convection zone might be due to diffusion.   The neutrino fluxes at one

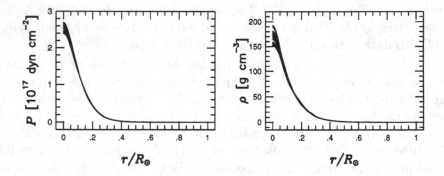

Figure 2.   Pressure profile (left) and density profile (right) obtained from the various combinations of SP/HLH/BISON/LOWL.

astronomical unit can be estimated along with a calculation of the nuclear

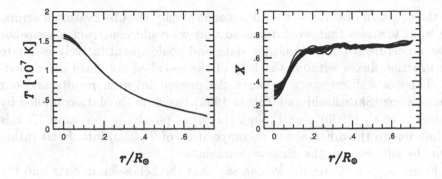

Figure 3. Temperature profile (left) and hydrogen abundance profile (right) obtained from the various combinations of SP/HLH/BISON/LOWL.

TABLE 1. Capture rates predicted by the seismic solar models

Helioseismic Data	Kamiokande	Cl (SNU)	Ga (SNU)
SP87/HLH/LOWL	$6.0\text{-}7.4\times10^{6}$ cm^{-2}s^{-1}	8.1-9.9	127.-135.
SP88/HLH/LOWL	6.7-9.4	9.0-12.3	130.-142.
SP90/HLH	4.0-6.6	5.7-8.6	116.-130.
SP87-90/HLH/BISON	6.1-7.2	8.2-9.5	127.-132.
LOWL/HLH	7.9-8.6	10.3-11.2	134.-139.

reaction rates from the estimated temperature and chemical composition distributions. Table 1 summarizes the estimated neutrino fluxes based on the present seismic solar models.

5. Discussion

We found that, as far as we adopt the most likely values of observed frequencies and the most likely micro-physics, the resulting luminosity of the model is smaller than $1L_{\odot}$. This was noted in our previous work (Shibahashi and Takata 1996) as well as by Roxburgh (1996) and Antia and Chitre (1996). A seismic solar model satisfying $L = L_{\odot}$ can be obtained taking account of uncertainties of either the seismic data or micro-physics. Indeed, Roxburgh (1996) and Antia and Chitre (1996) reproduced the solar luminosity by modifying the nuclear reaction rates. Our policy is to take account of the uncertainties of both the seismic data and micro-physics and to find the model satisfying $L = L_{\odot}$ with the least deviation from the most likely values in the multi-dimensional space of uncertainties. The present work is the first step of our attempt: we adopted the most likely micro-physics (but for the equation of state) and took account of only the observational errors. We want to stress that we did this so that we could construct a snapshot solar model from the helioseismic data and could quantitatively estimate the neutrino fluxes without the help of the so-called standard solar models. There is a discrepancy between the present inverted results and our previous one (Shibahashi and Takata 1996) based on the data compiled by Libbrecht et al. (1990) even though the input physics is the same. This is mainly due to the difference in decomposition of $\mathcal{T}(\tilde{w}, \omega)$ into $F(\tilde{w})$ rather than the difference in the observational data.

From the present result, we can say that the helioseismic data and the measured neutrino fluxes are inconsistent if we accept the most likely microphysics. It may seem that this supports the particle physics solution of the solar neutrino problem. However, we should further examine the neutrino fluxes of seismic models, taking account of uncertainties in micro-physics, before reaching any conclusions. Such approach is now in progress.

References

Antia, H. M. and Chitre, S. M. (1996) preprint.

Bachmann, K. T., Duvall, T. L., Jr, Harvey, J. W. and Hill, F. (1995) *Astrophys. J.*, **443**, 837

Bahcall, J. N., Pinsonneault, M. H. and Wasserburg, G. J. (1995) *Rev. Mod. Phys.*, **67**, 781

Basu, S. and Antia, H. M. (1995) *Monthly Notices Roy. Astron. Soc.*, **276**, 1402

Duvall, T. L., Jr. (1995) in *Proc. Fourth SOHO Workshop: Helioseismology, Vol. 1*, ed. J. T. Hoeksema, V. Domingo, B. Fleck and B. Battrick (ESA Publication Division, Noordwijk), pp. 107

Elsworth, Y., Howe, R., Isaak, G. R., McLeod, C. P., Miller, B. A., New, R., Speake, C.C. and Wheeler, S. J. (1994) *Astrophys. J.*, **434**, 801

Grevesse, N. (1984) *Physica Scripta*, **T8**, 49

Jefferies, S. M., Duvall, T. L., Jr., Harvey, J. W. and M. A. Pomerantz (1990) in *Progress of Seismology of the Sun and Stars* ed. Y. Osaki and H. Shibahashi (Springer-Verlag, Berlin), pp. 135

Libbrecht K. G., Woodard M. F. and Kaufman J. M. (1990) *Astrophys. J. Suppl.*, **74**, 1129

Rogers, F. J. and Iglesias, C. A. (1992) *Astrophys. J. Suppl.*, **79**, 507

Roxburgh, I. W. (1996) *Bull. Astron. Soc. India*, **24**, 89

Schou, J. and Tomczyk, S. (1995) in *GONG '94: Helio- and Asteroseismology from the Earth and Space*, ed. R. K. Ulrich, E. J. Rhodes, and W. Däppen (Astron. Soc. Pacific, San Francisco), pp. 448

Shibahashi, H. (1993) in *Frontiers of Neutrino Astrophysics*, ed. Y. Suzuki & K. Nakamura (Universal Academy Press, Tokyo), pp. 93

Shibahashi, H. and Takata, M. (1996) *Publ. Astron. Soc. Japan*, **48**, 377

Tomczyk, S., Streander, K., Card, G., Elmore, D., Hull, H. and Cacciani, A. (1995) *Solar Phys.*, **159**, 1

Vorontsov, S. S. (1990) in *Progress of Seismology of the Sun and Stars* ed. Y. Osaki and H. Shibahashi (Springer-Verlag, Berlin), pp. 67

Vorontsov, S. S. and Shibahashi, H. (1991) *Publ. Astron. Soc. Japan*, **43**, 739

STELLAR ROTATION AND MIXING

JEAN-PAUL ZAHN

Département d'Astrophysique Stellaire et Galactique,
Observatoire de Paris, 92195 Meudon, France

Abstract. Many observations indicate that some mixing occurs in stellar radiation zones: in massive stars, chemical elements resulting from nuclear burning in the core are detected at the surface, and in solar-type stars lithium is depleted with age. Since all mixing processes transport also momentum, the depletion of lithium should be linked with the loss of angular momentum through the stellar wind, and there are indeed signs of such a correlation in the behavior of tidally-locked binaries. Moreover, any transport process leaves its signature in the internal rotation profile, and this can help greatly in its identification. After reviewing the main transport mechanisms which have been considered so far, our present conclusion is that the uniform rotation observed in the radiative interior of the Sun is probably achieved by the action of internal waves emitted at the base of the convective envelope. It remains to be verified whether these waves contribute directly to the mixing, or whether they act only through the shaping of the rotation profile, which in turn determines the mixing through meridian circulation and turbulent diffusivity.

1. Mixing in stellar interiors: the observational evidence

At first sight, it seems that stars undergo no mixing at all during their main sequence phase, since most of them evolve towards the giant branch, which proves that they build up a helium rich core surrounded by a hydrogen rich envelope. This heterogeneous structure arises because a fraction of the stellar interior is in radiative equilibrium - if the star were fully convective, it would remain homogeneous.

At a closer look however, many stars show some signs of partial mixing occurring even in their radiative regions, and we shall briefly recall them.

175

J. Provost and F.-X. Schmider (eds.), Sounding Solar and Stellar Interiors, 175-188.
© 1997 IAU. Printed in the Netherlands.

Chemical elements which are produced through nuclear reactions in the innermost regions of stars are observed at their surface, such as ^{14}N in B stars (Lyubimkov 1984, 1991; Gies & Lambert 1992). An enhancement of ^{4}He is detected in some OB stars (Herrero et al. 1992), and there is an excess of ^{13}C in RGB stars (Lambert 1976). Likewise, the overabundance of ^{3}He in the solar wind suggests that this isotope originates from the deep interior of the Sun (Geiss & Reeves 1972).

On the other hand, the surface abundance of ^{7}Li decreases with age in solar-type stars (Herbig 1965; Soderblom et al. 1993), which impli that this fragile element has been carried some distance below the convective envelope, to be destroyed where the temperature exceeds ≈ 2.5 , 10^6K.

There is also indirect evidence that some mild mixing occurs in stars possessing a radiative envelope; otherwise, as was pointed out by Biermann (1937) and again by Schatzman (1969), microscopic processes such as atomic diffusion, gravitational settling and radiative acceleration would cause significant differences in their surface composition, which are not observed as a rule.

The purpose of this contribution is to examine the possible causes of such mixing, and to verify whether it is related to the rotation of the stars.

2. Rotational mixing in early-type stars

In the classical description, which we owe to Eddington (1925), Vogt (1925) and Sweet (1950), the centrifugal force causes a thermal imbalance in the radiative region of a rotating star, which in turn generates a large- scale meridian circulation whose turn-over time is

$$t_{ES} = t_{KH} \left(\frac{\Omega^2 R^3}{GM} \right)^{-1} , \qquad (1)$$

with $t_{KH} = GM^2/RL$ being the Kelvin-Helmholtz time (M, R and L are the mass, radius and luminosity of the star, Ω its angular velocity, and G the gravitational constant). This time is short enough for early-type stars to prevent them from ever becoming giants. Mestel (1953) resolved this paradox by explaining that the gradient of molecular weight which arises from hydrogen burning tends to choke the circulation in creating adverse μ-currents; according to him, mixing can only occur in very fast rotators, close to the centrifugal break-up.

But this appears in conflict with some observations, which indicate that helium is enhanced by 15 to 20% in rapid rotators (Herrero et al. 1992).

The weakness of the classical Eddington-Sweet theory is that it postulates a prescribed rotation law, most often uniform rotation. When one takes into the advection of angular momentum through the meridian circulation, which keeps modifying the internal rotation, the picture changes radically. Early-type stars lose little angular momentum during their main-sequence phase, as attested by their large rotation velocities. Hence, there is no need to carry this angular momentum from the interior to the surface, and therefore the circulation tends to vanish after an initial adjustment phase. The proof of this was brought by Busse (1981, 1982), but it was met with great scepticism, partly because his analysis was not carried to high enough order and therefore produced an asymptotic solution with uniform rotation.

In fact, it turns out that mixing is not fully suppressed, because even in the asymptotic regime predicted by Busse some differential rotation subsists, and gives rise to shear instabilities (Zahn 1992). The turbulence generated by the horizontal shear tends to restore uniform rotation in latitude, and to first approximation the rotation rate depends only on the radial coordinate: $\Omega = \Omega(r)$. The vertical shear $d\Omega/dr$ is stabilized by the (stable) entropy stratification, but only to some extent, because radiative leakage allows for turbulence at small enough scales (cf. Zahn 1974). Thus, a stationary solution is conceivable, in which the angular fluxes carried by the circulation and the turbulent motions cancel each other, and it would allow the star to be mixed.

But, the situation is more intricate, because the star evolves in a non-homologous way, and angular momentum keeps being redistributed, thus excluding an asymptotic regime. Moreover, the gradient of molecular weight $\nabla_\mu = d \ln \mu / d \ln P$ which arises from nuclear burning strongly inhibits the vertical shear instabilities, which play a key role in the mixing near the convective core (Meynet & Maeder 1996). According to the Richardson criterion, these are suppressed as soon as

$$\frac{g}{H_P} \nabla_\mu \geq Ri_c \left(\frac{r \sin \theta \, d\Omega}{dr} \right)^2, \tag{2}$$

provided no other instability is present (H_P is the pressure scale-height, g the local gravity, θ the colatitude and $Ri_c \approx 1/4$ the critical Richardson number). However, we have seen that the horizontal shear produces some turbulence which tends to restore uniform rotation in latitude; although the turbulent motions are highly anisotropic in this stable stratification, they will erode the density fluctuations and thus weaken the buoyancy force. The resulting turbulent transport in the radial direction is characterized by a diffusion coefficient (Talon & Zahn 1996)

$$D_v = 2 Ri_c \frac{H_P}{g} \frac{(r \sin \theta \, d\Omega/dr)^2}{(\nabla_{ad} - \nabla)/(K + D_h) + \nabla_\mu / D_h}, \tag{3}$$

where K is the radiative diffusivity and $D_h (\gg D_v)$ the horizontal turbulent diffusivity.

Presently, the weak point in the theory is the coefficient D_h, which we are unable as yet to derive from first principles. It forces us to settle for a substitute, in the form of a parametric relation which may be established as follows. Since the horizontal shear is sustained by the meridian advection of angular momentum, the strength of the turbulence must be related with the amplitude U of the circulation velocity, in order to keep the differential rotation in latitude at some moderate level. This leads to the following prescription, which we shall use in lack of something better (see Zahn 1992):

$$ D_h = \frac{rU}{C_h} \left[\frac{1}{3} \frac{d\ln \rho r^2 U}{d\ln r} - \frac{1}{2} \frac{d\ln r^2 \Omega}{d\ln r} \right] \tag{4} $$

with ρ being the density and C_h a parameter of order 1.

This description for the transport of matter and angular momentum has been implemented in the Geneva stellar evolution code, and the first calculations which have been performed for the main-sequence evolution of a $9 M_\odot$ star are rather encouraging. They allow indeed for partial mixing, the main effects being observable in the surface abundances of ^{12}C, ^{14}N and ^{16}O, accompanied by a slight increase of ^4He (Talon et al. 1996; see also the poster presented at this symposium by Talon & Zahn).

3. Rotational mixing in late-type stars

It has been noticed from the start that the depletion of lithium in late-type stars could well be related with their spin-down (see for instance Skumanich 1972). Both the abundance of ^7Li and the rotation rate decrease with age; in the Pleiades, where one observes a large spread of these parameters, the fastest rotators display the highest lithium content and the slow rotators are generally lithium depleted (Soderblom et al. 1993).

As was mentioned in §1, lithium is destroyed at some depth below the convective envelope of these stars, and therefore it has to be carried there by a not too efficient transport process, whose effect is felt in a few 10^8 years: from the age of the Pleiades to that of the Hyades, the lithium abundance decreases roughly by 0.5 dex in a 1 $M_\odot$ star.

Likewise, angular momentum must be extracted from the radiative interior of the star, since the torque exerted by the stellar wind is applied on its magnetosphere (Schatzman 1962), and thus on its convective envelope.

It is therefore natural to examine first whether the same physical mechanism may be responsible for both of these transports, and we shall call this possibility *rotational mixing of type I*.

Two physical processes are known to transport both angular momentum and matter, namely turbulent diffusion and advection through a large-scale circulation. The effect of turbulent mixing has been studied thoroughly at Yale by Endal and Sofia (1978), and more recently by Pinsonneault et al. (1989, 1990). In their picture, the magnetic torque slows down the convective envelope, and thereby induces in the radiative interior a differential rotation which is liable to various instabilities, thus leading to turbulent transport. The contribution of the meridian circulation is also included, but for simplicity it too is handled as a diffusive process. As mentioned above, the differential rotation in latitude is smoothed out by horizontal shear instabilities, which permits to consider the angular velocity Ω as a function of depth only. Its evolution in time is governed by the diffusion equation

$$\rho \frac{\partial}{\partial t} \left[r^2\, \Omega \right] = \frac{1}{r^2} \frac{\partial}{\partial r} \left[\rho \nu_{rot}\, r^4 \frac{\partial \Omega}{\partial r} \right], \tag{5}$$

with a prescription for the turbulent viscosity ν_{rot} inspired by the instabilities which are most likely to occur.

Equation (5) is integrated with a surface boundary condition specifying the loss of angular momentum dJ/dt through the wind as a function of the angular velocity, which never departs much from the power law given originally by Skumanich (1972):

$$\frac{dJ}{dt} = -f_w \Omega^3. \tag{6}$$

The additional parameter f_w involved in that prescription is adjusted so as to yield the present rotation rate.

The transport of lithium (or any other element) is modeled by a similar diffusion equation

$$\rho \frac{\partial c_i}{\partial t} = \dot{c}_i + \frac{1}{r^2} \frac{\partial}{\partial r} \left\{ r^2 \rho \left[D_{m,1} c_i + (D_{m,2} + f_c\, \nu_{rot}) \frac{\partial c_i}{\partial r} \right] \right\}, \tag{7}$$

where c_i designates the concentration of the considered species, $\dot{c}_i$ its creation or destruction rate through nuclear reactions; the $D_{m,i}$ account for the microscopic diffusion and $f_c \nu_{rot}$ is the turbulent diffusion coefficient.

To yield the observed ^{7}Li abundance, the parameter f_c must take a rather small value (≈ 0.035), whereas the turbulent diffusivity and viscosity should be of the same order.

This disparity, which has been pointed out by Law et al. (1984), has no physical reason: it is simply due to the crude treatment of the meridian circulation. When its transport is properly accounted for, the rotation rate obeys an advection/diffusion equation:

$$\rho \frac{\partial}{\partial t} \left[r^2\, \Omega \right] = \frac{1}{5r^2} \frac{\partial}{\partial r} \left[\rho r^4\, \Omega\, U \right] + \frac{1}{r^2} \frac{\partial}{\partial r} \left[\rho \nu_v\, r^4 \frac{\partial \Omega}{\partial r} \right], \tag{8}$$

where ν_v is the turbulent viscosity in the vertical direction, and $U(r)$ the amplitude of the meridian velocity).

The transport of lithium obeys an equation similar to (7)

$$\rho\frac{\partial c_i}{\partial t} = \dot{c}_i + \frac{1}{r^2}\frac{\partial}{\partial r}\left\{r^2\rho\left[D_{m,1}c_i + (D_{m,2} + \nu_v + D_{eff})\frac{\partial c_i}{\partial r}\right]\right\}; \qquad (9)$$

but it includes an extra diffusivity given by

$$D_{eff} = \frac{|rU(r)|^2}{30\,D_h}, \qquad (10)$$

where D_h is the horizontal diffusivity (4) which we have encountered in §2. This term represents the contribution of the meridian circulation, which is changed into a vertical diffusion because the turbulent motions associated with the latitudinal shear instabilities tend to erode the chemical composition on horizontal surfaces, hence reducing the advective transport (Chaboyer & Zahn 1992).

In contrast, the advection of angular momentum is hardly affected by this horizontal homogeneization, because a net momentum flux subsists even when Ω is constant on horizontal surfaces. We have thus a physical explanation for the different efficiencies characterizing respectively the transport of angular momentum and that of lithium, as observed in the Sun and solar-type stars.

An important property of this type of rotational mixing is that the transport of lithium is tightly linked with that of angular momentum. Therefore, if no other transport process is present, the depletion of lithium is correlated with the loss of angular momentum through the wind. To first approximation, the theory predicts a linear relation between the relative variation of the lithium concentration c_{Li} and that of angular momentum:

$$\frac{d}{dt}\ln c_{Li} \propto \frac{d}{dt}\ln J\Big|_{wind}, \qquad (11)$$

once the asymptotic regime has been reached (Zahn 1992, 1994).

But, there is an alternate possibility, which we shall call *rotational mixing of type II*: the extraction of angular momentum may be due to another, more powerful process, which enforces nearly uniform rotation in the radiative interior, whereas the transport of lithium is achieved mainly through the meridian circulation. In that case, the lithium depletion is related directly with the rotation rate:

$$\frac{d}{dt}\ln c_{Li} \propto -\Omega^2. \qquad (12)$$

With single stars, one cannot distinguish between these two types of mixing, since during their spin-down the angular velocity decreases with time roughly as $\Omega \propto t^{-1/2}$ (Skumanich 1972), and this means that

$$\frac{d}{dt} \ln J \propto -\Omega^2.\qquad (13)$$

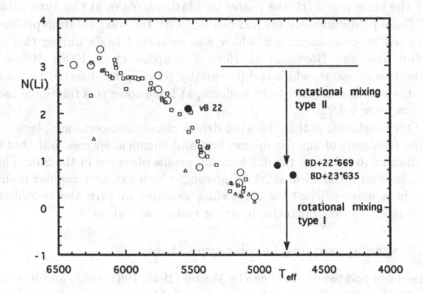

Figure 1. The lithium abundance in the Hyades (on logarithmic scale), as a function of effective temperature (Thorburn *et al.* 1993). The binaries with known orbital elements are indicated by circles, which are filled for the three tidally-locked binaries (Zahn 1994). Those binaries permit to distinguish between the two types of rotational mixing, as explained in the text.

In *tidally locked binaries*, however, the angular momentum carried away by the wind is drawn from the orbit, not from the interior of the stars (Ryan & Deliyannis 1995; Zahn 1994). Therefore, if they are subjected to what we have called rotational mixing of type I, they should deplete less lithium than single stars of the same mass. This is indeed confirmed by the observations, at least to some degree: in the Hyades, the three tidally-locked binaries all show higher lithium abundances than single stars of the same effective temperature (Thorburn *et al.* 1993).

However, these binaries do not display the original lithium abundance of $N(Li) \approx 3$, which indicates that they experience in addition some rotational mixing of type II. The depletion associated with each type of mixing is sketched in Fig. 1.

The presence of another, more efficient process for the transport of angular momentum is confirmed by the internal rotation of the Sun. If this momentum were transported only by meridian circulation and related turbulent diffusion, as in type I mixing, the solar rotation profile would be rather steep, with the core spinning about 20 times faster than the surface. This was shown already by Pinsonneault *et al.* (1989), and it has been confirmed in the most recent work by Chaboyer *et al.* (1995). The calculations we have performed with the improved treatment of the meridian circulation give the same result (cf. the poster by Matias & Zahn at this symposium).

These predictions are in clear conflict with the solar rotation profile determined by helioseismology, which was presented again during this symposium (see also Brown *et al.* 1989; Thompson *et al.* 1996). Below the convective envelope, which is differentially rotating as observed at the surface the rotation appears to be uniform, at least down to a fractional radius of about $r \approx 0.4 R_\odot$.

Our conclusion is that the wind-driven circulation certainly plays a role in the transport of angular momentum and chemical species, but that it is insufficient to establish the flat rotation profile observed in the Sun. Therefore, another mechanism must be present, which extracts angular momentum in a more efficient way. We shall examine in turn the two obvious candidates, namely magnetic torquing and wave transport.

4. Magnetic torquing in solar-type stars

It has been pointed out long ago by Mestel (1953, 1961, 1965) and Roxburgh (1963) that even a very weak magnetic field is able to enforce uniform rotation in the radiative interior of a star. The reason is clear: differential rotation winds up any existing meridian (or poloidal) field $\vec{B}_p$ into a toroidal component $\vec{B}_\phi$, and the resulting Laplace-Lorentz force reacts back on the rotation rate.

The magnetic diffusivity is so small that the toroidal field reaches easily the level where the magnetic torque balances the loss of angular momentum through the wind, which occurs when

$$ r_c^3 < B_p\, B_\phi > \approx \left. \frac{dJ}{dt} \right|_{wind} \tag{14} $$

at the base of the convection zone ($r = r_c$). That torque then enforces nearly uniform rotation along the field-lines of the poloidal field $\vec{B}_p$, a law first predicted by Ferraro (1937). The rapid evolution towards this state has been described by Mestel, Moss and Tayler (1988), and more recently by Charbonneau and MacGregor (1993), who considered various configurations of the poloidal field.

However, if the magnetic field is responsible for the extraction of angular momentum from the radiative interior, its poloidal component must be deeply rooted in the radiation zone and it must be connected to the convective envelope. The alternating dynamo field does not satisfy the first condition, since it does not penetrate much into the radiation zone.

A slowly decaying fossil field would comply with both conditions, but it would pose another problem: along each field-line of $\vec{B}_p$ the angular velocity would be that of its entry point into the differentially rotating convection zone, which means that the rotation would be non-uniform in the radiative interior, unless all its field-lines converge into the convection zone within a narrow latitude band in each hemisphere. Such a situation cannot be excluded a priori, but no coherent model has been built yet to substantiate it.

Furthermore, the question of the stability of stellar magnetic fields has still not been answered in a satisfactory way, and the subject has not advanced significantly since the excellent review by Spruit (1987). Purely poloidal and purely toroidal fields are known to be dynamically unstable, although rotation may reduce the growth-rate of the instability. Moreover, when radiative damping and magnetic diffusion is taken into account, additional instabilities occur, of the double diffusive type.

We thus conclude that magnetic torquing is a very efficient process to put the radiation zone of the Sun into corotation along the field-lines of the poloidal field, but that the outcome would probably be a differentially rotating interior, contrary to the observations. Furthermore, it remains to be verified whether an internal field of this type would last long enough to play the expected role.

5. Internal waves in the solar interior

Gravity waves (also called internal waves) play an important role in the tidal braking of massive binary stars, as was shown by Zahn (1975) and confirmed by Goldreich and Nicholson (1989). But, they have been invoked only recently as a process which could extract angular momentum from the radiative interior of the Sun (Zahn 1990; Schatzman 1993), and the efficiency of this transport has just been evaluated (Kumar & Quataert 1996; Zahn, Talon & Matias 1996).

These waves are characterized by their cyclic frequency σ and their horizontal wavenumber k_h (or equivalently by their spherical order ℓ), and these determine the vertical wavenumber k_v through the dispersion relation

$$k_v^2 = \left(\frac{N^2}{\sigma^2} - 1\right) k_h^2 \equiv \left(\frac{N^2}{\sigma^2} - 1\right) \frac{\ell(\ell+1)}{r^2}, \qquad (15)$$

with N being the Brunt-Väisälä frequency ($\sigma < N$). In a differentially rotating star, the local frequency σ is Doppler-shifted through the horizontal motion, and it varies with depth according to:

$$\sigma(r) = \sigma_0 - m\Omega(r), \tag{16}$$

with σ_0 being the frequency in the inertial frame. It is straightforward to evaluate the flux of angular momentum $\mathcal{F}_J$ carried by a travelling wave, which is related to the flux of kinetic energy by

$$\mathcal{F}_J = 2\frac{m}{\sigma}\mathcal{F}_K, \tag{17}$$

with mr being the azimuthal component of the horizontal wavenumber. In the adiabatic case, the angular momentum is conserved, i.e.

$$4\pi r^2 \mathcal{F}_J(r) \equiv \mathcal{L}_J = \text{cst}, \tag{18}$$

but when radiative damping is taken into account, the angular momentum "luminosity" decreases as

$$\mathcal{L}_J(r) = \mathcal{L}_J(r_c) \exp -\tau(r), \tag{19}$$

where τ, akin to an optical depth, is given by

$$\tau(r) = [\ell(\ell+1)]^{3/2} \int_r^{r_c} K \frac{N^3}{\sigma^4} \left(\frac{N^2}{N^2 - \sigma^2}\right)^{\frac{1}{2}} \frac{dr}{r^3}, \tag{20}$$

with r_c being the coordinate of the top of the radiation zone, K the radiative diffusivity. (The assumption has been made here that the star is homogeneous; for the general case where composition gradients are present, see Zahn *et al.* 1996). Note that the damping is tremendously enhanced when $\sigma \to 0$, which in a differentially rotating star arises at the depth where $m\Omega = \sigma(r) - \sigma_0$.

It remains to integrate the angular momentum luminosity $\mathcal{L}_J(r)$ over the whole spectrum of the internal waves emitted at the base of the convective envelope. We describe here the approach taken by Zahn et al (1996). Following García López and Spruit (1991), we match the pressure fluctuation in the wave with that of the turbulent convection, and allow the convective eddies of spherical order ℓ_ω to generate, by stochastic excitation, waves of lower order ℓ. Furthermore, we assume that the energy spectrum of the convective motions is adequately represented by the Kolmogorov law. This yields the following expression for the flux of internal wave energy at the top of the radiative interior:

$$\mathcal{F}_K(r_c) = \rho_c v_c^3 \frac{\omega_c}{N_c^2} \int_{\omega_c}^{N_c} \frac{d\omega}{\omega} (N_c^2 - \omega^2)^{\frac{1}{2}} \left[\frac{\omega}{\omega_c}\right]^{-2} \int_0^\ell \frac{d\ell}{\ell_c}, \tag{21}$$

with ω_c, ℓ_c and the velocity v_c corresponding to the largest convective eddies, and N_c being the Brunt-Väisälä frequency, which takes a finite value here due to convective penetration (see Zahn 1991). To derive the flux of angular momentum, we need to know how the energy is distributed over the azimuthal order m. We settle for the simplest assumption, which seems to be substantiated by the solar p-modes, namely that this distribution is uniform. Making use of (17), of (19) and of (21), we then get the following expression for the angular momentum luminosity:

$$\mathcal{L}_J(r) = 4\pi r^2 \frac{\rho_c v_c^3}{N_c \ell_c} \int_{\omega_c}^{N_c} \frac{d\omega}{\omega} \left(1 - \frac{\omega^2}{N_c^2}\right)^{\frac{1}{2}} \left[\frac{\omega}{\omega_c}\right]^{-3} \int_0^{\ell_\omega} \frac{d\ell}{\ell} \int_{-\ell}^{\ell} \exp(-\tau) \, m \, dm.$$

(22)

Obviously, the damping factor $\exp(-\tau)$ plays a crucial role in this integral: in order to extract angular momentum from the deep interior of the Sun, a wave must be able to reach that far, which requires its spherical order ℓ to be as small as possible. A drastic simplification consists in retaining only those waves for which the Sun would be transparent, if their frequency were not Doppler-shifted to zero. For moderate differential rotation, this leads to

$$\mathcal{L}_J(r) = \mathcal{L}_J(r_c) - \frac{4\pi r^2}{3} \frac{\rho_c v_c^3}{N_c \ell_c} \left(\frac{\omega_c{}^4}{I}\right) \frac{\Omega(r) - \Omega(r_c)}{\omega_c},$$

(23)

with $I(r)$ being the damping integral deduced from (20)

$$I(r) = \int_r^{r_c} K \, N^3 \frac{dr}{r^3}.$$

(24)

If the wave transport were the only present, the angular velocity would evolve according to

$$\frac{\partial}{\partial t}\left(\rho r^4 \Omega\right) = \frac{1}{2} \frac{\rho_c v_c^3}{N_c \ell_c} \frac{\partial}{\partial r}\left[r^2 \left(\frac{\omega_c{}^4}{I}\right) \frac{[\Omega(r) - \Omega(r_c)]}{\omega_c}\right].$$

(25)

From this equation one readily derives the timescale characterizing this transport process; approximating $\rho_c v_c^3$ by $1/10$ of the convective flux, and this flux by $L_\odot/4\pi r_c^2$, one obtains an estimate for the synchronization time t_{sync}, with $\bar{\rho}$ being the mean density (see the poster by Zahn $et\ al.$):

$$t_{sync} \approx 60 \frac{M_\odot R_\odot^2}{L_\odot} \frac{\rho}{\bar{\rho}} \left(\frac{r}{R_\odot}\right)^3 \left(\frac{r_c}{R_\odot}\right)^2 N_c \omega_c \ell_c \left(\frac{I}{\omega_c{}^4}\right) \approx 10^7 \text{ years.}$$

(26)

The transport efficiency evaluated independently by Kumar and Quataert (1996), using a somewhat different approach, is of the same order.

Those estimates are obtained with a number of simplifying assumptions, and more work is needed to refine the theory. For instance, the Coriolis force has been ignored in the description of the internal waves; it may well introduce an unbalance between waves of opposite azimuthal order m, leading to an extra momentum transport. Moreover, the recipes used to couple the internal waves with the convective motions are rather crude, and these are assumed to obey Kolmogorov's law. Fortunately, the result quoted above is not too sensitive to the slope of the power spectrum.

6. Conclusion

We have seen that the rotational mixing due to the meridian circulation and related shear instabilities seems to explain rather well the composition anomalies which are observed in rotating early-type stars (Talon et al. 1996). But, the picture may still be incomplete, and other physical processes may well compete in the transport of angular momentum, as it has become obvious recently for late-type stars. Furthermore, semi-convection contributes also to the mixing in the more massive stars, and it has still not received a satisfactory treatment.

For late-type stars, the situation is also more complex than it was thought before. Although the rotational mixing which we designated by type I is rather successful in modeling the lithium abundance as a function of age (see Pinsonneault et al. 1989, 1990; Chaboyer et al. 1995), it fails to predict the correct internal rotation of the Sun. Having reviewed the few mechanisms which have been proposed so far for the transport of angular momentum in the solar interior, we conclude that the process which seems to prevail, at least in the later stages of evolution, is the transport by internal waves. Its efficiency is so high that it can easily enforce the flat rotation profile revealed by helioseismology (Zahn et al. 1996). This result has been anticipated by Schatzman (1993), and it is confirmed also by the estimates of Kumar and Quataert (1996).

Magnetic torquing is another possibility, although its action depends crucially on the topology of the field, as discussed in §4. To our best knowledge, no consistent model has been published yet, with realistic boundary conditions at the base of the convection zone and with meridian circulation advecting the decaying poloidal field. But, there is no need to invoke magnetic fields. Most observational facts can be explained by the conjugated action of the rotation-induced meridian circulation and of the gravity waves. Let us sketch what we hold for the most plausible scenario.

During the early phases of the evolution of a solar-type star, while its rotation speed is large, the transport of angular momentum is ensured mainly through the meridian circulation, and the depletion of lithium is correlated with the loss of angular momentum by the wind. These processes combine in what we have called the rotational mixing of type I. Their strength scales as the deceleration rate; it thus decreases as the spin-down proceeds. At some point, to be determined by the more detailed calculations which are in progress, the transport of angular momentum by internal waves takes the leading role, and it tends to flatten the rotation profile. This will not stop the meridian circulation, which subsists even when rotation is uniform (the classical Eddington-Sweet case), but the depletion of lithium is then no longer correlated with the loss of angular momentum: instead, it will scale as Ω^2, and we have called this process rotational mixing of type II. Referring back to the tidally-locked binaries in the Hyades which have been discussed above, type I mixing is responsible for the difference in lithium abundance between these binaries and single stars of the same mass, whereas the depletion from the original value log $(N_{Li}) \approx 3$ to the present $\approx$ 1.5 occurs during the later phase, which is dominated by rotational mixing of type II (see Fig. 1).

Further work is required to confirm this scenario, and to adapt it to early-type stars. As top priorities, I would rank a realistic description for the anisotropic shear turbulence which is generated by differential rotation, and an improved treatment for the excitation of internal waves at the boundary of a convection zone. But, it is also very important to verify whether internal waves can contribute directly to the mixing, as it has been suggested by Press (1981), García López and Spruit (1991), and Schatzman (1993, 1996). Finally, we look forward for even tighter observational constraints, and hope that those will be provided soon thanks to helio- and asteroseismic techniques which have been discussed during this symposium.

References

Biermann, L. (1937) *Astron. Nachr.*, **263**, 185

Brown, T.M, Christensen-Dalsgaard, J., Dziembowski, W.A., Goode, P., Gough, D.O. and Morrow, C.A. (1989) *ApJ*, **343**, 526

Busse, F.H. (1981) *Geophys. Astrophys. Fluid Dynamics*, **17** , 215

Busse, F.H. (1982) *ApJ*, **259**, 759

Chaboyer, B., Demarque, P. and Pinsonneault, M.H. (1995) *ApJ*, **441**, 865

Charbonneau, P. and MacGregor, K.B. (1993) *ApJ*, **417**, 762

Eddington A.S. (1925) *Observatory*, **48**, 78

Endal A.S. and Sofia, S. (1978) *ApJ*, **220**, 279

Ferraro, V.C.A. (1937) *MNRAS*, **97**, 458

García López, R.J. and Spruit, H.C. (1991) *ApJ*, **377**, 268

Geiss, J. and Reeves, H. (1972) *A&A*, **18**, 126

Giess, D.R. and Lambert, D.L. (1992) *ApJ*, **387**, 673

Goldreich, P. and Nicholson, P.D. (1989) *ApJ*, **342**, 1079

Herbig, G.H. (1965) *ApJ*, **141**, 588

Herrero, A., Kudritzski, R.P., Vilchez, J.M., Kunze, D., Butler, K. and Haser, S. (1992) *A&A*, **261**, 209

Kumar, P. and Quataert, E.J. (1996) *ApJ*, (in press)

Lambert, D.L. (1976) *ApJ*, **210**, 684

Law, W.Y., Knobloch, E. and Spruit, H.C. (1984) *Observational Tests of Stellar Evolution Theory* (A. Maeder & A. Renzini ed.; Reidel)

Lyubimkov, L.S. (1984) *Astrofisika*, **20**, 475

Lyubimkov, L.S. (1991) *Evolution of Stars: The Photospheric Abundance Connection* (G. Michaud & A. Tutukov ed.; Kluwer), 125

Meynet, G. and Maeder, A. (1996) *A&A*, (in press)

Mestel, L. (1953) *MNRAS*, **113**, 716

Mestel, L. (1961) *MNRAS*, **122**, 473

Mestel, L. (1965) *Stellar Structure, in Stars and Stellar Systems* (G.P. Kuiper & B.M. Middlehurst; Univ. Chicago Press), **8**, 465

Mestel, L., Moss, D. and Taylor, R.J. (1988) *MNRAS*, **231**, 873

Pinsonneault, M.H., Kawaler, S.D., Sofia, S. and Demarque, P. (1989) *ApJ*, **338**, 424

Pinsonneault, M.H., Kawaler, S.D. and Demarque, P. (1990) *ApJS*, **74**, 501

Press, W.H. (1981) *ApJ*, **245**, 286

Roxburgh, I. (1963) *MNRAS*, **126**, 157

Ryan, S.G. and Deliyannis, C.P. (1995) *ApJ*, **453**, 819

Schatzman, E. (1962) *Ann. Astrophys.*, **25**, 18

Schatzman, E. (1969) *A&A*, **3**, 331

Schatzman, E. (1993) *A&A*, **279**, 431

Schatzman, E. (1996) *J. Fluid Mech.*, **322**, 355

Skumanich, A. (1972) *ApJ*, **171**, 563

Soderblom, D.R., Burton, F.J., Balachandran, S., Stauffer, J.R., Duncan, D.K., Fedele, S.B. and Hudon, J.D. (1993) *AJ*, **106**, 1059

Spruit, H.C. (1987) *The Internal Solar Angular Velocity* (B.R. Durney & S. Sofia ed.; Reidel) 185

Sweet, P.A. (1950) *MNRAS*, **110**, 548

Talon, S. and Zahn, J.-P. (1996) *A&A*, (in press)

Talon, S., Zahn, J.-P., Maeder, A. and Meynet, G. (1996) *A&A*, (in press)

Thompson, M.J., Toomre, J., Anderson, E.R., Antia, H.M., Berthomieu, G., Burtonclay, D., Chitre, S.M., Christensen-Dalsgaard, J., Corbard, T., DeRosa, M., Genovese, C.R., Gough, D.O., Haber, D.A., Harvey, J.W., Hill, F., Howe, R., Korzennik, S.G., Kosovichev, A.G., Leibacher, J.W., Pijpers, F.P., Provost, J., Rhodes Jr., E.J., Schou, J., Sekii, T., Stark, P.B. and Wilson, P.R. (1996) *Science*, **272**, 1300

Thorburn, J.A., Hobbs, L.M., Deliyannis, C.P. and Pinsonneault, M.H. (1993) *ApJ*, **415**, 150

Vogt, H. (1925) *Astron. Nachr.*, **223**, 229

Zahn, J.-P. (1974) *Stellar Instability and Evolution* (P. Ledoux, A. Noels & R.W. Rogers ed.; Reidel, Dordrecht), 185

Zahn, J.-P. (1975) *A&A*, **41**, 329

Zahn, J.-P. (1990) *Inside the Sun* (G. Berthomieu & M. Cribier ed.; Kluwer) 425

Zahn, J.-P. (1991) *A&A*, **252**, 179

Zahn, J.-P. (1992) *A&A*, **265**, 115

Zahn, J.-P. (1994) *A&A* **288**, 829

Zahn, J.-P., Talon, S. and Matias, J. (1996) *A&A* (in press)

INTERNAL SOLAR ROTATION

TAKASHI SEKII

Institute of Astronomy, University of Cambridge
Madingley Road, Cambridge CB3 0HA, UK

1. Introduction

The rotation of the sun affects the wave propagation in the solar cavity. The measurement of the effect can be used to infer the internal rotation of the sun, and indeed much of what we know today about the dynamics of the solar interior has been derived from observation and interpretation of the rotationally induced splitting of solar p-mode frequencies. Inversion has proven to be a very powerful tool in such investigations, and will remain at the centre of our effort in studying solar rotation from helioseimic data. Study of the flow inside the sun is not only important in its own right but is vital for improving our understanding of solar activity and its cycle. Now that both GONG and SOHO are operational and are accumulating data, there is no doubt that we are about to learn a great deal more about the dynamical structure of our own star.

In this review various inversion techniques, specifically in the context of the rotation inversion, are discussed. Then, what we have learned so far is summarized, and results from the latest observations are discussed, particularly those from the GONG experiment (Harvey *et al.* 1996).

2. Rotational splitting

Normal modes of the sun, a three-dimensional body, are identified by three indices: radial order n, spherical harmonic degree l and azimuthal order m. In the absence of any symmetry-breaking agent, the eigenfrequency has $(2l + 1)$-fold degeneracy in m. The solar rotation breaks the spherical symmetry and lifts this degeneracy, causing the frequency to split into $2l + 1$ different values. Let us describe rotation by the two-dimensional distribution of angular velocity $\Omega(r, \mu)$, where r is the radial coordinate and μ is the cosine of the colatitude θ. By treating rotation as a perturbation to

189

J. Provost and F.-X. Schmider (eds.), Sounding Solar and Stellar Interiors, 189-202.
© 1997 *IAU. Printed in the Netherlands.*

a spherically symmetric sun, and by applying linear perturbation theory, we obtain

$$\Delta\omega_{nlm} \equiv \omega_{nlm} - \omega_{nl} = m\langle\Omega\rangle_{nlm} \ , \tag{1}$$

where ω_{nlm} is the frequency in the presence of the rotation, ω_{nl} is the unperturbed degenerate frequency and $\langle\Omega\rangle_{nlm}$ denotes a weighted integral of angular velocity $\Omega(r,\mu)$:

$$\langle\Omega\rangle_{nlm} = \frac{\Delta\Omega_{nlm}}{m} = \int_0^R dr \int_{-1}^1 d\mu K_{nlm}(r,\mu)\Omega(r,\mu) \ , \tag{2}$$

where $K_{nlm}(r,\mu)$ is the splitting kernel for the mode under consideration, representing the extent to which $\Omega(r,\mu)$ influences this particular mode. Observationally, $\langle\Omega\rangle_{nlm}/\omega_{nl} \sim 0.5\mu\text{Hz}/3\text{mHz} \sim 10^{-4}$, and this justifies the application of linear perturbation theory. Our aim is, given a set of observations of rotational splitting, $\{\Delta\omega_{nlm}\}$, to solve the corresponding set of linear integral equations (2) for the angular velocity $\Omega(r,\mu)$.

There is more than one way to write down $K_{nlm}(r,\mu)$ explicitly (e.g., Hansen, Cox and van Horn 1971), but the following is convenient in presenting the mathematical structure:

$$K_{nlm}(r,\mu) = K_{nl}(r)W_{lm}(\mu) + L_{nl}(r)X_{lm}(\mu) \ , \tag{3}$$

where

$$\begin{align}
K_{ln}(r) &= (\xi_r^2 + \{l(l+1) - 1\}\xi_h^2 - 2\xi_r\xi_h)\rho r^2/I_{ln} \tag{4}\\
L_{ln}(r) &= \xi_h^2 \rho r^2/I_{ln} \tag{5}\\
W_{lm}(\mu) &= P_l^m(\mu)^2 \tag{6}\\
X_{lm}(\mu) &= \frac{d}{d\mu}\left[\frac{1-\mu^2}{2}\frac{d}{d\mu}W_{lm}(\mu) + \mu W_{lm}(\mu)\right] \tag{7}\\
I_{ln} &= \int[\xi_r^2 + l(l+1)\xi_h^2]\rho r^2 dr \ , \tag{8}
\end{align}$$

ρ is density, P_l^m is a normalized associated Legendre function and ξ_r and ξ_h determine the radial and horizontal components of the radial part of the displacement vector $\boldsymbol{\xi}_{nlm}$ according to

$$\boldsymbol{\xi}_{nlm} = \left[\ \xi_r(r),\ \xi_h(r)\frac{\partial}{\partial\theta},\ \xi_h(r)\frac{1}{\sin\theta}\frac{\partial}{\partial\phi}\ \right]P_l^m(\cos\theta)e^{im\phi - \omega_{nlm}t} \ , \tag{9}$$

with respect to spherical polar coordinates (r,θ,ϕ). Some examples of the splitting kernels are shown in Figure 1.

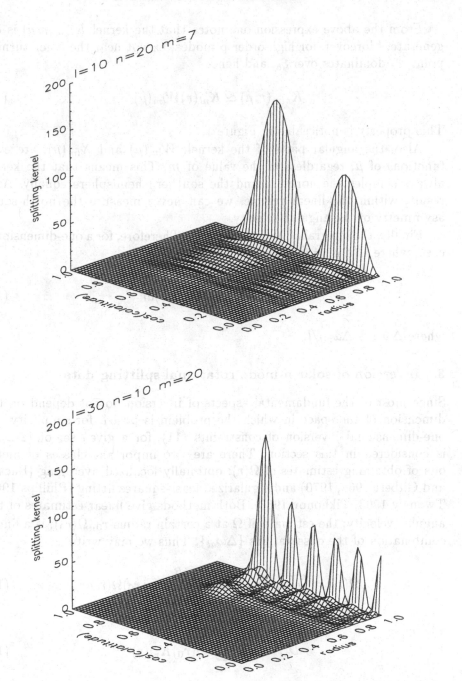

Figure 1.   Some examples of the splitting kernels, represented as surface plots against r, the radial coordinate, and μ, the cosine of the colatitude. $(l, n, m) = (10, 20, 7)$ [top] and $(30, 10, 20)$ [bottom].

From the above expression one notes that the kernel $K_{nlm}(r, \mu)$ is degenerate. Moreover, for high-order p modes, except near the inner turning point, ξ_r dominates over ξ_h, and hence

$$K_{nlm}(r, \mu) \simeq K_{nl}(r)W_{lm}(\mu) . \tag{10}$$

This property is noticeable in Figure 1.

Also, the angular parts of the kernel, $W_{lm}(\mu)$ and $X_{lm}(\mu)$, are even functions of μ, regardless of the value of m. This means that the kernel always samples the northern and the southern hemisphere equally. As a result, within the linear regime, we can never measure the north-south asymmetry of the angular velocity.

Finally, the integral of $X_{lm}(\mu)$ vanishes. Therefore, for a one-dimensional case, where $\Omega(r, \mu) = \Omega(r)$, we have

$$\Delta\omega_{nl} = \int_0^R K_{nl}(r)\Omega(r)dr , \tag{11}$$

where $\Delta\omega_{nl} \equiv \Delta\omega_{nll}/l$.

3. Inversion of solar p-mode rotational-splitting data

Since most of the fundamental aspects of inversion do not depend on the dimension of the space in which the problem is posed, for simplicity the one-dimensional inversion of constraints (11), for a given set of $\{\Delta\omega_{nl}\}$, is considered in this section. There are two important classes of methods of obtaining estimates of $\Omega(r)$: optimally localized averaging (Backus and Gilbert 1968, 1970) and regularized least-squares fitting (Phillips 1962, Twomey 1963, Tikhonov 1963). Both methods give linear estimates of the angular velocity; the estimate of Ω at a certain radius r_0, $\hat{\Omega}(r_0)$, is a linear combination of the observations $\{\Delta\omega_{nl}\}$. Thus we may write

$$\hat{\Omega}(r_0) = \sum_{nl} c_{nl}(r_0)\Delta\omega_{nl} = \int_0^R D(r; r_0)\Omega(r)dr , \tag{12}$$

where

$$D(r; r_0) \equiv \sum_{nl} c_{nl}(r_0)K_{nl}(r) \tag{13}$$

is the *averaging kernel*. The coefficients $c_{nl}(r_0)$ are called inversion coefficients at the target radius r_0. To discuss the error in the estimate, in general we should consider the the error covariance matrix of $\Delta\omega_{nl}$ (*e.g.*, Gough 1996). Once again for simplicity, however, let us assume that the errors are independent and the matrix is diagonal. Then the standard deviation in the estimate $\hat{\Omega}(r_0)$ is given by

$$\delta\hat{\Omega}(r_0) = \left(\sum_{nl} c_{nl}(r_0)^2 \sigma_{nl}^2 \right)^{1/2} , \qquad (14)$$

where σ_{nl} is the standard deviation of $\Delta\omega_{nl}$.

3.1. OPTIMALLY LOCALIZED AVERAGING (OLA)

The basic idea of optimally localized averaging (OLA) is readily explained by equations (12) – (14). Since we would like $\hat{\Omega}(r_0)$ to be as close as possible to $\Omega(r)$, we would like the averaging kernel $D(r; r_0)$ to be well localized. However, the more one localizes $D(r; r_0)$, usually the greater is the amplitude of the coefficients $c_{nl}(r_0)$, and hence the greater error in the estimate: a trade-off between resolution and error must be sought.

In OLA one seeks a balance by minimizing, e.g.,

$$S \equiv \int D(r; r_0)^2 (r - r_0)^2 dr + \alpha \sum_{nl} c_{nl}^2 \sigma_{nl}^2 , \qquad (15)$$

subject to the unimodular condition

$$\int_0^R D(r; r_0) dr = 1 , \qquad (16)$$

where α is a regularization parameter which controls the trade-off.

An important variant of the standard OLA described above is the subtractive optimally localized averaging (SOLA) in the form considered by Pijpers and Thompson (1992). In SOLA one attempts to fit $D(r; r_0)$ to a prescribed function such as a Gaussian centred at a target radius with an appropriate width.

3.2. REGULARIZED LEAST-SQUARES FITTING (RLS)

In regularized least-squares fitting (RLS), one seeks to fit the data by minimizing

$$S = \sum_{nl} \frac{1}{\sigma_{nl}^2} \left[\Delta\omega_{nl} - \int_0^R K_{nl}(r)\Omega(r)dr \right]^2 + \alpha \int_0^R [\mathcal{L}\Omega(r)]^2 dr , \qquad (17)$$

where α is again a regularization parameter and $\mathcal{L}$ is, say, a differential operator that measures the 'smoothness' of $\Omega(r)$, or some other property that one might wish to optimize. Minimization of S leads to another integral

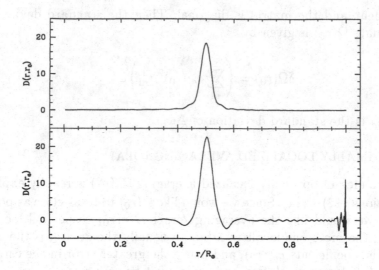

Figure 2. Examples of averaging kernel for $r_0 = 0.5$, obtained by an OLA inversion (top) and an RLS inversion (bottom). The mode set is identical to the GONG Hare-and-Hound exercise (Gough and Toomre 1993).

equation that involves the adjoint of $\mathcal{L}$, which one might solve by expanding $\Omega(r)$ in a series of functions. Usually, however, the expansion is carried out before the minimization. Either way, it can be shown that we can still write the estimate of $\Omega(r)$ at a certain radius r_0 as a linear combination of the data in the manner of equation (12).

Had the regularization term not been included, in the case of solar p modes the minimization of S would require the inversion of a nearly singular matrix producing an 'ill-posed problem', and the solution would typically contain spurious highly oscillatory components which are generated by numerical and data errors. This is due to the fact that all the splitting kernels have large amplitude near the surface with a large amount of redundancy in the information that the datasets carry.

3.3. COMPARISON BETWEEN OLA AND RLS

Helioseismologists have been using both OLA and RLS for their inversions. Between the two methods, OLA gives better localization, because the method is designed that way. Figure 2 compares typical averaging kernels produced from an OLA inversion and from a RLS inversion. The OLA averaging kernel has almost no sidelobes and no surface contamination, both of which are undesirable. Depending on the mode set, even OLA kernels can have these features, but to a much less extent than the RLS kernels. As was pointed out by Christensen-Dalsgaard, Schou and Thompson (1990), however, the negative sidelobes can contribute to improve the estimate. For example, if $\Omega(r)$ is convex (or concave) in the region where the averaging is taking place, the localized average by a positive definite kernel would underestimate (or overestimate) $\Omega(r)$, but negative sidelobes push up (or down) the estimate, provided that the averaging kernel stays unimodular. This is what the regularization term does: interpolation and extrapolation. Part of the side lobes therefore maybe accounted for as a designed property of RLS methods. Still, OLA gives better averaging kernels overall, and estimates that are more robust and more easily interpretable.

On the other hand, RLS is more responsive to internal inconsistency of the data, if it exists. Since an OLA method is concerned only about localization, it is left to helioseismologists to check the consistency of the data, but a RLS method directly looks at the misfit, and therefore such inconsistency, including ones arising from underestimating random errors, will show up immediately.

RLS is computationally cheaper, too. An OLA inversion involves inverting a matrix of size M, where M is the number of modes, per target point. If one needs to estimate $\Omega(r)$ on N_T target points, this takes $\sim M^3 N_T$ flops. RLS involves inverting a matrix of size N, where N is the number of functions in the expansion, or, in many cases, the number of grid points. So this takes $\sim N^3$ flops. If $M = 1000$, $N = 500$ and $N_T = 100$, then the ratio of the operation count is ~ 800. For a two-dimensional case, this ratio increases. For $M = 10^5$, $N = 5000$ and $N_T = 1000$, the ratio is $\sim 8 \times 10^6$.

In spite of these differences, in the real cases in helioseismology, OLA and RLS are usually in general agreement (see Figure 4, next section). However, this does not mean that the difference between the two is merely an operational one. To make this point, let us consider a linear equation

$$A x = b \,, \tag{18}$$

where A is an $m \times n$ matrix, and x and b are vectors of size n and m, respectively. From this one obtains $\hat{x}$, an estimate of x, as

$$\hat{x} = R b \,. \tag{19}$$

The resolution matrix R depends on the method one chooses to obtain the estimate. The estimate $\hat{x}$ is related to the real solution x through the following formula (hence the name 'resolution' matrix)

$$\hat{x} = R b = R A x \,, \tag{20}$$

while the vector of misfit is given by

$$b - A \hat{x} = (I_m - A R) b \,, \tag{21}$$

where I_m is the identity matrix of size m. In the absence of data errors, OLA seeks to choose R in such a way that RA is as close as possible to I_n, the identity matrix of size n:

$$RA \to I_n \ . \tag{22}$$

On the other hand, RLS seeks to reduce the misfit and therefore aim to achieve

$$AR \to I_m \ . \tag{23}$$

We may write, therefore,

$$R_{OLA} \simeq A_L^{-1} \ , \quad R_{RLS} \simeq A_R^{-1} \ , \tag{24}$$

where R_{OLA} and R_{RLS} are the resolution matrices obtained from an OLA and an RLS inversion, respectively, and A_L^{-1} and A_R^{-1} denote the left and the right inverses. This suggests the fundamental difference and the complementary nature of OLA methods and RLS methods. It also means that these two classes of methods are not simply the ones we *happen* to know, but that they are *the* two fundamental bases.

We know from linear algebra that if one of either the right or the left inverse exists, then the other inverse exists, too, and they are identical. In this sense, the general agreement between the two inverses in helioseismic inversions is another indication that what we have been doing is not too far off the mark.

4. Inversions in two dimensions

Before GONG, measurement of individual splittings ω_{nlm} were not available. Instead, it was customary to expand the splittings in the form

$$\frac{\Delta \omega_{nlm}}{2\pi} = \sum_k a_k(n, l) \mathcal{P}_k(m; l) \ ,$$

where $\mathcal{P}_k(m; l)$ is some polynomial and a_k is an expansion coefficient. Ritzwoller and Lavely (1991) pointed out that it is advantageous to choose polynomials $\mathcal{P}_k(m; l)$ that are orthogonal in the discrete space of m. In any case, inversions of expansion coefficients normally reduce to inverting each set of expansion coefficients (at fixed k), which are one-dimensional inversions, and then combining them together.

Mathematically, inference of two-dimensional rotation, $\Omega(r, \theta)$, from individual splitting frequencies $\Delta \omega_{nlm}$, is no different from inference in one dimension. Computationally, however, it is much harder. With the number of modes of the order of 10^5, a straightforward application of OLA is

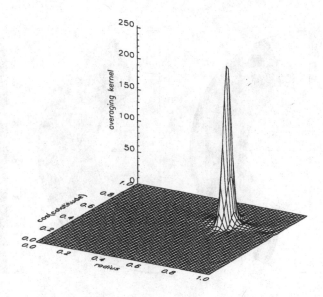

Figure 3. Surface plot of an averaging kernel obtained by a 1D⊗1D OLA inversion, centred at $r = 0.7R_\odot$, $\mu = 1/\sqrt{2}$.

prohibitively expensive. This was the reason the first two-dimensional inversions were RLS inversions (*e.g.* Sekii 1990, 1991; Schou 1991, also Corbard *et al.* 1995)

A fully two-dimensional OLA was presented by Christensen-Dalsgaard *et al.* (1995). To invert a huge matrix they carried out singular-value decomposition and then truncated the spectral expansion to reduce the size of the problem. However, singular-value decomposition itself is a computationally intensive procedure, and therefore a fully two-dimensional OLA remains an extremely expensive method.

Sekii (1993a) pointed out that a property of the splitting kernels of the solar p-modes (equation 10) can be exploited to decompose the two-dimensional problem into successive one-dimensional inversions (1D⊗1D inversions). The sun's spherical geometry introduces a complexity, as was discussed by Sekii (1993b), but this decomposition still reduces the computational labour substantially, and it is possible to formulate OLA methods on this basis (Sekii 1993b, 1995; Pijpers & Thompson 1996). An example of an averaging kernel obtained by this method is shown in Figure 3.

5. What we have learned

What we know today about the internal rotation of the sun, from observations of the past and the present, and from various analysis of them, are

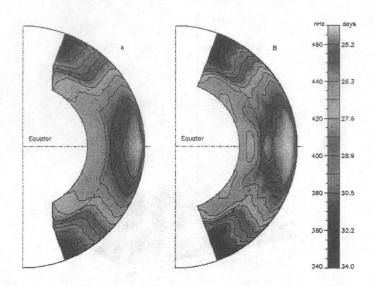

Figure 4.   Rotation rate in the sun inferred from GONG data, by a 1D⊗1D OLA inversion (left) and by a 2D RLS inversion (right). From Thompson *et al.* (1996).

summarized as follows:

– the latitudinal variation of the angular velocity observed at the surface more-or-less persists throughout the convection zone,
– there appears to be weak differential rotation in the radiative interior,
– the maximum of the angular velocity occurs at $r/R_\odot \simeq 0.95$ (unless we have a rapidly spinning core!),
– there is suspected a shear layer immediately beneath the surface,
– there is a shear layer beneath the base of the convection zone.

Such features are seen in the inversions shown in Figure 4 (Thompson *et al.* 1996), obtained from 4 months of GONG data, and are in agreement with the LOWL result (Tomczyk, Schou and Thompson 1995), the latest GONG result (Korzennik *et al.* 1996) and the MDI result (Kosovichev *et al.* 1996).

A 1D⊗1D OLA inversion of the latest GONG data (Korzennik *et al.* 1996) is shown in Figures 5 and 6. The result is consistent with those shown in Figure 4. To investigate the structure below the base of the convection zone specifically, Douglas Gough and I have carried out a nonlinear least-squares fitting. (Figure 7). An analytical model of the rotation rate, comprising a rigidly rotating interior and a differentially rotating outer layer (latitudinal dependence is independent of radius), with a transition zone between them, is fitted in the least-squares sense to the GONG data. One

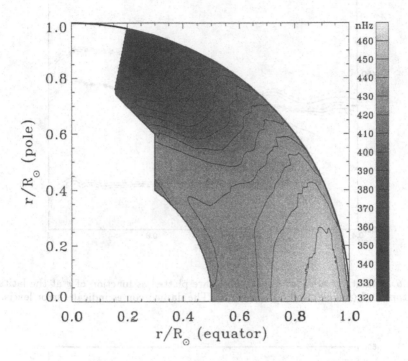

Figure 5. Contour diagram of a 1D⊗1D OLA inversion of the latest GONG data (Korzennik *et al.* 1996).

might note that the positions of the upper and the lower boundaries of the transition zone are nonlinearly related to the measurements, and that the fitting is accomplished by iteration. Please compare the result with that shown in Figure 6. The result indicates that the zone is centred at $r = 0.696R_\odot$, and the thickness of the zone is $0.064R_\odot$. Such a transition zone, known as a tachocline, was suggested to be a site of turbulent mixing by Spiegel and Zahn (1992) (cf. Gough and Sekii 1996).

Finally, an inversion in deeper layers was attempted (Figure 8). The averaging kernel is well localized, and the result suggests a rate slower than the surface value. This agrees with BiSON (Chaplin *et al.* 1996) and LOWL but not IRIS (Lazrek *et al.* 1996).

6. Future problems

In spite of our long efforts, the kinematics in the core remains largely uncertain.

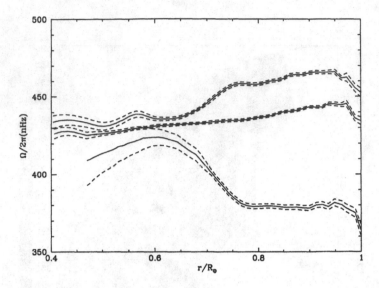

Figure 6. Rotation rate shown in Figure 5 are plotted as function of r at the latitudes of $0°$ (top), $30°$ (middle) and $60°$ (bottom). The dashed curves indicate error levels.

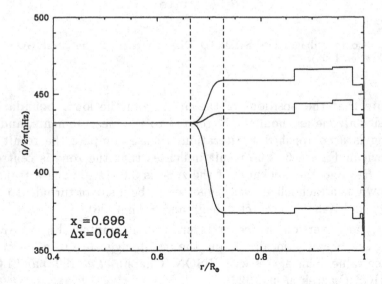

Figure 7. Result of a nonlinear least-squares fitting to the latest GONG data (Korzennik *et al.* 1996). Rotation rates at the latitudes of $0°$ (top), $30°$ (middle) and $60°$ (bottom) are obtained from fitting an analytic expression to the data. The transition zone is found to be centred at $r = 0.696R_\odot$ with a thickness $\Delta r = 0.064R_\odot$. The vertical dashed lines indicate the position of the upper and the lower boundaries.

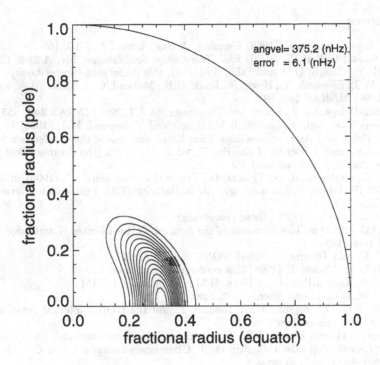

Figure 8. An averaging kernel obtained from a 1D⊗1D OLA inversion of the latest GONG data (Korzennik *et al.* 1996. The estimated rotation rate is $\Omega/2\pi = 375.2 \pm 6.1(\mathrm{nHz})$.

The low-degree splitting measurement is a key to the issue, but observations are still not in mutual agreement, as was discussed by Pallé (1996). The disagreement may be coming from the difference in data analysis procedures, which shows how difficult such tasks indeed are (cf. Chang, Gough and Sekii 1996). G-modes, if found, would play an extremely important role in our investigation of the core. The helioseismic instruments on SOHO, as they accumulate more data, might solve both problems.

A new method of investigating the dynamics of solar interior is now developing. Local-helioseismological approaches do not rely on frequency splitting as such, but on measuring the travel times of acoustic waves in the sun between various parts of the solar surface. Inferences concerning the internal flow in three dimensions have been made (Kosovichev and Duvall 1996, Duvall *et al.* 1996). It is still unclear how deep down into the solar interior such an approach is effective, but this is certainly an exciting new tool in helioseismology that enables us to investigate the dynamics of the outer layer in great detail.

References

Backus, G. and Gilbert, F. (1968) *Geophys. J. Roy. Astron. Soc.* **16**, 169

Backus, G. and Gilbert, F. (1970) *Phil. Trans. Roy. Soc. London, Ser. A* **266**, 123

Chang, H.-Y., Gough, D.O. and Sekii, T. (1996), this conference (poster book)

Chaplin, W.J., Elsworth, Y., Howe, R., Isaak, G.R., McLeod, C.P., Miller, B.A. and New, R. (1996) *MNRAS* **280**, 849

Christensen-Dalsgaard, J., Schou and Thompson, M.J. (1990) *MNRAS* **242**, 353

Christensen-Dalsgaard, J., Larsen, R.M., Schou and Thompson, M.J. (1995) in *GONG 1994: Helio- and Astero-Seismology from Earth and Space*, the Conference Series of the Astronomical Society of the Pacific, ed R. K. Ulrich (the Astronomical Society of the Pacific, San Francisco), 70

Corbard, T., Berthomieu, G., Gonczi, G., Provost, J. and Morel, P. (1995) in *The 4th SOHO Workshop, Helioseismology*, ed. B. Battrick (ESA Publication Division, Noordwijk), 289

Duvall, T.L., Jr. *et al.* (1996) these proceedings

Gough, D.O. (1996) in *The Structure of the Sun*, ed T. Roca Cortés (Cambridge University Press), 141

Gough, D.O. and Toomre, J. (1993) *GONG Report No. 11*

Gough, D.O. and Sekii, T. (1996) this conference (poster book)

Hansen, C.J., Cox, J.P. and van Horn, H.M. (1977) *ApJ* **217**, 151

Harvey, J.W. *et al.* (1996) *Science* **272**, 1284

Korzennik, S.G., Thompson, M.J., Toomre, J. and the GONG internal rotation team (1996) these proceedings

Kosovichev, A.G. and Duvall, T.L., Jr. (1996) in *Solar Convection and Oscillations and their Relationship* eds. F.P. Pijpers, J. Christensen-Dalsgaard and C.S. Rosenthal (Kluwer Academic), in press

Kosovichev, A.G. *et al.* (1996) *Sol. Phys.*, submitted

Lazrek, M *et al.* (1996) *Sol. Phys.*, **166**, 1

Pallé, P.L. (1996) these proceedings

Phillips, D.L. (1962) *J.Ass.Comput. Mach.* **9**, 84

Pijpers, F.P. and Thompson, M.J. (1992) *A&A* **262**, L33

Pijpers, F.P. and Thompson, M.J. (1996) *MNRAS* **279**, 498

Ritzwoller, M.H. and Lavely, E.M. (1991) *ApJ* **369**, 557

Schou, J. (1991) in *Challenges to theories of the structure of moderate-mass stars*, eds. D.O. Gough and J. Toomre (Springer, Heidelberg)

Sekii, (T. 1990) in *Progress of Seismology of the Sun and Stars*, eds. Y. Osaki and H. Shibahashi (Springer-Verlag, Berlin), 337

Sekii, T. (1991) *PASJ*, **43**, 381

Sekii, T. (1993a) in *GONG1992: Seismic Investigation of the Sun and Stars*, the Conference Series of the Astronomical Society of the Pacific, ed T.M. Brown (the Astronomical Society of the Pacific, San Francisco), 237

Sekii, T. (1993b) *MNRAS* **264**, 1018

Sekii, T. (1995) in *The 4th SOHO Workshop, Helioseismology*, ed. B. Battrick (ESA Publication Division, Noordwijk), 285

Spiegel, E.A. and Zahn, J.-P. (1992) *A&A*, **265**, 106

Thompson, M.J. *el al* (1996) *Science* **272**, 1300

Tomczyk, S., Schou, J. and Thompson, M.J. (1995) *ApJ*, L57

Tikhonov, A.N. (1963) *Sov. Maths-Dokl.* **4**, 1035

Twomey, S. (1963) *J. Ass. Comput. Mach.* **10**, 97

INTERNAL STRUCTURE AND ROTATION OF THE SUN: FIRST RESULTS FROM THE MDI DATA

A.G. KOSOVICHEV, J. SCHOU, P.H. SCHERRER , R.S. BOGART,
R.I. BUSH, J.T. HOEKSEMA, J. ALOISE , L. BACON,
A. BURNETTE, C. DE FOREST, P.M. GILES , K. LEIBRAND,
R. NIGAM, M. RUBIN , K. SCOTT AND S.D. WILLIAMS
W.W. Hansen Experimental Physics Laboratory,
Stanford University, Stanford, CA 94305, USA

SARBANI BASU AND J. CHRISTENSEN-DALSGAARD
Theoretical Astrophysics Center, Danish National Research
Foundation, and Institute of Physics and Astronomy,
Aarhus University, DK-8000 Aarhus C, Denmark

W. DÄPPEN AND E.J. RHODES, JR.
Department of Physics and Astronomy, University of Southern
California, Los Angeles, CA 90089, USA

T.L. DUVALL, JR.
Laboratory for Astronomy and Solar Physics, NASA Goddard
Space Flight Center, Greenbelt, MD 20771, USA

R. HOWE AND M.J. THOMPSON
Astronomy Unit, Queen Mary and Westfield College, London,
E1 4NS, UK

D.O. GOUGH AND T. SEKII
Institute of Astronomy, and Department of Applied Mathematics
and Theoretical Physics, Madingley Road, Cambridge, CB3
0HA, UK

J. TOOMRE
JILA, University of Colorado, CO 80309, USA

T.D. TARBELL, A.M. TITLE, D. MATHUR , M. MORRISON,
J.L.R. SABA , C.J. WOLFSON AND I. ZAYER
Lockheed-Martin Advanced Technology Center, 91-30/252, 3251
Hanover St., Palo Alto, CA 94304.

P.N. MILFORD
Parallel Rules, Inc. 41 Manzanita Ave., Los Gatos, CA 95030

J. Provost and F.-X. Schmider (eds.), Sounding Solar and Stellar Interiors, 203-210.
© 1997 IAU. Printed in the Netherlands.

Abstract. The Medium-l Program of the Michelson Doppler Imager (MDI) instrument on board SOHO provides continuous observations of oscillation modes of angular degree, l, from 0 to ~ 300. The initial results show that the noise in the Medium-l oscillation power spectrum is substantially lower than in ground-based measurements. This enables us to detect lower amplitude modes and, thus, to extend the range of measured mode frequencies. The MDI observations also reveal the asymmetry of oscillation spectral lines. The line asymmetries agree with the theory of mode excitation by acoustic sources localized in the upper convective boundary layer. The sound-speed profile inferred from the mean frequencies gives evidence for a sharp variation at the edge of the energy-generating core. In a thin layer just beneath the convection zone, helium appears to be less abundant than predicted by theory. Inverting the multiplet frequency splittings from MDI, we detect significant rotational shear in this thin layer.

1. MDI Medium-l Program

MDI has three basic helioseismology programs: Medium-l, Low-l, and Dynamics (Scherrer *et al.*, 1996). The Medium-l data are spatial averages of the full-disk Doppler velocity out to 90% of the disk's radius measured each minute. This results in 26,000 bins of approximately 10 arcsecond resolution that provide sensitivity to solar p modes up to $l = 300$. The Low-l observables are velocity and continuum intensity images summed into 180 bins, with the intent of detecting oscillations up to $l = 20$. The Dynamics Program provides 1024×1024 images of the whole disk, thus covering all of the p modes up to $l = 1500$. However, the Dynamics Program can run continuously for only 2 months each year when the high-rate telemetry channel is available.

The Medium-l and Low-l Programs have been run with almost no interruptions since 18 April 1996. The first Dynamics observations have been carried out from 23 May to 24 July, 1996. In this paper, we present first results from the Medium-l Program.

To optimize the Medium-l observing program, we performed simulations of several vector-weighted binning schemes, using a 20-hour series of full-disk Dopplergrams obtained on 25 January 1996. An optimal set of Gaussian weights has been found that substantially reduces spatial aliasing in the angular degree range from 0 to 300 while preserving most of the power of the modes (see Kosovichev *et al.*, 1997).

2. Medium-*l* Power Spectrum

A standard helioseismic data analysis procedure has been applied to obtain the oscillation power spectra from the Medium-*l* data. 60 days of data have been processed.

The *m*-averaged power spectrum, obtained from 10 days of the data is shown in Figure 1. The ridges in the diagrams correspond to modes of different radial order *n*. The lowest weak ridge is the f mode.

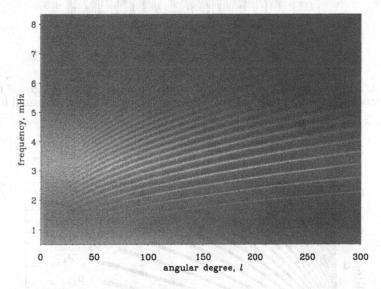

Figure 1. Power spectrum ($l - \nu$ diagram) obtained from the MDI Medium-*l* data for the modes averaged over *m*.

An important difference between the MDI Medium-*l* Program and the ground-based networks is the ability to detect the low-*n* low-frequency modes that carry substantial independent information about the solar structure and dynamics. However, the amplitude of the low-*n* modes is very small, so long stable time series are required to detect them. One example of low-frequency modes: the f mode of $l = 95$, is shown in Figure 2. Parameters of this modes have been measured for the first time. With longer time series, we should be able to observe more low-frequency modes, with even lower amplitudes. So far, the smallest mode amplitude we have been able to detect was about 1 mm/s.

The frequencies of the solar oscillations have been measured as described by Schou (1992). For the initial frequency measurements, we approximated the line profiles by Lorentzians and used a maximum likelihood method to determine the parameters of the Lorentzians, taking into account possible

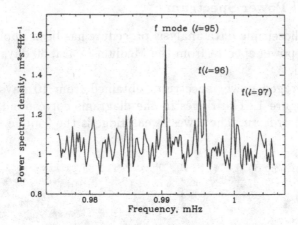

Figure 2. Power spectrum of f mode of $l = 95$ obtained from 2 months of the MDI data. Modes of adjacent l leak into the spectrum because of imperfect spatial filtering.

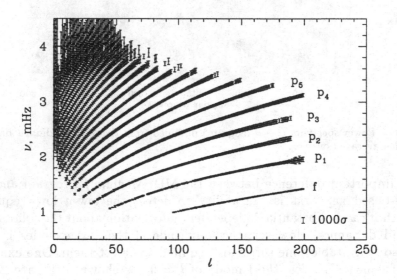

Figure 3. Mean frequencies of the mode multiplets obtained from 2 months of the Medium-l data. The error bars show the standard errors multiplied by 1000.

overlaps of lines. The mean frequencies of mode multiplets are shown in Figure 3. The error bars in this figure indicate estimated errors multiplied by 1000.

3. Line Asymmetry

Because of the significant reduction in the noise, the Medium-l data have revealed interesting characteristics of the line profiles of the oscillation power spectra. Figure 4 shows the power spectra of the modes for $l = 200$.

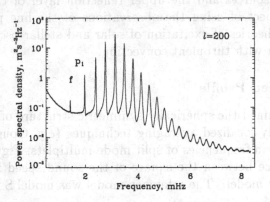

Figure 4. Power spectrum of $l = 200$ modes obtained from the MDI medium-l data.

The most interesting feature is the asymmetry of the line profiles. Though the asymmetry has been noticed in the ground-based data (Duvall *et al.*, 1993), its properties have not reliably measured. Several authors have studied this problem theoretically and have found that there is an inherent asymmetry whenever the waves are excited by a localized source (*e.g.* Gabriel, 1992; Kumar *et al.* 1994).

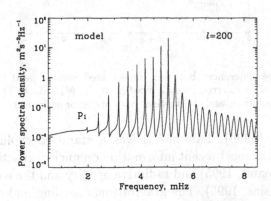

Figure 5. Theoretical power spectrum of $l = 200$ modes for a model with a source of the uniform spectral density localized 80 km beneath the photosphere. The background noise is assumed to be uniform with the power spectral density 10^{-2} m^{2}s^{-2}Hz^{-1}. (Nigam and Kosovichev, 1996)

Physically, the asymmetry is an effect of interference between an outward direct wave from the source and a corresponding inward wave that passes through the region of wave propagation. Figure 5 shows a theoretical power spectrum of p modes of $l = 200$ obtained by Nigam and Kosovichev (1996). The degree of the asymmetry depends on the relative locations of the acoustic sources and the upper reflection layer of the modes. This opens the prospect of using the observations of the line profiles of solar modes to test theories of excitation of solar and stellar oscillations and of their interaction with turbulent convection.

4. Sound-Speed Profile

We have determined the spherically symmetric structure of the Sun by using the optimally localized averaging techniques (*e.g.* Gough *et al.*, 1996) to invert the mean frequencies of split mode multiplets. Figure 6 shows the relative difference between the square of the sound speed in the Sun and a reference solar model. The reference model was model S of Christensen-

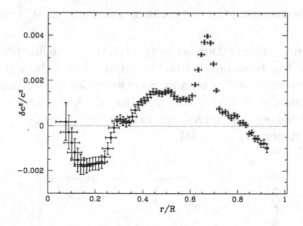

Figure 6.   Relative differences between the squared sound speed in the Sun and a standard solar model as inferred from 2 months of the MDI data. The horizontal bars show the spatial resolution, and the vertical bars are error estimates.

Dalsgaard *et al.* (1996). This model was a standard evolutionary model computed using the most recent information on nuclear reaction rates (Bahcall and Pinsonneault, 1995) and radiative opacity and the equation of state (Rogers and Iglesias, 1996). The gravitational settling and diffusion of helium and heavier elements were taken into account following the theory by Michaud and Proffitt (1993).

    The inversion results show that the maximum difference in the square of the sound speed between the model and the Sun is only 0.4%. Two features of the sound-speed profile are particularly notable. The first is a narrow peak centered at 0.67 R, just beneath the convection zone. This peak was previously detected in the LOWL (Basu *et al.*, 1996) and GONG data (Gough *et al.*, 1996) and is most likely due to a deficit of helium in this narrow region.

Another interesting feature is the sharp decrease of the sound speed compared to the model at the boundary of the energy-generating core, at 0.25 R. It is quite possible that the drop in the sound speed results from an overabundance of helium at the edge of the solar core. The steep increase of the sound speed towards the solar center can be explained if helium is less abundant than in the standard solar model. This indicates that the helium abundance profile seems to be more flat in the solar core than it is in the standard solar model. Gough *et al.* (1996) came to a similar conclusion from the GONG data. However, if the transition at the edge of the core is really as sharp as we have found from the initial MDI data, then it strongly suggests material redistribution by macroscopic motion in the core, possibly induced by the instability of ^{3}He burning, as first suggested by Dilke and Gough (1972).

5. Rotation Rate

The internal rotation rate at three latitudes, 0 (equator), 30 and 60 degrees, is shown in Figure 7.

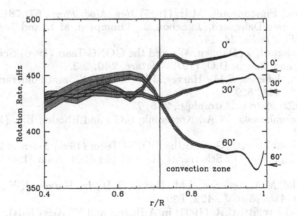

Figure 7. Solar rotation rate inferred from 2 months of MDI Medium-*l* data as a function of radius at three latitudes, 0°, 30°, and 60°. The formal errors are indicated by the shaded regions. The arrows indicate the Doppler rotation rate directly measured on the surface.

The rotation is inferred using a regularized least-squares technique. The inversion results confirm the previous findings that latitudinal differential rotation occurs only in the convection zone, that the radiative interior rotates almost rigidly, and that there is a thin shear layer near the surface (*e.g.* Goode *et al.*, 1991; Thompson *et al.*, 1996). The most interesting is that the transition layer (tachocline) between the radiative and convection zone is mostly located in the radiative zone, at least at the equator where it is also fairly thin, certainly less than $0.1\ R$. The layer seems to be wider at high latitudes.

The sharp radial gradient of the angular velocity beneath the convection zone gives strong evidence that the sharp narrow peak of the sound speed at $0.67\ R$ seen in the structure inversion (Figure 6) is due to rotationally-induced turbulent mixing in the tachocline (Spiegel and Zahn, 1992). The location and the width of the tachocline are also consistent with the requirement of recent dynamo theories of the solar cycle.

Acknowledgments

The authors acknowledge many years of effort by the engineering and support staff of the MDI development team at the Lockheed Palo Alto Research Laboratory (now Lockheed-Martin Advanced Technology Center) and the SOI development team at Stanford University. SOHO is a project of international cooperation between ESA and NASA. This research is supported by the SOI-MDI NASA contract NAG5-3077 at Stanford University.

References

Bahcall, J.N., and Pinsonneault, M.H. (1995) *Rev. Mod. Phys.*, **67**, 781.
Basu, S., Christensen-Dalsgaard, J., Schou, J., Thompson, M.J., and Tomczyk, S. (1996) *Bull. Astr. Soc. India*, **24**, 147.
Christensen-Daalsgard, J., Däppen, W., and the GONG Team (1996) *Science*, **272**, 1286.
Dilke, F.W.W., and Gough, D.O. (1972) *Nature*, **240**, 262.
Duvall, T.L., Jr., Jefferies, S.M., Harvey, J.W., Osaki, Y., and Pomerantz, M.A. (1993) *Astrophys. J.*, **410**, 829.
Gabriel, M. (1992) *Astron. Astrophys.*, **265**, 771.
Goode, P.R., Dziembowski, W.A., Korzennik, S.G., and Rhodes, E.J. (1991) *Astrophys. J.*, **367**, 649.
Gough, D.O., Kosovichev, A.G., and the GONG Team (1996) *Science*, **272**, 1296.
Kosovichev, A.G., Schou, J., Scherrer, P.H., and the MDI Team (1997) *Solar Phys.*, in press.
Kumar, P., Fardal, M.A., Jefferies, S.M., Duvall, T.L., Jr., Harvey, J.W., and Pomerantz, M.A. (1994) *Astrophys. J.*, **422**, L29.
Michaud, G., and Proffitt, C.R. (1993) in A.Baglin and W.Weiss (eds), *Inside the stars*, ASP Conf. Series, vol. 40, San Francisco, p.246.
Nigam, R., and Kosovichev, A.G. (1996) in J. Provost and F.X. Schmider (eds), *Sounding Solar and Stellar Interiors*, Kluwer Acad. Publ., in press.
Rogers, F.J., and Iglesias, C.A. (1996) *Astrophys. J.*, **456**, 902.
Scherrer, P.H., and the MDI Team (1996) *Solar Phys.*, **162**, 129.
Schou, J. (1992) *On the Analysis of Helioseismic Data*, Thesis, Aarhus University .
Spiegel, E.A., and Zahn, J.-P. (1992) *Astron. Astrophys.*, **265**, 106.
Thompson, M.J., Toomre, J., and the GONG Team (1996) *Science*, **272**, 1300.

INTERNAL ROTATION AND DYNAMICS OF THE SUN FROM GONG DATA

S. G. KORZENNIK
Harvard-Smithsonian Center for Astrophysics
60 Garden St, Cambridge MA 02138, USA

M. J. THOMPSON
Astronomy Unit, Queen Mary & Westfield College
Mile End Rd, London E1 4NS, UK

J. TOOMRE
Joint Institute for Laboratory Astrophysics
University of Colorado, Boulder CO 80309, USA

AND

THE GONG INTERNAL ROTATION TEAM

1. Introduction

We report inferences for the Sun's internal rotation from GONG months 4–10 averaged power spectra.[1] In keeping with the international collaborative nature of the GONG project, the results presented here are based on the work of several groups around the world inverting the GONG data and sharing their results via the world-wide web. These groups are at the Observatoire de la Côte d'Azur, Nice (T. Corbard, G. Berthomieu, J. Provost); Theoretical Astrophysics Center, Aarhus (J. Christensen-Dalsgaard, F. Pijpers); Center for Astrophysics, Cambridge MA (A. Eff-Darwich, S. Korzennik); QMW, London (R. Howe, M. Thompson, in collaboration with J. Schou, Stanford); Institute of Astronomy, Cambridge (T. Sekii, D. Gough);

[1]This work utilizes data obtained by the Global Oscillation Network Group (GONG) project, managed by the National Solar Observatory, a Division of the National Optical Astronomy Observatories, which is operated by AURA, Inc. under a cooperative agreement with the National Science Foundation. The data were acquired by instruments operated by the Big Bear Solar Observatory, High Altitude Obseratory, Learmonth Solor Observatory, Udaipur Solor Observatory, Instituto de Astrofísico de Canarias, and Cerro Tololo Interamerican Observatory.

J. Provost and F.-X. Schmider (eds.), Sounding Solar and Stellar Interiors, 211-218.
© *1997 IAU. Printed in the Netherlands.*

University of Sydney (D. Burtonclay, Li Yan, P. Wilson); and Tata Institute of Fundamental Research, Bombay (H. Antia, S. Chitre).

2. Key issues

The first inferences made by the team from GONG data were presented in the special GONG issue of Science (Thompson *et al.* 1996). The present paper is a report on work in progress towards a more mature understanding of the inferences that can be drawn from the GONG data at the present time. Two principal datasets have been used, both based on averaged 7-month (GONG months 4–10) power spectra. One set comprises individual m frequencies from the GONG project pipeline (Hill *et al.* 1996), in the ranges $\nu < 5000\,\mu Hz$ and $0 \leq l \leq 150$: this set contains 109 483 frequencies with IERR and BAD flags[2] both zero, though individual inverters will generally have made smaller selections and/or fitted low-order polynomials in m to these frequencies. The second set consists of 7 910 Clebsch-Gordon a coefficients in approximately the same ranges of frequency and degree $(1111\,\mu Hz \leq \nu \leq 4563\,\mu Hz,\ 5 \leq l \leq 150)$, obtained by S. Korzennik using a completely different frequency estimation procedure.

In addition to the global picture of the internal rotation of the Sun, the particular aspects of the solar rotation that we have identified as likely targets for study with these data are:

- near-surface shear layer
- shear layer at the base of the convection zone (tachocline)
- variations within the convection zone
- latitudinal constancy in the radiative interior
- rotation of the core

Another important issue that must be addressed is the reliability of the inferences that we make, and the sources of uncertainty. Thus we wish to assess:

- uncertainties coming from the data reduction (peak finding)
- discrepancies from inverting different observations
- discrepancies between different inversions of the same data

Then we might hope to give an answer to the question posed by John Leibacher during this conference: "How can we render more certain our inferences from the data?".

---

[2]IERR$\neq$0 indicates a problem within the peak-find itself; BAD$\neq$0 indicates that the estimated mode parameters fail some post-processing criterion

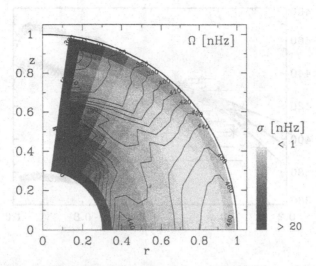

Figure 1. Inversion of GONG project frequencies using $1 \otimes 1$ SOLA (Aarhus group). Contours of isorotation are shown, superimposed on a grey-scale plot of the formal errors. A very dark background means a less reliable determination.

3. Results

Figure 1 shows the solar rotation profile inferred from individual m splittings using a $1 \otimes 1$ SOLA inversion. Similar results are shown by Sekii (these proceedings). Results from two other methods – RLS and 2D SOLA – applied to these same data are illustrated in Fig. 2. All of the inversions show the same overall behaviour: the persistence of surface-like differential rotation through much of the convection zone; enhanced rotation around $r = 0.95R$; a transition to essentially latitudinally-independent rotation beneath the convection zone. The inversions compared in Fig. 2 are in quite good agreement, even though the RLS used individual m splittings and the 2D SOLA used a coefficients only up to a_7. The small but systematic differences in the convection zone (in particular at 60° latitude) need to be investigated but probably can be understood in terms of the differences in averaging kernels: specifically, the RLS kernels have structure near the surface, so that the interior solution may be biased by the near-surface rotation rate. The averaging kernels differ even more in the deeper interior, where the present data constrain the rotation rate rather poorly. The large asymmetric horizontal bars on the high-latitude SOLA inversion at around $0.4R$ indicate that the method failed to localize a kernel at this target location.

The sensitivity to different peak-bagging reductions applied to the same

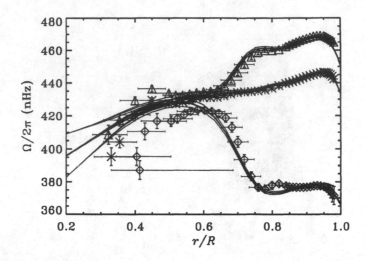

Figure 2. Inversions of GONG project frequencies using 2D RLS (solid curves; QMW group) and 2D SOLA (symbols; Aarhus group) at latitudes 0°, 30°, 60°. ±1-σ formal errors are indicated by extra curves and by vertical error bars on the symbols; horizontal bars represent radial resolution of the averaging kernels.

GONG power spectra is illustrated in Fig. 3, where the results of applying the same 1.5D RLS inversion to a coefficients from the GONG project and from Sylvain Korzennik are compared. The two panels show similar comparisons with similar methods, but from two groups of collaborators. The inferred differences from the two data reductions are slight in the convection zone and outer radiative interior. Only in the deep interior do the two datasets produce strikingly different results (panel a), with the project data indicating a slow rotation while the Korzennik data favour a roughly uniform rotation profile. Because the rotation in the deep interior is poorly constrained, the solution in $r < 0.4R$ is largely a result of extrapolation according to the regularization used. For this reason, in the inversion in panel b, conditions of zero radial and latitudinal gradient have been imposed on the solution at $r = 0.3R$, which effectively enforces a uniform rotation in the deep interior. While this may avoid giving a misleading impression caused by a wild extrapolation in the core, it too is an a priori prejudice and in this case serves to mask the discrepancy between the two datasets.

To investigate the effect of inverting data from different experiments, we have compared inversions of GONG project frequencies with inversions of splittings derived from two-year averaged LOWL data ($5 \leq l \leq 95$). Results of a 1.5D RLS inversion of both datasets are shown in Fig. 4. The

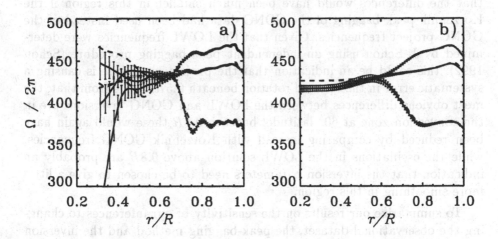

Figure 3. Comparison of inferences from project (continuous curves, with error bars) and SGK a coefficients (broken curves) using 1.5D RLS. The two panels show (a) results from CfA group; (b) results from TIFR group. The solution is depicted at latitudes 0° (equator), 30° and 60°.

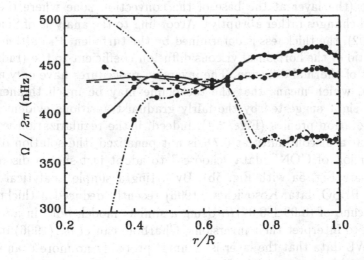

Figure 4. Comparison of inferences from GONG project frequencies and from LOWL frequencies, using a 1.5D RLS inversion (CfA group). The curves with solid dots are the GONG results (as in Fig. 3a). The solution is shown for latitudes 0° (solid curves), 30° (dashed) and 60° (triple-dot dashed).

main difference is beneath $r = 0.5R$: comparing with Fig. 3a, it is clear
that the differences would have been much smaller in this region if the
Korzennik peak-bagging of the GONG data had been used instead of the
GONG project frequencies. Given that the LOWL frequencies were determined by J. Schou using an independent peak-bagging procedure (Schou
1992), this could be an indication that the project procedure is causing a
systematic error in the inferred rotation beneath $0.5R$. Aside from that, the
most obvious differences between the LOWL and GONG inversions are in
the convection zone at 60° latitude: beneath $0.8R$ these would again have
been reduced by comparing instead with Korzennik GONG frequencies,
while the oscillations in the LOWL solution above $0.8R$ are probably an
indication that the inversion parameters need to be chosen to give a little
more smoothing in this region.

To summarize our results on the sensitivity of our inferences to changing the observational dataset, the peak-bagging method and the inversion
technique, all three lead to some generally small but noticeable differences
in the solutions for the internal rotation. Judging from our findings, the uncertainties from all of these are similar in magnitude. In the deep interior
(say beneath $0.4R$) the solutions are more poorly constrained by the data
and the differences are larger.

A new measurement of some interest to solar dynamo theorists and
those interested in the dynamics of the solar interior is the thickness of the
tachocline (the layer at the base of the convection zone where the rotation speed changes rather abruptly). According to the analysis of Spiegel &
Zahn (1992), the thickness is determined by the turbulent Prandtl number,
i.e. the ratio of the horizontal viscous diffusion coefficient to the (radiative)
coefficient of thermal diffusion. The inversion procedures have only a finite
resolution, which means that the tachocline may be much thinner than
is at first sight suggested by the fairly gradual transitions evident in the
inferred rotation profiles (Figs. 2-4). Indeed, if the regularization is modified so that a discontinuity at $0.7R$ is not penalized, the solution of a 2D
RLS inversion of GONG data "chooses" to adopt just such a discontinuity (compare Fig. 5a with Fig. 5b). By fitting a simple analytical model
profile to BBSO data, Kosovichev (1996) recently deduced a thickness of
the tachocline of $(0.09 \pm 0.04)R$. Using a similar model, both in a forward
sense and to interpret their inversions, Charbonneau et al. (1996) inferred
from LOWL data that the layer is thinner, probably no more than $0.06R$,
with a hint also that the tachocline is prolate. By making a nonlinear least-
squares fit to the GONG data of a rotation profile incorporating a transition
of adjustable position and width, Sekii (these proceedings) infers that the
tachocline is centred on $r = 0.696R$ and has width $0.064R$. This value is
not very different from that considered by Spiegel & Zahn. It is interesting

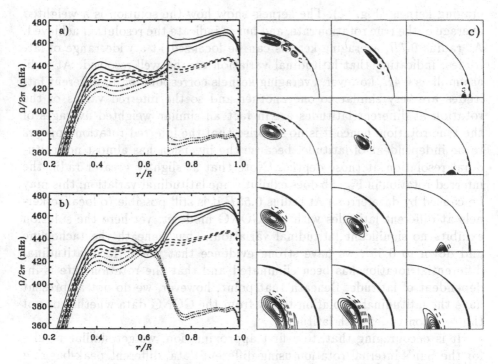

Figure 5. (a) Solution at latitudes 0° (solid), 30° (dashed) and 60° (triple-dot dashed) for a 2D RLS of GONG data, allowing the solution to adopt a discontinuity at the base of the convection zone. Adjacent curves indicate ±1-σ error limits. (b) As (a) but not allowing a discontinuity ('standard' 2D RLS). (c) Averaging kernels corresponding to the solution in panel b, at target latitudes (left to right) 90°, 45° and 0° and radii (top to bottom) 0.9R, 0.7R, 0.55R, 0.4R. (Nice group)

to note also that the value of the almost uniform angular velocity Ω_c in the radiative interior, about 0.94 of the equatorial value Ω_0 in the convection zone, is another indication of the stress in the tachocline. The purely horizontal viscous stress based on two-dimensionally isotropic turbulence assumed by Spiegel & Zahn leads to $\Omega_c = 0.91\Omega_0$, whereas if the turbulence were isotropic in three dimensions, and large-scale advection were unimportant, Ω_c would be 0.96Ω_0 (Gough 1985). The intermediate observed value suggests perhaps that reality lies between these two extremes. It is important to recognise also that perhaps a more radical modification to our theoretical ideas should be entertained, such as transport by anisotropic two-dimensional turbulence or by Lorentz forces.

With only slight differences, all our inversions show that the inferred rotation rates at latitudes 0°, 30° and 60° converge below the base of the convection zone. How firmly can we therefore say that the rotation in the radiative interior is independent of latitude? To assess this, we inspect averaging kernels (Fig. 5c). The kernels show how the solution is a weighted average of the true rotation rate, and hence indicate the resolution achieved. At radius $0.7R$, averaging kernels can be localized at a wide range of latitudes, indicating that latitudinal variation can be well-resolved. At radii as small as $0.4R$, however, averaging kernels corresponding to different latitudes are very similar to one another and so the inferred values of the rotation at different latitudes are in fact all similar weighted averages of the true rotation: hence it is no surprise that the inferred rotation appears to be independent of latitude, because the inversion has almost no latitudinal resolution at those depths. (Note that at slightly greater radii, the inferred rotation in Fig. 5b does exhibit some latitudinal variation: this may be caused by data error.) At radius $0.55R$ it is still possible to localize kernels at different latitudes with the GONG modeset; yet here the solution exhibits no significant latitudinal variation. Thus beneath the tachocline and down to $0.55R$ we have strong evidence that the surface latitudinal differential rotation has been eliminated and that the rotation rate is independent of latitude. Beneath that point, however, we do not at present have the latitudinal resolution to say from the GONG data whether or not the rotation varies with latitude.

It is encouraging that, to a first approximation, we get similar results for the Sun's internal rotation using different data, different peak-bagging procedures and different inversion methods. However, to address subtler questions we need to understand the differences that arise from all these three sources: these differences are fairly small throughout much of the Sun, but are substantial in the deep interior. As the random errors are reduced and we push our inversions harder, the systematic errors will become even more important. Clearly, more work is still required both in peak-bagging and in the inversions to get better agreement.

References

Charbonneau, P. *et al.* (1996) Observational Constraints on the Dynamical Properties of the Shear Layer at the Base of the Solar Convection Zone, in *Sounding Solar and Stellar Interiors*, Proc. Symposium IAU 181, poster volume, in press

Gough, D.O. (1985) Theory of solar oscillations, in *Future missions in solar, heliospheric and space plasma physics*, eds Rolfe, E. & Battrick, B., ESA SP-235, ESTEC, Noordwijk, 183 – 197

Hill, F. *et al.* (1996) The Solar Acoustic Spectrum and Eigenmode Parameters, *Science*, **272**, 1292

Kosovichev, A.G. (1996) Helioseismic Constraints on the Gradient of Angular Velocity at the Base of the Solar Convection Zone, *Astrophys. J. Letters*, **469**, L61

Schou, J. (1992) On the Analysis of Helioseismic Data, PhD Thesis, Aarhus University.

Spiegel, E.A. & Zahn, J.-P. (1992) The solar tachocline, *Astron. Astrophys.*, **265**, 106

Thompson, M.J. *et al.* (1996) Differential Rotation and Dynamics of the Solar Interior, *Science*, **272**, 1300

SPECIAL SESSION
DEDICATED TO PHILIPPE DELACHE

THE SCIENTIFIC ACHIEVEMENTS OF P. DELACHE

R. M. BONNET

Agence Spatiale Européenne

8-10 rue Mario Nikis, 75738 Paris CEDEX 15, France

Abstract. An overview of the multiple facets of Philippe Delache's scientific achievements is presented, from his early involvement in experimental work and space research, through his re-orientation towards radiative transfer, astrophysics, and then the study of the global Sun and its internal structure, and the study of the interior of Jupiter. The constant concern of P. Delache for a proper and correct mathematical analysis of the problems and for modelling, are outlined in this brief overview. The generous character, the profound and complex personality, as well as the human qualities of this very talented scientist, among which modesty is very prominent, are presented in this description of a remarkable career which ended much too early.

1. Introduction

Describing Philippe Delache's scientific achievements in a few pages is a real challenge. It is both simple and very difficult. It is simple because he is the author or co-author of an abundant scientific literature and has written several excellent reports which describe how he himself perceived his own work. On the other hand, it is difficult because of the diversity of the topics he addressed in his too short scientific life, always merging a rigorous mathematical analysis with a precise and clear physical judgement. As a mark for a long-lasting friendship over the years and of my deep appreciation for his work and that of his students, I have accepted to take up this challenge.

Delache was born on 8 October 1937 at Semur-en-Auxois in Burgundy, some two- and-a-half months before me. After his secondary studies at the Lycée Carnot in Dijon, he joined the Ecole Normale Supérieure in Paris where he stayed from 1956 until 1960. A very important aspect of his stay

221

J. Provost and F.-X. Schmider (eds.), Sounding Solar and Stellar Interiors, 221-234.
© 1997 *IAU. Printed in the Netherlands.*

at the Ecole Normale Supérieure was his strong inclination towards mathematics, a discipline which he enjoyed more than any other at that time. This explains his constant inclination for the analytical approach, for modelling, and his rigour in the understanding of physical phenomena.

At the same time, like many of that generation, including P. Léna and myself, he was influenced by the enormous possibilities offered by space research, following the successful launch of Sputnik-1, the first artificial satellite of the Earth, by the Soviets in October 1957. This is how he became involved in experimental work and rejoined the newly created Service d'Aéronomie of J.E. Blamont, still hosted at that time at the Ecole Normale Supérieure.

I have known Delache since 1961 when I joined the Service d'Aéronomie myself. He was then preparing his thesis based on an experimental type of work. During the four years that we cohabited at the Service d'Aéronomie and until he rejoined the Observatory of Nice, then directed by J-C. Pecker, we had frequent and friendly contacts. After his departure to Nice, these contacts became less frequent, due to the distance, but remained excellent and always profound and very productive. We worked again closely in 1980-1983, when he joined the study team of the DISCO mission, the ancestor of SOHO. We also interacted frequently during the time he was a member of the Comité de Direction of my own laboratory, LPSP, at Verrières-le-Buisson.

I can identify four main phases in the scientific career of Delache: the experimenter phase, the return to theory and the thesis work, the radiative transfer phase, and the study of the global Sun. This enumeration defines a broad frame which I will now detail, trying to project an illumination on the extremely rich life and on the achievements of this bright brain. I will not address however, any of the long periods in Delache's career devoted to administration and teaching, as well as to the directorship of the Observatory at Nice (he was Director in 1969-1972, in 1975, and in 1989-1994), neither his frequent involvement in committees as an advisor or as an expert in the many areas where he was able to contribute wise and needed advice, in particular in solar physics and space research. These duties certainly absorbed a substantial portion of his time, which makes a posteriori his very productive scientific career even more impressive.

2. The Experimenter Phase

Any student at the Ecole Normale Supérieure willing to become a Professor had to prepare a Diplôme d'Etudes Supérieures, a sort of small thesis representing about one year of scientific work. Delache prepared it under the directorship of Blamont in the group of A. Kastler and P. Brossel, where an important effort was devoted to the study of the optical resonance of hydrogen and of alkaline vapours, in view of studying the Earth's atmosphere and various astrophysical objects.

The topic of his work was the study of an infrared transition of oxygen involving a level excited by the absorption of Lyman beta photons from hydrogen at 102.5nm, a phenomenon which had been invoked to explain an anomaly in the intensity of this line in various planetary nebulae. His work, of an experimental nature, consisted in simulating the process experimentally, using laboratory hydrogen and oxygen cells. The experimental set-up was not simple but Delache could confirm the reality of the process. After his Diplôme (Delache,1960), he was offered the possibility to continue his experimental work, studying the solar Lyman alpha radiation of hydrogen at 121.6 nm and its absorption by the Earth's upper atmosphere by means of rockets or artificial satellites.

His main thesis work consisted in developing the hydrogen cells, a delicate experimental work based on the development of sophisticated technologies, first to create atomic hydrogen from the dissociation of the molecule, then to seal the cells so that the optical thickness could be precisely controlled (Blamont, Delache and Stober,1964). The tightness of the metal-to-glass sealing was particularly challenging and Delache spent long days working in semi-darkness in the noisy ambiance created by an army of oil vapour pumps.

We often met in the laboratory where I was also preparing an experimental thesis on the ultraviolet imagery and spectroscopy of the Sun using rocket- borne spectrographs and telescopes of my own making. These encounters offered opportunities for scientific discussions and for semi-sour comments on the difficulties of space research where results are only achieved once in space, after many months or years of hard laboratory work with, unfortunately, little to publish, and the slogan "publish or perish", even though he and I could invoke being involved in pioneering work, applied already. That scarcity of scientific production, although not of his responsibility, concerned Delache deeply. It is also fair to say that he was more inclined toward the theoretical implications of his work than by its experimental character, as pioneering as it might be, and for which substantial fractions of his days, and of his nights, were absorbed.

In the course of our discussions I advised him to talk to Pecker, then my thesis advisor for the astrophysical implications of my work. Delache was particularly appreciative of the type of balance through which, between Blamont and Pecker, I could control the progress of my research, in spite of some spectacular failures of the Véronique rocket which carried my instruments. Torn apart as he was between staying at the Service d'Aéronomie and joining Pecker at Nice, he decided on the latter, however not without sorrow and regrets. His relations with his former director certainly cooled down substantially after that departure, but they always remained courteous. Delache, as the manifestation perhaps of a hidden regret, never did break the umbilical chord which tied him to his former research career, and he always felt inclined to return, while distancing himself nevertheless from too deep an involvement.

A posteriori, there was a lot for him to be proud of, which his experimental work led to in the study of the Sun and of the geocorona. Using hydrogen absorption cells and sounding rockets, his student, J. Quessette, measured the vertical distribution of hydrogen in the Earth's upper atmosphere. Using similar devices, Blamont's student, A. Vidal-Madjar, obtained the best, at the time, solar Lyman alpha profile from the American OSO-5 satellite. Similarly, the distribution of atmospheric hydrogen and its temperature were measured extensively on board the French D2-A satellite as well as on NASA's OSO-5. Finally, the distribution of hydrogen around comets was determined for the first time by J-L. Bertaux, another student of Blamont, using the same technique. In fact, hydrogen cells are still used today by Bertaux on SOHO to measure the penetration of interstellar hydrogen in the heliosphere (SWAN experiment). All these experiments are essentially the legacy of Delache's early experimental work at the Service d'Aéronomie.

3. The Return to Theory and the Thesis Work

At Meudon, under Pecker's guidance, Delache started purely theoretical work, far away from his early experimental involvement. His work was to study and interpret the apparent over-abundances of heavy ions such as nickel and iron, as compared with their photospheric values. The problem is non-stationary and must be treated as such. Delache applied irreversible thermodynamic processes, taking into account thermal diffusion between the cold photosphere and the hot corona, as well as gravitational sorting in the expanding solar wind. He resolved simultaneously the equations of diffusion and of ionisation for all states of ionisation and for different velocities of coronal expansion (Figure 1 and 2).

He presented his thesis in 1967 (Delache, 1967), some two years after he started this work totally new for him, a time record! Later, it was shown that the introduction of new and more precise oscillator strengths considerably reduces the amplitude of the abundance anomalies. However, Delache's basic treatment of the problem, in particular the effect of gravitational sorting, is correct. Today, his work has a direct application in the understanding of the abundance variations measured between the photosphere and the corona for elements with low first ionisation potential (10eV), as measured in particular by Ulysses and SOHO (Geiss et al., 1995).

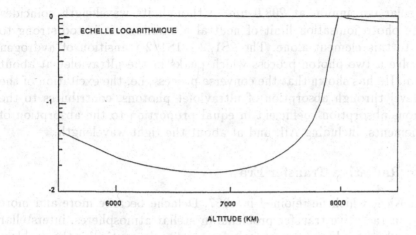

Figure 1. Variation of the total abundance of nickel for a given model of the corona and an initial wind velocity of 5 km/s (from Delache, 1967).

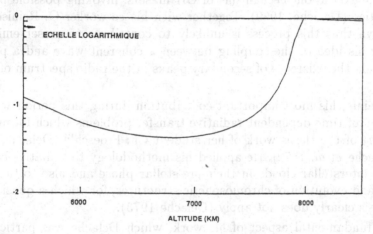

Figure 2. Same as Fig. 1 but with an initial wind velocity of 10 km/s (from Delache, 1967).

Although we were no longer in the same laboratory, we continued to interact closely, and it is worth mentioning at this stage his contribution to the inversion of the Laplace Transform, an essential problem for me in attempting to derive a source function from my solar limb-darkening data. Unfortunately, he never published a paper on that issue. He could have done so, in particular on the occasion of the Bilderberg conference which he and I attended in April 1967, which aimed at establishing a new model of the solar atmosphere (de Jager and Gingerich, 1968), and where his participation and contribution to this conference was very active.

Like me, he was puzzled by the presence of a prominent discontinuity in the solar continuum at 208.0 nm. Although its wavelength coincides with the photo ionisation limit of neutral aluminium, it is too strong to be due to this element alone. The 2S1/2 - 1S1/2 transition of hydrogen can involve a two-photon process which peaks in the ultraviolet at about 208.0 nm. He has shown that the converse process, i.e. the excitation of the upper level through absorption of ultraviolet photons, contributes to the continuous absorption coefficient in equal proportion to the absorption of other elements, including AlI, and at about the right wavelength.

4. The Radiative Transfer Era

Once at Nice, which he rejoined in 1967, Delache became more and more involved in radiative transfer problems in stellar atmospheres, interstellar matter, and kept a keen interest for solar physics, in particular the problem of coronal heating. Amazingly, he also became interested in the understanding in the excitation mechanisms of OH masers, invoking possible plasma instabilities (Delache, 1969). Together with his co-worker G. Reinisch, he has shown that this process is unlikely to contribute to maser emission. However his idea of the coupling between a coherent wave and a plasma can explain the existence of secondary peaks in the radio spectrum of these sources.

Certainly, his most important contribution during this period was the resolution of time dependent radiative transfer problems which formed the main part of the thesis work of her student Ch. Froeschlé (Delache, 1968, and Delache *et al.* 1972). He applied his methodology to radiative cooling of dense interstellar clouds in their pre-stellar phase and also to the equilibrium and evolution of chromospheric structures, for which a quasi static hypothesis clearly does not apply (Delache 1973).

The fundamental aspect of his work, which Delache was particularly proud of, was the demonstration that an approximate analytical method, as opposed to lengthy and heavy numerical simulations using the big - and then expensive - computers, could represent fairly accurately the source functions of lines emitted in non LTE conditions: through a rigorous and simple physical reasoning, he follows the time evolution of a photon and evaluates its probability of escape from an optically thick and scattering medium at various frequencies in the line profile. Figure 3 reproduces his results which have been published in the Astrophysical Journal (Delache 1974).

His ideas have been exploited and refined by several scientists, in particular R.G Athay and A. Skumanich in Boulder, and by his colleagues at Nice , Uriel and Hélène Frisch. They are often quoted and considered as a reference in the domain. Generalising the escape probability method, he

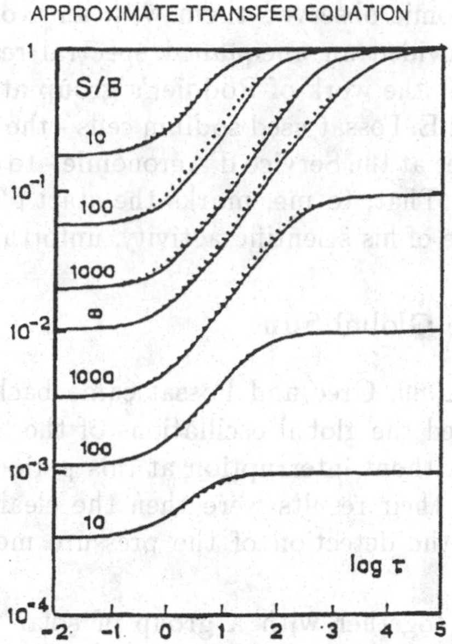

APPROXIMATE TRANSFER EQUATION

Figure 3. full lines, ordinary iteration, dotted lines, solution of the approximate transfer equation. Curves are labeled with the number of iterations, corresponding also to the nondimensional time T. Below the limiting solution ($T = \infty$), they correspond to the initial condition $S = 0$; above, to S = B. (From Delache, 1974).

applied them to the interpretation of quasar spectra and, together with S. Collin-Souffrin, S. Dumont and H. Frisch (Collin-Souffrin et al., 1981) he then proposed observing programmes for the Hubble Space Telescope, in a sense: a return to space research!

Space research in fact was always present just beneath the surface of his interests. I could test it directly on the occasion of the preparation of the IAU symposium which the two of us organised at Nice in the summer of 1976 which was devoted to the Energy Balance and Hydrodynamics of the Solar Chromosphere and Transition Region and was essentially based on the availability of new space data from 0SO-8 and Saliut-6 (Bonnet and Delache 1977).

His involvement in radiative transfer problems gradually faded out in the late seventies. He was still interested, however, in solar physics and in such fundamental issues as coronal heating where he recognised the limits imposed by the insufficient quality of the data. The then fashionable assumption that the dissipation of acoustic waves in the upper solar atmosphere might cause the temperature to rise in the chromosphere and the corona, led him to look more closely at solar oscillations. He had always been interested in these oscillations, as far back as 1964 when F. Roddier at the Service d'Aéronomie observed them with an atomic strontium jet spectrometer which provided an unequalled spectral resolution. He then looked more carefully at the work of Roddier's group at the University of Nice where G. Grec and E. Fossat used sodium cells - the same which M.-L. Chanin developed earlier at the Service d'Aéronomie - to observe the global oscillations of the Sun. That, to me, marks the start of the most intense and of the richest phase of his scientific activity, unfortunately the last.

5. The Study of the Global Sun

In the winter of 1979-1980, Grec and Fossat came back from Antarctica where they had observed the global oscillations of the Sun for more than four consecutive days without interruption at this period of the year. Due to the long time span, their results were then the cleanest and certainly the most accurate for the detection of the pressure modes. Delache was enthusiastic.

At the same time, together with a group of solar phycisists, among whom was C. Fröhlich, I was involved in an ESA assessment study for a small satellite to measure the solar constant and its variation with the solar cycle, as well as the spectral irradiance of the Sun in the visible and the ultraviolet. The "Dual Irradiance and Solar Constant Observatory", DISCO, was to be placed in orbit around Lagrangian point L1 located between the Sun and the Earth. Already, C. Fröhlich using radiometric data obtained with ACRIM on the Solar Maximum Mission, had provided evidence of p-modes, and DISCO, with its L1 orbit, would clearly lead to an improvement. Delache proposed that we should also consider the possibility of observing the solar velocity oscillations, using sodium cells. His main interest was to observe the gravity modes which probe the solar interior down to the very core, in view of addressing such fundamental questions as the solar neutrinos deficit. He was invited to join the DISCO study team.

For more than two years we worked very closely on the definition of DISCO, the ancestor of SOHO. One of the most promising experiments was indeed the sodium cell, similar in its working principle to that of Grec and Fossat, but which had to be miniaturised from a size of a fraction of a cubic metre to a few tens of cubic centimetres. That required a complete re-definition of the instrumentation together with its space qualification. That was the first opportunity for Delache to become again intensively involved, at least for the duration of the DISCO study, in an experimental type of work.

DISCO was not selected by ESA, having lost in March 1983 a fierce competition with ISO, the Infrared Space Observatory. Our study work was therefore put on hold, but SOHO was born (Huber *et al.* 1996) and, naturally, Delache became involved in the SOHO study. His interests switched back more to theory and in particular to the possibility of detecting the global and gravity modes, an effort which required a thorough mathematical and physical treatment of the noise problem, which Delache was very motivated to address.

Most intriguing in the 1980's was the presence in the power spectrum of solar oscillations of a 160 mn mode, detected both at Stanford University and at Crimea Observatory. Delache was keen to understand the nature of this prominent feature which might be either a gravity mode or simply an atmospheric artefact. He invested considerable time and energy in its interpretation (Delache, 1981), and started a close collaboration with P. Scherrer at Stanford. As P. Scherrer describes in his contribution to this special session (Scherrer, this issue), Delache became a real "gravity mode hunter". Both co-signed a paper in Nature on the detection of gravity modes (Delache and Scherrer, 1983). The SOHO data have shown how difficult their detection is, and how optimistic it was then to think that they could be detected from the ground. However, that work was very useful in assessing the limits and the difficulties of their detection. A close collaboration also started with C. Fröhlich, whom he knew from his earlier involvement in the DISCO team, on the detection of solar luminosity pressure and gravity modes (Fröhlich and Delache, 1984).

During our DISCO study, the possibility that long-term luminosity variations might result in similar variations in the solar diameter was often raised. Assuming that thermal energy is converted into potential energy and vice versa, if the solar constant varies with solar activity, as proven by SMM and other space data, the diameter should also vary. Quietly, and with his usual curiosity, Delache looked in more depth into the astrometric measurements of the solar diameter made with the astrolabe at CERGA, near Nice, by F. Laclare. The two of them showed the existence of an apparent anticorrelation between the length of the solar diameter and the level of solar activity (Delache, Laclare and Sadsaoud 1985). Not content with this observation alone, he discussed in more depth and for the first time the validity of the astrolabe measurements, taking into account the effect of active regions at the limb.

In the last period of his activity, he never abandoned his permanent inclination for analytical and mathematical problems. With A. Vigouroux he started a thorough and critical analysis of the Fourier transform as applied to long term solar variations. They showed that the wavelet transform, which makes no implicit assumptions as to the standing character of the waves, offered a better representation of the long-term variations of the solar diameter and of its irradiance (Vigouroux and Delache, 1993). Having in mind to reach the physical phenomena at play in the solar core, he

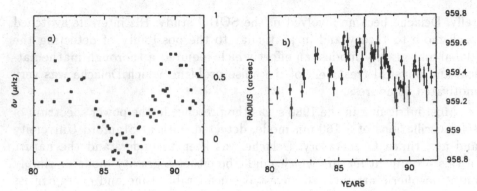

Figure 4. (a) Frequency shifts of low-degree solar p-modes vs . time. The reference is taken as the value at solar activity minimum (see text). (b) Radius measurements averaged over identical time intervals vs . time. (From Delache *et al.*, 1993).

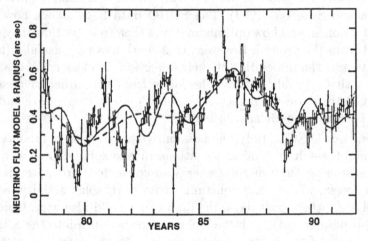

Figure 5. Combinations "a" (dashed line) and "d" (solid line) of significant harmonics in neutrino spectrum compared with radius ($R_\odot - 959, 0$ arcsec) (points with error bars). (From Gavryusev *et al.*, 1994).

analysed possible correlations between the flux of neutrinos, as measured by Davis in the US, and the solar diameter. Together with an international group of collaborators, he discussed possible time correlations between the long term variations of the frequencies of p-modes, the solar radius and the neutrino flux (Delache *et al.* 1993), paying due and careful attention to the problem of extracting a meaningful signal from a high level of noise (Delache *et al.* 1992), Figure 4 and 5.

Having tested the usefulness and power of this new mathematical instrument, he then exploited fully the wavelet analysis, applying it with A. Vigouroux to all possible manifestations of solar long term variability, having permanently in mind to properly understand the solar machinery and more particularly the solar dynamo, and possibly to predict the intensity of solar activity, based on a correct understanding of the past. That was the subject of the last publication he wrote (Vigouroux and Delache, 1994). This work also led to a series of posthumous articles with A. Vigouroux and J.M. Pap (Pap, Vigouroux and Delache, 1995; Vigouroux, Pap and Delache, 1996).

This description of Delache's achievements would be incomplete if I were not to mention his work on Jupiter. This apparently curious derivation from his previous work is a natural consequence of what he did for the study of the solar interior. Knowing the discontinuities which exist in the internal structure of the giant planet, he proposed to B. Mosser to compute the spectrum of its possible oscillations (Mosser, Delache and Gautier, 1988-a and 1988-b). This led him, together with Schmider and Fossat, to propose an observation of the planet at the Observatoire de Haute Provence and then to use the infrared Fourier spectrometer of the large Canadian French and Hawaïan telescope, a work which has been conducted by J. Gay, J.P Maillard and D. Mekarnia (Mosser *et al.*, 1991; and 1993); Figure 6.

He deeply regretted that his too intense involvement in administrative tasks prevented him from being more directly involved in this research. He gave up the Directorship of the Observatoire de la Côte d'Azur in Spring 1994, just a few months before he disappeared and, unfortunately, he could never fully enjoy the exploitation of his remarkable ideas. He was without any doubt the discreet conductor of an activity which had its centre of gravity at Nice and which, for more than 10 years, using new theoretical and observatory methods, led to remarkable and determining progress in our understanding of the solar, stellar, and planetary interiors. Naturally, he became a co-investigator on all the seismology experiments of SOHO: GOLF, VIRGO, and MDI. Knowing the extraordinary quality of the data that these instruments provide, I cannot but deeply regret that he is not here to see them and to work on them. His diagnostic capability would certainly have created an explosion of new ideas. In fact it is the whole set of SOHO data which would have enthused him, from seismology to coronal observations, with the newest and the most accurate abundance determinations of the solar wind. SOHO's success would have offered him the crowning of his constant involvement and interest in solar physics.

Our last scientific discussion dates back to July 1994, just a few days after Comet Shoemaker-Levy hit Jupiter, and he did not yet know whether the impact had any signature in the power spectrum of the oscillations of the planet. This was in Austria where he gave the introductory lecture at the Alpbach Summer School which that year was devoted to the Sun. This is the last time I saw him and talked to him. His lecture was the

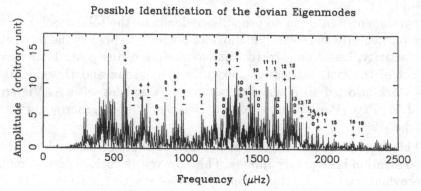

Figure 6. Identification of the $\ell = 0$ and $\ell = 1$ p modes in the observed Fourier spectrum. The values of the radial order n, the degree ℓ and the azimuthal order m ($\pm$ signifies $m = \pm 1$) are given for each identified eigenmode (from Mosser *et al.*, 1991).

most remarkable description of our present knowledge of the Sun and of its interior, a beautiful and exhaustive review of solar physics. It is a pity that the proceedings of the Summer School are no longer published, because we would certainly possess today one of the most complete and one of the clearest descriptions of the solar machinery, one which would prove once again how great an astrophysicist Philippe was.

6. Conclusion

These few pages written by someone who has followed Philippe Delache in his scientific career, sometimes closely, too often from a distance, cannot pay tribute to the achievements of such an outstanding and rich personality, who contributed so much to our understanding of the Sun and to astrophysics. To characterise and summarise such a scientific life, the best might be to use Philippe's own words, proving how modest but also how complex his personality was:

> "I am not the man of a long lasting kind of work, methodic, based on developing in depth a unique theme of research or a complex formalism. I rather like to put my fingers on those physical phenomena which come in support of, and help explaining such and such observations yet unexplained, even letting others taking care of deepening, or of extending the field on which my own ideas apply."

I think that if there is one parameter which can without any doubt be invoked to describe Philippe's attitude and ethics, it is t, time. The time he introduced in radiative transfer. The time it takes an atom to become ionised in the solar wind. The time of the oscillations of the Sun, of stars, of Jupiter. The time of correlations between p-modes frequency shifts, the neutrino flux, the solar diameter and the activity cycle. The time which is too long to conduct an experiment to completion. The time which passes too fast and forced his thirst of knowledge and his insatiable curiosity. The following paragraph is a beautiful text that he wrote about time.

I have not addressed at all here the other and numerous activities he developed in non purely scientific tasks, which absorbed a lot of his time, as Director of the Observatory at Nice, or as a professor at the University, or when he was sitting on committees, as advisor to CNRS, CNES, ESA and others. He was also an excellent communicator and he wrote several articles and gave several exemplary lectures for the general public (Delache and Amiot, 1977; Delache, 1980; Delache, 1993). He possessed a wonderful gift for explaining in simple terms the most arduous physical problems, and I can only regret that he did not spend enough time on that particular activity. He played a key role as chief editor of "Le Journal des Astronomes Français" in 1977-1980 in placing that review at the high level it has reached today.

He has been and has done all that and much more. He was the man who helped so many young, and less young, scientists, - J.P. Lafon kindly reminded me that Delache put considerable effort in 1990-1991 into the setting-up of the CNRS Research Group on circumstellar envelopes. Always with enthusiasm, always providing an invaluable moral support, never imposing his leadership by force, always letting the initiative of the young take over after his initial impulsion had been given.

He felt the permanent need to interact with others, another reason why he touched on so many issues. He felt the need to be in osmosis with his colleagues, and in nearly all fields of astrophysics, without however pretending he knew everything, having the frankness to confess his ignorance in those fields where he considered himself a novice. He felt the need to be appreciated and loved, the need to maintain his friendship, even in the case of adverse circumstances. He was a man of no rupture, and at the same time in permanent transition. He was an artist and a humorist, a magician in many senses. He was such a generous character and such a profound and sensitive personality. Above all he was a friend. The numerous seeds and trees he left will grow and blossom and give fruit for a very long time.

References

Blamont, J.E., Delache, P., Stobber, A.K. (1964) NASA Tech. Note $n°$ 654.
Bonnet, R.M. and Delache, P. (1977) Proceedings of IAU Colloquium $n°$ 36, *The Energy Balance and Hydrodynamics of the Solar Chromosphere and Corona*, Nice, 1976.
Collin-Souffrin, S., Delache, P., Dumont, S. and Frisch, H. (1981) *Astron. Astrophys.* **104**, 264.
De Jager, C. and Gingerich, O. (1968) *Solar Phys.* **3**, 5
Delache, P. (1960) Diplôme d'Etudes Supérieures, Paris.

Delache, P. (1967) *Ann. Astrophys.* **30**, 827.

Delache, P. (1968) *J. Quant. Spect. Radiat. Transfer* **8**, 317.

Delache, P. (1969) *J. Phys. C.* **3**, 21.

Delache, P. *et al.* (1972) *Astron. Astrophys.* **16**, 348.

Delache, P. (1973) Proceedings of IAU Colloquium n° 19, *Stellar Chomospheres*, A. Underhill and S. Jordan ed., NASA, Scientific and Technical Information Office.

Delache, P. (1974) *Astrophys. J.* **192**, 475.

Delache, P. and Amiot, M. (1977) "Neptune et la Gloire de Le Verrier", *Le Monde*, 30 Nov. 1977.

Delache, P. (1980) *La Science Spatiale*, Encyclopédie Hachette.

Delache, P. (1981) *Compt. R. Acad. Sci.* **293**, 949.

Delache, P. and Scherrer, P. (1983) *Nature* **306**, 651.

Delache, P., Laclare, F. and Sadsaoud, H. (1985) *Nature* **317**, 416.

Delache, P., Laclare, F., Gavryusev, V. and Gavryuseva, E. (1992) *Mem. Soc. Astr. Ital.* **64**, 237.

Delache, P. (1993) *Sciences Aujourd'hui*, Encyclopedia Universalis.

Delache, P., Gavryusev, V., Gavryuseva, E., Laclare, F., Regulo, C. and Roca-Cortès, T. (1993) *Astrophys. J.* **407**, 801.

Fröhlich, C. and Delache, P. (1984) Proceedings of EPS Colloquium *Oscillations as a Probe of the Sun's Interior, Mem. Soc. Astron. Ital.* **55**, 99.

Gavryusev, V., Gavryuseva, E., Delache, P. and Laclare, F. (1994) *Astron. Astrophys.* **286**, 305.

Geiss, J., Gloeckler, G., von Steiger, R., Balsiger, H., Fisk, L.A., Galvin, A. B., Ipavich, F.M., Livi, S., McKenzie, J.F., Ogilvie, K.W. and Wilken, B. (1995) *Science* **268**, 1033.

Huber, M.C.E., Bonnet, R.M., Dale, D. C., Arduini, M., Fröhlich, C., Domingo, V. and Whitcomb, G. (1996) ESA Bulletin **86**, 25.

Mosser, B., Delache, P. and Gautier, D. (1988-a) *B.A.A.S.* **20**, 870.

Mosser, B., Delache, P. and Gautier, D. (1988-b) in *Seismology of the Sun and Sun-like Stars* ESA-SP 286, 671.

Mosser, B., Schmider, F.-X., Delache, P. and Gautier, D. (1991) *Astron. Astrophys.* **251**, 356.

Mosser, B., Mekarnia, D., Maillard, J.P., Gay, J. and Gautier, D. (1993) *Astron. Astrophys.* **267**, 604.

Pap, J.M., Vigouroux, A. and Delache, P. (1996) *Solar Phys.* **167**, 125.

Vigouroux, A. and Delache, P. (1993) *Astron. Astrophys.* **278**, 607.

Vigouroux, A. and Delache, P. (1994) *Solar Phys.* **151**, 267.

Vigouroux, A., Pap, J.M. and Delache, P. (1996) *Solar Phys.*, in press.

LONG-TERM SOLAR IRRADIANCE VARIABILITY

JUDIT M. PAP

*Department of Physics and Astronomy, University of
California, Los Angeles, CA 90095-1562 and
Jet Propulsion Laboratory, California Institute of Technology,
Pasadena, CA 91109, USA*

Abstract. Measurements of the solar energy throughout the solar spectrum
and understanding its variability provide important information about the
physical processes and structural changes in the solar interior and in the
solar atmosphere. Solar irradiance measurements (both bolometric and at
various wavelengths) over the last two decades have demonstrated that the
solar radiative output changes with time as an effect of the waxing and
waning solar activity. Although the overall pattern of the long-term variations is similar in the entire spectrum and at various wavelengths, being
higher during high solar activity conditions, remarkable differences exist
between the magnitude and shape of the observed changes. These differences arise from the different physical conditions in the solar atmosphere
where the irradiances are emitted. The aim of this paper is to discuss the
solar-cycle-related long-term changes in solar total and UV irradiances. The
space-borne irradiance observations are compared to ground-based indices
of solar magnetic activity, such as the Photometric Sunspot Index, full disk
magnetic flux, and the Mt. Wilson Magnetic Plage Strength Index. Considerable part of the research described in this paper was stimulated by the
discussions with the late Philippe Delache, who will always remain in the
heart and memory of the author of this paper.

1. Introduction

The Sun, a fairly typical star, dominates the physical conditions throughout the solar system due to its influence on planetary atmospheres and the
interplanetary medium. The total radiative output of the Sun establishes
the Earth's radiative environment and controls its temperature and atmospheric composition. This mostly continuum radiation originates in the

235

J. Provost and F.-X. Schmider (eds.), Sounding Solar and Stellar Interiors, 235-250.
© 1997 *IAU. Printed in the Netherlands.*

photosphere and its major part reaches the troposphere and the Earth's surface and oceans. Consequently, small but persistent variations in the solar energy received on the top of the Earth's atmosphere ("solar constant") may be responsible for slow climate changes such as produced the Little Ice Ages, which was accompanied with an unexceptionally low magnetic activity of the Sun, known as the "Maunder Minimum" (Nesme-Ribes *et al.*, 1994). Therefore, the accurate knowledge of the solar radiation received by the Earth and its temporal variations, especially over decades, is critical for an understanding of the role of solar variability in climate change and the climate response to increasing greenhouse gas concentrations.

Although the existence of possible global climate changes based on the changing solar constant had been doubted and debated for a long time, the results of various space experiments for monitoring solar irradiance opened an exciting new era in both atmospheric and solar physics. Various space-borne observations of total solar irradiance over the last two decades established conclusively that total solar irradiance is not constant and the changes on time scales from minutes to the 11-year solar activity cycle are related to changes in the Sun's interior and atmosphere (Willson and Hudson, 1988; Fröhlich *et al.*, 1991). High precision photometric observations of solar-type stars clearly show that year-to-year variations connected with magnetic activity is a widespread phenomenon among such stars (Radick, 1994). As the nearest star, the Sun is the only star where we can observe and identify a variety of structures and processes which lead to irradiance variability on time scales of minutes to decades. On the other hand, the observable characteristics of other stars expand our knowledge on how the Sun works, simply by enlarging the sample to a larger set of conditions.

To establish the possible link between solar variability and climate change, it is necessary to analyze long-term data sets representing solar activity. Fortunately, ever since the earliest telescopic observations, the Sun's variability in the form of sunspots and related magnetic activity has been the subject of careful study. Considerable efforts have been made to develop irradiance models to help identify their physical causes and to provide irradiance estimates when no satellite observations exist. The ultimate goal is to uncover how and why the Sun is changing in order to reconstruct and predict the solar-induced climate changes.

2. Variations Observed in Solar Irradiance

The value of the integrated solar energy flux over the entire solar spectrum, hence total irradiance, arriving at the top of the Earth's atmosphere at 1 AU is called the "solar constant". Continuous observations of total solar irradiance to detect its variability started at the beginning of this century at the

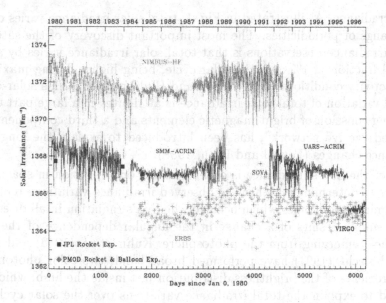

Figure 1.   The time series of various space-borne irradiance experiments are presented (from Fröhlich, 1997). Fig. 1 has been provided for this paper by courtesy of the SOHO/VIRGO Science Team.

Smithsonian Institute, first from high altitude mountain stations and later on from balloons and aircraft. These early measurements, however, could not reveal variations in total solar irradiance related to solar effects because of the lack of sufficient radiometric precision and the selective absorption of the terrestrial atmosphere (Fröhlich *et al.*, 1991).

The first continuous and high precision observations of total solar irradiance from space started in the late seventies from various satellite platforms. The time series of the space-borne experiments are plotted in Fig. 1 (from Fröhlich, 1997). The different scale of these measurements is related to the absolute accuracy (±0.2%) of the calibration of the individual instruments (Fröhlich, 1997). Because of the low absolute accuracy of the current total irradiance measurements, it is extremely difficult to maintain long-term homogeneous irradiance data sets, especially when there are interruptions in the observed data. The UARS/ACRIM II data presented in Fig. 1 have been scaled to the SMM/ACRIM I irradiance level by Fröhlich (1997) via the intercomparison of the two ACRIM data sets with the Nimbus-7/ERB and ERBS measurements. The used scaling factor is 1.00123 (Fröhlich, 1997).

Although the absolute accuracy of the total irradiance measurements is limited to about ±0.2%, the precision and stability of the instruments is much better, which makes it possible to study the relative variations in total irradiance. As illustrated in Fig. 1, total solar irradiance varies over a wide range of periodicities. The most important discovery of the satellite-based irradiance observations is that total solar irradiance varies by about a small fraction of 1% over the solar cycle, being higher during maximum solar activity conditions (e.g. Willson and Hudson, 1988). This solar-cycle-related variation of total solar irradiance is attributed in a large part to the changing emission of bright magnetic elements and a third component, the so-called "active network", has been introduced to explain the long-term irradiance changes (Foukal and Lean, 1988).

Since the Sun's irradiance is observed from one direction in space, it is difficult to determine whether the observed irradiance variations represent true luminosity changes which occur in the Sun's radiation in all directions or are simply results of a change in the angular dependence of the radiation field emerging from the photosphere. Kuhn et al. (1988) and Kuhn and Libbrecht (1991) have performed broad-band, two-color photometric measurements of the brightness distribution just inside the limb, which can be used to explain the total irradiance variations over the solar cycle. By integrating the limb brightness measurements, one can compute the solar luminosity. It has been demonstrated that the observed ERB and ACRIM total irradiance and the computed luminosity change in phase and relative amplitude, (e.g., Kuhn et al., 1988) and the active regions faculae alone fail by more than a factor of two in explaining the solar cycle related long-term changes (Kuhn et al., 1988). These brightness observations indicate that the long-term irradiance changes may also be related to variations in the photospheric temperature (Kuhn et al., 1988), although it is not yet clear whether this change can be linked with the bright network component. Note that former comparisons of the Swiss infrared measurements (Müller et al., 1975) with visible data have also indicated that there is a latitudinal dependence of the effective temperature of the Sun which may explain the long-term irradiance observations (Pecker, 1994).

The conclusion we can derive from the current measurements is that the solar radiation flux is anisotropic, a function of latitude, and naturally a function of the time during the migration of solar activity (Pecker, 1994). Note that additional global effects, such as changes in the solar diameter (Delache et al., 1986; Ulrich and Bertello, 1995), large scale convective cells (Ribes et al., 1985), the differential rotation of the Sun's interior and solar dynamo magnetic fields near the bottom of the convective zone (Kuhn, 1996) may also produce variations in total irradiance. These results demonstrate that the long-term changes are really luminosity changes, which may play an important role in global changes of the Earth's climate.

Short-term changes on time scales of days to months are superimposed on the long-term irradiance variations, which are primarily associated with the evolution of active regions via the combined effect of dark sunspots and bright faculae (Chapman, 1987). The most striking events in the short-term irradiance changes are the sunspot-related dips in total irradiance with an amplitude up to 0.3% (Willson et al., 1981). The effect of sunspots on total solar irradiance has been modeled with the "Photometric Sunspot Index" (PSI), which relates the area, position, and contrast of sunspots to a net effect on the radiative output of the observed solar hemisphere and it is corrected for the limb darkening (Hudson et al., 1982). The relationship between the PSI model and total solar irradiance is presented in Fig. 2 for the time interval of 1980 to 1994, where the PSI model has been calculated by Fröhlich et al. (1994). The dashed line in Fig. 2a indicates the daily values of total irradiance as measured by UARS/ACRIM II (Willson, 1994), whereas the solid line gives its adjusted value to the SMM/ACRIM I scale (Fröhlich, 1997).

As can be seen from Fig. 2, dips in total irradiance always correspond to the peaks in the PSI model, convincing the skeptics that the darkening effect of sunspots on total irradiance has indeed been detected and that the strong magnetic fields of sunspots can cause negative excursions in the total flux. It is interesting to note that in spite of the irradiance deficit due to sunspots, which are the most pronounced at solar maximum, total solar irradiance varies in phase with the solar cycle (Fröhlich et al., 1991). This indicates that the bright magnetic features overcompensate the effect of sunspots and give the primary source of long-term irradiance variations. The total irradiance corrected for sunspot darkening (thereafter S_c) is shown in Fig. 3a, indicating that the solar cycle variability of total irradiance would be considerably larger without the effect of sunspots. It has been found that the long-term variation of S_c is very similar to that of the solar UV irradiance (Foukal and Lean, 1990; Pap et al., 1991) and about one fifth (Lean, 1989) to one third (London et al., 1989) of the long-term change in total irradiance is related to the variability in the integrated 200 – 300 nm UV flux. To compare the variation of the UV irradiance with S_c, the combined Nimbus-7/SBUV1 and NOAA9/SBUV2 Mg II h & k core-to-wing ratio (Mg c/w) is plotted in Fig. 3b. Although the SBUV experiments suffer from a significant degradation in their diffuser reflectivity, the ratio of the irradiance in the core of the Mg 280 nm line to the irradiance at the neighboring continuum wavelengths can be used as a good index of solar chromospheric variability (Heath and Schlesinger, 1986). The combined Nimbus-7/SBUV1 and NOAA/SBUV2 Mg c/w ratio provides a long-term UV irradiance data set covering two solar cycles, which makes it possible to study the long-term UV irradiance changes in a great detail. As can be seen from Fig. 3., Sc and the Mg c/w ratio vary in parallel over the solar cycle. The formation of the Mg II line is very similar to that of the

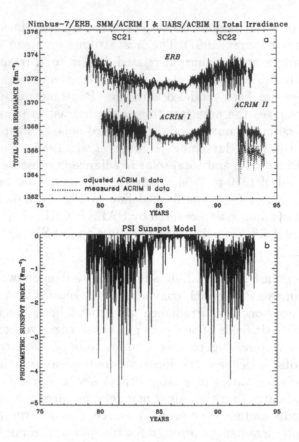

Figure 2. The Nimbus-7/ERB, SMM/ACRIM I and UARS/ACRIM II total irradiance
data are shown in Fig. 2a. Fig. 2b shows the PSI model calculated by Fröhlich *et al.*
(1994).

Ca II K line, and the two time series correlate very well (Donnelly *et al.*,
1994). This demonstrates that the long-term variations in total solar and
UV irradiances are primarily caused by the same events, i.e., the changing emission of bright magnetic features, such as plages and the magnetic
network (Lean, 1988).

It is interesting to note that the variation of the Mg c/w ratio over solar
cycles 21 and 22 is very symmetrical. As can be seen from Fig. 3, the Mg
c/w ratio shows about a 3-year long "flat" maximum during solar cycles 21
and 22. In contrast to the similar variation of the Mg c/w over the two solar
maxima, there is a substantial difference between the decline of the Mg c/w
during solar cycles 21 and 22. As can be seen from Fig. 3, the Mg c/w ratio
decreased steadily from 1982 to 1986. Note that the decline of solar UV
irradiance at Lyman-α was very similar and both the Mg c/w ratio and

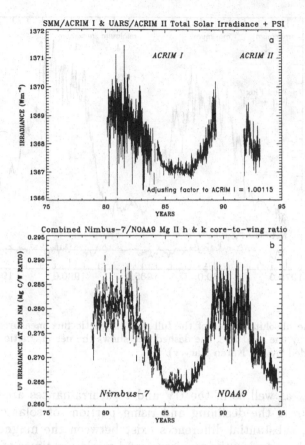

Figure 3. Total solar irradiance corrected for sunspot darkening is presented in Fig. 3a. The combined Nimbus-7/SBUV1 and NOAA9/SBUV2 Mg c/w ratio is shown in Fig. 3b.

Lyman-α irradiances showed a pronounced 300-day periodicity during the descending phase of solar cycle 21 (Pap *et al.*, 1990). In contrast, a sharp decrease was observed in both the Mg c/w ratio and Lyman-α between February and June 1992 (White *et al.*, 1994), which was seen in additional solar indices, e.g. in the PSI function, 10.7 cm radio flux and the full disk magnetic flux.

3. Solar irradiance variability and magnetic activity

It has been assumed that the observed long-term changes in total solar and UV irradiances are related to the evolution of the magnetic fields over the solar cycle (e.g. Harvey, 1994). The absolute value of the full disk integrated magnetic flux measured at the National Solar Observatory at Kitt Peak is plotted in Fig. 4. As can be seen, the long-term variations in the full disk

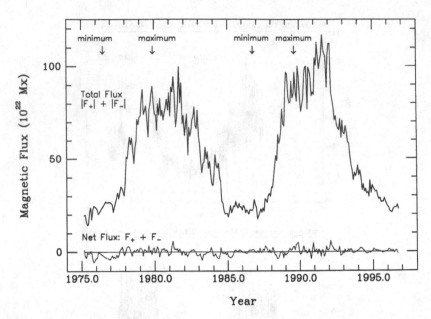

Figure 4.   The absolute values of the full disk magnetic flux measured at NSO/Kitt Peak is shown by the solid line, the dashed line shows the net magnetic flux. (The plot has been provided by Dr. Karen Harvey).

magnetic flux as well as in the UV and total irradiances are very similar, especially during the declining and rising portions of solar cycles 21 and 22. However, substantial differences exist between the magnetic field and irradiance values during the maximum and minimum times of the solar cycle. During the maximum of solar cycle 21, total solar irradiance decreased steadily from at least 1980, while the magnetic field peaked almost two years later, in 1982. In addition, as shown in Fig. 3b, the UV irradiance reached the maximum level of solar cycle 22 in late 1989, whereas the maximum of the magnetic flux occurred only in late 1991.

The relationship between total solar and UV irradiances and the magnetic flux during solar minimum was studied by Pap *et al.* (1996) in detail, using the so-called "dispersion diagrams". Any changes shorter than the solar rotational period were considered as noise and removed from the data by calculating monthly averages. The beginning and the end of the minimum time of solar cycle 21 was established from the distribution of the data in the dispersion diagrams. It was shown that the length of solar minimum was much shorter in the case of total and UV irradiances than in the case of solar indices representing strong magnetic fields, such as the full disk magnetic flux and PSI. These results indicate, as also seen in Figs. 2-4, that total solar and UV irradiances started to increase about 10 months prior to the rise of the magnetic flux (and PSI) at the beginning of the ascending phase of solar cycle 22.

Since the full disk magnetic flux includes the magnetic field of both sunspots and plages, the "Magnetic Plage Strength Index" (MPSI) has been used to study the relation between the changes of the UV irradiance and the magnetic flux related to plages. MPSI has been derived from magnetograms taken at the Mt. Wilson Observatory in the Fe I line at 525.0 nm (Ulrich, 1991). MPSI is defined as the sum of the absolute magnetic fields of all pixels with magnetic strength between 10 and 100 gauss divided by the total number of pixels in the image. Chapman and Boyden (1986) showed that pixels with magnetic fields of 10 to 100 gauss are associated with faculae and plages, whereas pixels above 100 gauss represent sunspots. The time series of the Mg c/w ratio (solid line) and MPSI (dotted line) are presented in the upper panel of Fig. 5 for the time interval of November 1978 to September 1986 (from maximum to minimum of solar cycle 21). The lower panel gives the scatter plot diagram between the two time series, which shows their high correlation ($r = 0.95$). However, the linear relation between the two time series breaks down at the time of solar minimum.

To study the nonlinear relationship between the Mg c/w ratio and MPSI, a relatively new technique called "Singular Spectrum Analysis" (SSA) has been used. SSA has been developed to study nonlinear and chaotic dynamical systems (Vautard et al., 1992). SSA is based on Principal Component Analysis in the time domain. The examined time series is augmented into a number of shifted time series. The cornerstone of SSA is the spectral, i.e., eigenvalue – eigenvector, decomposition of the lagged covariance matrix which is composed of the covariances determined between the shifted time series. The eigenvalues of the lagged covariance matrix compose the Singular Spectrum, where the eigenvalues are arranged in a monotonically decreasing order. The eigenvalues cut-off at a certain order, forming a "tail" in the spectrum which is considered as the noise floor of the data. The number of eigenvalues above the noise floor represents the degree of freedom of the variability, or in other words, the statistical dimension of the data, which is associated with the number of oscillatory components in the signal. The highest eigenvalues represent the fundamental oscillations in the data and in many cases they are related to the trend.

Fig. 6. shows the Singular Spectra of the Mg c/w ratio and MPSI. As can be seen, the two Singular Spectra are very similar, identifying about 50 oscillatory components in the two time series. The most interesting aspect of SSA is the reconstruction of the examined time series above the noise level and/or part of interest. The various oscillatory components can be reconstructed as a projection to the eigenvectors of the lagged covariance matrix. The Reconstructed Components (RCs) related to the first eigenvalues of the Mg c/w ratio and MPSI give the solar cycle related trends,

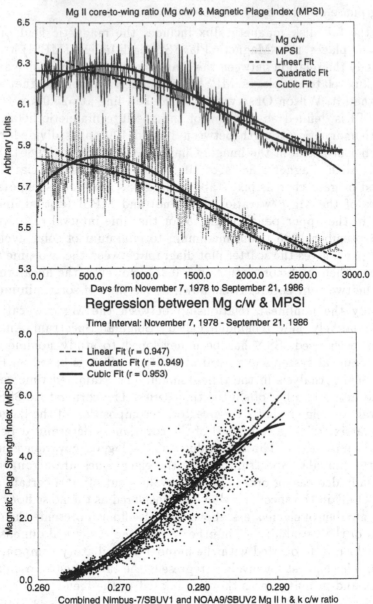

Figure 5. The upper panel shows the Mg c/w ratio (solid line) and MPSI (dotted line) for the time interval of November 7, 1978 to August 31, 1994. The lower panel shows the scatter plot between the two time series.

plotted in the upper panel of Fig. 7. The lower panel shows the correlation between the two trends, which clearly indicates the nonlinear relation between the solar cycle related trends in the Mg c/w ratio and MPSI. As

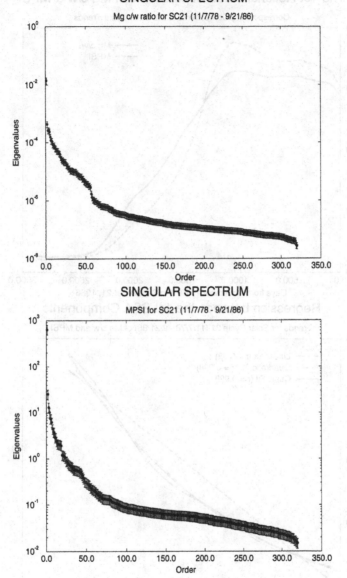

Figure 6. The Singular Spectra of the Mg c/w ratio (upper panel) and MPSI (lower panel) for solar cycle 21.

Fig. 7 shows, using a quadratic or a cubic fit, the long-term change in UV irradiance can be predicted by MPSI with a correlation coefficient: r = 0.998.

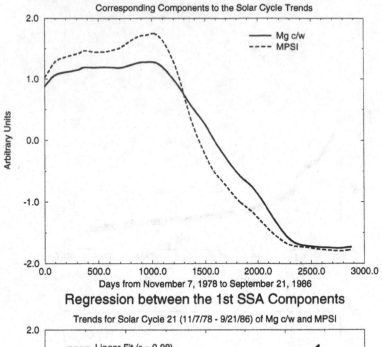

The 1st Reconstructed Components of Mg c/w & MPSI

Corresponding Components to the Solar Cycle Trends

Days from November 7, 1978 to September 21, 1986

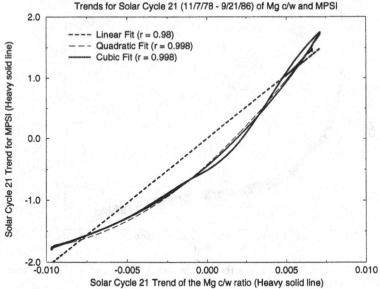

Regression between the 1st SSA Components

Trends for Solar Cycle 21 (11/7/78 - 9/21/86) of Mg c/w and MPSI

Figure 7. The upper panel shows the 1st RCs of the Mg c/w ratio (heavy solid line) and MPSI (heavy-dashed line) for solar cycle 21. The lower panel shows the correlation between the two RCs.

Finally, the 27-day variability in the Mg c/w ratio and MPSI has been reconstructed as well and the results are presented in Fig. 8. As can be seen, the amplitude of the 27-day variability is the strongest at the time of the

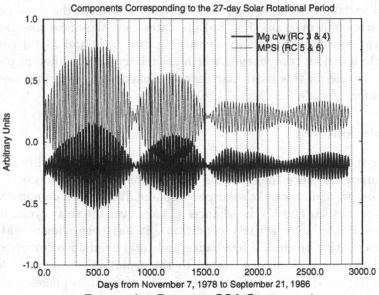

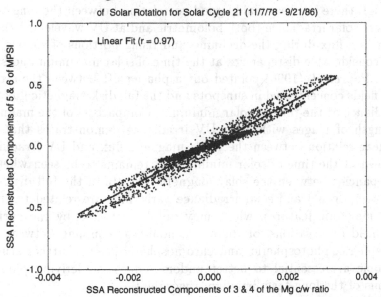

Figure 8. The 3rd and 4th Reconstructed Components of the Mg c/w ratio (heavy line) for solar cycle 21 are presented on the upper panel. These RCs represent the 27-day solar rotational component in the signal. The correlation between these components is shown on the lower panel.

solar maximum in both indices and it decreases towards solar minimum.
It is interesting to note that there is about a two year modulation in the
rotational variability of both indices. The distribution of the data points
on the scatter plot diagram (Fig. 8b) indicates that the linear association
between the rotational variability of the Mg c/w ratio and MPSI can be
divided into 3 various components, which are related to the different phases
of the solar cycle, in this particular case, the maximum, declining portion
and minimum of cycle 21.

4. Conclusions

Measuring the solar energy output and understanding its variability are extremely important since they provide information about the physical processes in, below, and above the solar photosphere. It has been demonstrated
that both total solar and UV irradiances change over the solar cycle, being
higher during maximum activity conditions. It has been shown that the
long-term variation of the UV irradiance and total irradiance corrected for
sunspot darkening is primarily related to the evolution of magnetic fields
over the solar cycle via the changing emission of faculae, plages and the
magnetic network (Lean, 1988).

Although there is a reasonably good correlation between the long-term
variations of solar irradiance (both bolometric and at UV wavelengths) and
solar magnetic flux during the declining and rising portions of solar cycle,
there are considerable discrepancies at the time of solar maximum and solar
minimum. Pap *et al.* (1996) pointed out a phase shift between the strong
magnetic fields concentrated in sunspots and the full disk magnetic flux and
solar irradiance at the time of solar minimum. Comparison of the magnetic
field strength of plages with solar UV irradiance demonstrates that the
strong linear relation between the plage magnetic flux and UV irradiance
breaks down at the time of solar minimum. It remains to be seen whether
the discrepancies between the solar magnetic flux (both the full disk and
its plage component) and solar irradiance variability are related to small
and faint magnetic features, which may not be detected by the current
magnetic field observations, or there is a nonlinear coupling between the
subphotospheric, photospheric, and chromospheric layers. Further studies
on this topic are essential to better understand the underlying physical
mechanisms of the long-term irradiance variabilities.

Acknowledgements: The research described in this paper was carried out by
the University of California, Los Angeles and the Jet Propulsion Laboratory, California Institute of Technology under a contract with the National
Aeronautics and Space Administration. The author expresses her gratitude
to the entire SOHO/VIRGO Science Team for providing the first VIRGO

results for this paper. The author expresses her gratitude to Dr. K. Harvey for providing the plot of the full disk magnetic flux measured at the National Solar Observatory at Kitt Peak and produced co-operatively by NSF/NOAO, NASA/GSFC, and NOAA/SEL. Useful comments by Drs. L. Bertello, C. Fröhlich, D. Parker, R. Ulrich and F. Varadi are highly acknowledged.

References

Chapman, G.A. (1987) *Ann. Rev. Astron. Astrophys.* **25**, 633.
Chapman, G.A. and Boyden, J. (1986) *Astrophys. J.* **302**, L71.
Delache, Ph., Lacrare, F., and Sadsaoud, H. (1986) in J. Christensen Dalsgaard and S. Frandsen (eds.), *Advance in Helio- and Astroseismology*, IAU Press, p. 223.
Donnelly, R.F., White, O.R., and Livingston, W.C. (1994) *Solar Physics* **152**, 69.
Foukal, P. and Lean, J. (1988) *Ap.J.* **328**, 347.
Foukal, P. and Lean, J. (1990) *Science* **247**, 505.
Fröhlich, C. (1997) in J.M. Pap, C. Fröhlich, and R.K. Ulrich (eds.), *The Proceedings of the SOLERS22 1996 Workshop*, Kluwer Academic Publishers, Dordrecht, in press.
Fröhlich, C., Foukal, P., Hickey, J.R., Hudson, H.S., and Willson, R.C. (1991) in C.P. Sonnett, M.S. Giampapa, and M.S. Matthews (eds.), *The Sun in Time*, Univ. Arizona Press, Tucson, p. 11.
Fröhlich, C., Pap, J., and Hudson, H.S. (1994) *Solar Physics* **152**, 111.
Fröhlich, C., Andersen, B., Appourchaux, T., Berthomieu,G., Crommelynck, D., Domingo, V., Fichot, A, Finsterle, W., Gómez, M., Gough, D., Jiménez, A., Jeifsen, T., Lombaerts, M., Pap, J., Provost, J., Roca Cortés. T., Romero, J., Roth, H., Sekii, T., Telljohann, S., Toutain, T., Wehrli, Ch. (1997) *Solar Physics* **170**, 1.
Harvey, K. (1994) in J.M. Pap, C. Fröhlich, H.S. Hudson, and S.K. Solanki (eds.), *The Sun as a Variable Star: Solar and Stellar Irradiance Variations*, Cambridge University Press, Cambridge, p. 217.
Heath, D.F. and Schlesinger, B.M. (1986) *J. Geophys, Res* **91**, 8672.
Hudson, H.S., Silva, S., Woodard, M. and Willson, R.C. (1982) *Solar Physics* **76**, 211.
Kuhn, J. (1996) in T. Roca Cortes (ed.), *Global Changes in the Sun*, Cambridge University Press, Cambridge, in press.
Kuhn, J. and Libbrecht, K.G. (1991) *Ap.J.* **381**, L35-L37.
Kuhn, J., Libbrecht, K.G. and Dicke, R.H. (1988) *Science* **242**, 908.
Lean, J. (1988) *Adv. Space Res.* **Vol 8** (5), 263.
Lean, J. (1989) *Science* **244**, 197.
London, J., Pap, J., and Rottman, G. (1989) in J. Lastovicka, T. Miles, and O. Neill (eds.), *Middle Atmosphere Program Handbook* **Vol 29**, p. 9.
Müller, E.A., Stettler, P., Rast, J., Kneubühl, F.K., and Huguenin, D. (1975) in Chiuderi, Landini, and Righini (eds.), *First European Solar Meeting*, Firenze **105**.
Nesme-Ribes, E., Sokoloff, D., and Sadourney, R. (1994) in J.M. Pap, C. Fröhlich, H.S. Hudson, and S.K. Solanki (eds.), *The Sun as a Variable Star: Solar and Stellar Irradiance Variations*, Cambridge University Press, Cambridge, p. 244.
Pap, J., Tobiska, W.K., and Bouwer, D. (1990) *Solar Physics* **129**, 165.
Pap, J., London, J., and Rottman, G. (1991) *Astron. Astrophys.* **245**, 648.
Pap, J., Vigouroux, A. and Delache, Ph. (1996) *Solar Physics* **167**, 125.
Pecker, J.-C. (1994) *Vistas in Astronomy* **Vol. 38**, 111.
Radick, R. (1994) in J.M. Pap, C. Fröhlich, H.S. Hudson, and S.K. Solanki (eds.), *The Sun as a Variable Star: Solar and Stellar Irradiance Variations*, Cambridge University Press, Cambridge, p. 109.
Ribes, E., Mein, P., and Manganey, A. (1985) *Nature* **318**, 170.

Ulrich, R. (1991) *Adv. Space Res.* **Vol. 11** (4), 217.

Ulrich, R. and Bertello, E. (1995) *Nature* **377**, 214.

Vautard, R., Yiou, P., and Ghil, M. (1992) *Physica D* **58**, 95.

White, O.R., Rottman, G.J., Woods, T.N., Knapp, B.G., Keil, S.L., Livingston, W.C., Tapping, K.F. Donnelly, R.F., and Puga, L. (1994) *J. Geophys. Res.* **89**, 369.

Willson, R.C. (1994) in J.M. Pap, C. Fröhlich, H.S. Hudson, and S.K. Solanki (eds.), *The Sun as a Variable Star: Solar and Stellar Irradiance Variations*, Cambridge University Press, Cambridge, p. 54.

Willson, R.C. and Hudson, H.S, (1988) *Nature* **332**, 810.

Willson, R.C., Gulkis, S., Janssen, M., Hudson, H.S., and Chapman, G.A. (1981) *Science* **211**, 700.

GIANT PLANETS SEISMOLOGY

B. MOSSER
Institut d'Astrophysique de Paris
98b, bld Arago, F-75014 Paris
mosser@iap.fr

Abstract. The giant planets Jupiter and Saturn belong to the interesting category of possible goals for remote seismic analysis. Their first seismic observations and their analysis were attempted in 1987 and 1991 respectively, under Philippe Delache's initiative. The theoretical analysis of giant planets seismology reveals the strong signature of the dense planetary core and the tiny one of the hydrogen plasma phase transition. The asymptotic formalism makes possible to obtain pertinent information for the observation of planetary oscillations and for their analysis. Specific observational techniques were developed to detect the seismic signature of giant planets. However, the first observations (Schmider *et al.* 1991, Mosser *et al.* 1993) of Jovian oscillations remain tentative. Even if the Jovian origin of the signal is beyond doubt, the interpretation in terms of Jovian global modes remains speculative. The collision of comet SL9 onto Jupiter provided an unexpected and unique opportunity to search for oscillations excited by the cometary impacts (Mosser *et al.* 1996). Seismic observations of Saturn remain negative so far. Therefore, this review focuses on Jupiter. Finally, the almost 10-years long experience of seismic observations of Jupiter and Saturn has not yet provided new constraints for planetary interior models. However, guidelines for future observational projects dedicated to Jovian seismology can be drawn. The different techniques of observation are compared, and observational requirements are precisely described.

1. Introduction

A complete review of giant planets seismology was already given by Mosser (1994). The theoretical aspects of giant planets oscillation were described exhaustively, as well as the current status of what is known about the

251

J. Provost and F.-X. Schmider (eds.), Sounding Solar and Stellar Interiors, 251-264.

Jovian interior. The current review focuses on observational aspects, and try to determine the manner to optimize further seismic observations of Jupiter.

A survey of the papers dealing with giant planets seismology is presented in Section 2. The Jovian standard interior model is described in Section 3. Properties of the planetary pressure modes oscillation pattern estimated with the asymptotic theory are briefly presented, in order to provide information for observational work. Section 4 presents past observations, with a tentative comparative study of their performance. Perspectives are proposed in Section 5.

2. Historical review

Giant planets seismology is a recent field of interest. The first paper devoted to Jovian oscillations was published in 1976 by Vorontsov *et al.* The first tentative observations were reported by Deming *et al.* (1989). It is possible to define three periods in the past twenty years.

- First, Vorontsov *et al.* provided a very complete set of theoretical papers dealing with the specific problems of Jovian seismology.
- Then, the first observations of Jovian oscillations were attempted. In parallel, various theoretical papers were published.
- In July 1994, the collision of comet Shoemaker-Levy 9 provided a unique opportunity to search for Jovian oscillations. One third of the papers in Jovian seismology are devoted to this event.

Figure 1 presents an histogram of all papers published between 1976 and 1996 in journals with an international audience and lecture committee. The three periods mentioned above appear clearly. The first theoretical papers (Vorontsov *et al.* 1976, 1989, Vorontsov & Zharkov 1981, Vorontsov 1981, Vorontsov 1984a,b), even if covering the main theoretical aspects of the subject, remained isolated (except Bercovici & Schubert 1987), and giant planets seismology was still a confidential field. However, the successes in helioseismology opened new horizons.

Observations made in 1987 (Deming *et al.* 1989, Schmider *et al.* 1991) correspond to beginning of the second period. Philippe Delache played a very important role, making possible theoretical, instrumental and observational progresses. Synergy between planetologists and seismologists was highly fruitful. Independently of the observational results, the potential obtention of new constraints on the planetary interior structure induced further works (Marley 1991, Marley & Porco 1993, Lee 1993).

The collision of comet Shoemaker-Levy 9 fragments with Jupiter opened a new period. Several groups proposed or achieved observations for monitoring the seismic consequences of the impacts (Kanamori 1993, Deming

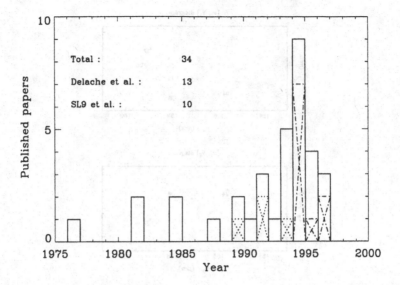

Figure 1. Histogram of papers dealing with Jovian seismology in the past two decades. Papers reporting new observational results are indicated by dotted lines. Philippe Delache, even if he does not explicitly appear in the author lists, was at the origin of 13 papers, among whose can be the first possible detection of Jovian oscillations.

1994, Gough 1994, Lee & Van Horn 1994, Lognonné *et al.* 1994, Marley 1994, Cacciani *et al.* 1995). Searching for the seismic signature of the cometary impacts was considered as a "high risk, high return" operation. This period will remain a short parenthesis, and giant planets seismology confined to very few groups, if future observations remain unable to provide new constraints on the Jovian interior structure.

3. Giant planets interior models and seismology

3.1. STANDARD MODEL

The Jovian standard interior model is based on the following assumptions:

- The gravitational moments, measured by the Voyager spacecraft (Campbell & Synnot 1985), imply a planetary core of heavy elements.
- Convection, measured in the upper troposphere, is needed to extract the interior heat of the planet. That induces an adiabatic structure.
- The composition is supposed to be homogeneous. In the fluid envelope, the mass fraction are respectively X = 74%, Y = 24% and Z = 2%, according to the Galileo probe (von Zahn & Hunten 1996).
- The transition between the molecular and metallic phases of hydrogen (plasma phase transition, PPT) occurs at the 1.2 Mbar pressure level (Chabrier *et al.* 1992).

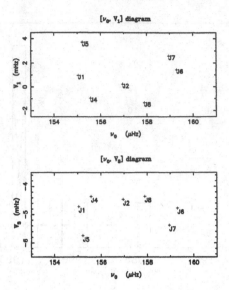

Figure 2. Asteroseismic HR diagram: comparison of different Jovian interior models, based on parameters of the oscillation pattern: ν_0 is the characteristic frequency, V_1 and V_3 the second-order frequencies (Provost *et al.* 1993). Despite the fact that all these standard models satisfy the gravitational constraints, their seismic properties are very different.

Different interior models were constructed following this scenario . When displayed in a seismic HR diagram (Christensen-Dalsgaard 1988), they prove to be very different (Fig. 2). The gravitational moments, integrated quantities of the mass distribution, do not provide a precise determination of the interior. Furthermore, the previous assumptions are denied by the non-standard point of view:

- The core may be more or less diluted in the envelope (Stevenson 1985)
- Condensation of heavy elements, or non-miscibility in the metallic phase imply an inhomogeneous composition.
- The PPT remains a theoretical construction.
- A possible radiative window, near the 2000 K temperature level, should modify completely the whole interior adiabat (Guillot *et al.* 1994).

The discrepancies between the standard Jovian interior models constrained by the gravitational moments, as well as the comparison of the standard and non-standard descriptions of the Jovian interior structure, defines the minimum output expected from giant planets seismology: constraints on the core size, existence and location of the PPT, mass distribution within the planet.

3.2. OSCILLATION PATTERN

3.2.1. *Asymptotic analysis*
The asymptotic analysis, derived from Tassoul (1980) and adapted to the Jovian interior with a dense core, is valid for low degree and high radial order modes of the non-rotating planet (Provost *et al.* 1993):

$$\nu_{n,\ell} = \left[n' + \frac{L^2 V_1 + V_2}{4\pi^2 \nu_{n,\ell}} - \frac{\varepsilon}{\pi} \sin \alpha_{n,\ell} - \frac{\varepsilon^2}{\pi} \frac{N-2}{2N} \sin 2\alpha_{n,\ell} \right] \nu_o$$

$$\text{with} \quad \alpha_{n,\ell} = 2\pi \left(\frac{n'}{N} - \frac{\ell}{2} - \frac{L^2 V_3 + V_4}{4\pi^2 \nu_{n,\ell}} \right) \quad \text{and} \quad n' = n + \frac{\ell}{2} + \frac{n_e}{2} + \frac{1}{4}$$

and where n and ℓ are the radial order and the degree of the degenerated mode; n_e is the polytropic index at the surface, ν_0 is the main characteristic frequency while $V_{1\rightarrow4}$ are the second order frequencies, ε and N represent respectively the amplitude and the period of the core modulation. The PPT occurs at a too-high level in the envelope to have any signature on low degree modes. On the other side, the modulation due to the dense core cancels the regularity of the pattern developed by the asymptotic theory (Tassoul 1980). All the conclusions based on the asymptotic results are confirmed by numerical calculations (Vorontsov *et al.* 1989, Lee 1993, Provost *et al.* 1993, Gudkova *et al.* 1995).

3.2.2. *Rotation*
For low degree high frequency modes, the planetary rotation can be treated as a perturbation (Mosser 1990), despite its relative importance: $\nu_{\rm rot}/\nu_0 \simeq 1/5$ for Jupiter, compared to about 1/300 for the Sun. The rotational removal of degeneracy is given by

$$\nu_{n,\ell,m} = \nu_{n,\ell} \left[1 + E(e, n, \ell) \left(\frac{1}{3} - \frac{m^2}{\ell(\ell+1)} \right) \right] - m \, \nu_{\rm rot}$$

This result agrees with the numerical analysis made by Lee (1993) taking into account the coupling between the modes. The small correction E is of the same order as the planetary oblateness $e = 6.5\%$. The additive term $- m \, \nu_{rot}$ is due to the rotating frame. The relative weight of the correction due to oblateness is high enough to completely cancel the regularity of the Zeeman-like multiplet. The frequency differences in a $\ell = 2$ multiplet with $\nu \simeq 2$ mHz vary between 14 and 42 μHz, whereas $\nu_{\rm rot} = 28$ μHz. Inside a given multiplet, the eigenfrequencies of opposite azimuthal orders satisfy to:

$$\nu_{n,\ell,-m} - \nu_{n,\ell,m} = 2 \, m \times \nu_{\rm rot}$$

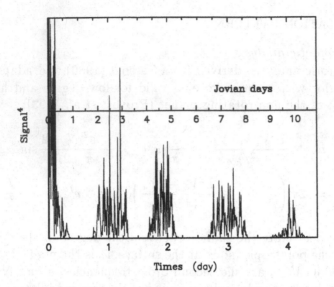

Figure 3. Spectrum of the spectrum recorded at CFHT in 1991, with the signature of the rotational splitting.

This equation, which expresses that the difference between two unknown eigenfrequencies remains a well known term, gives an important clue for the detection of an oscillating signal. The identification of doublets separated by an even multiple of the rotation frequency represents the signature of an oscillating phenomenon affected by the rotation. In the spectrum of the spectrum, this signature will consist of a series of peaks multiple of half the period of rotation (Fig. 3).

It must be emphasized that this signature is not a photometric effect, provided it appears at frequency much higher than the rotational frequency. The contamination of the spectrum by photometric inhomogeneities was examined by Lederer *et al.* (1995), and concluded that the spectrum range over 0.5 mHz is exempted from spurious photometric signal.

3.2.3. *Ray tracing*

The ray tracing theory was used by Bercovici & Schubert (1987) in order to obtain in a simple way the oscillation diagram, and by Mosser *et al.* (1988) to put in evidence the influence of the planetary core on the Jovian seismic pattern. However, this approach is too crude to describe precisely the core influence. A ray not propagating in the core simply ignores its existence, whereas the evanescent behaviour of a Jovian mode in the central region of the planet will seriously influence its eigenfrequency.

On the other hand, the ray tracing method was adequate for describing the propagation of very high frequency waves generated by the impacts of

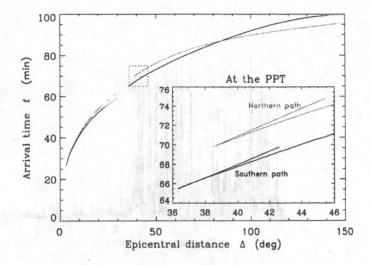

Figure 4. Arrival time versus epicentral distance for a Jovian interior model with PPT. The epicentral distance is the angular distance between the impact point and the region where the wave reaches the troposphere again. Due to the planetary oblateness, the north and south paths are not synchronized. The zoom shows the signature of the PPT, namely the shadow zone occurring about 1 hour after the impact.

comet SL9 fragments in the Jovian fluid envelope (Fig. 4). This event provided a unique opportunity to search for the signature of the PPT (Gudkova *et al.* 1995).

3.3. ENERGETICS

It is doubtless that Jupiter oscillates. The planet is fluid, mostly convective, and emits more energy than received from the Sun. Waves with frequency less than 3.1 mHz are trapped below the tropopause (Mosser 1995). However, the amplitude of the oscillations remains unknown. Theoretical predictions are currently unable to predict any amplitude level, just because the Jovian interior remains too mysterious. The mechanism proposed for the excitation of solar pressure modes by turbulent convection seems to be not efficient in Jupiter, since it gives very low amplitude (Deming *et al.* 1989). On the other side, the latent heat at the PPT should provide a powerful possible excitation. In fact, observations are the only way to solve the problem.

Jovian oscillations may be solar-like. In that case, Jovian and solar oscillation pattern should be similar, due to the equality between the free fall characteristic frequency $\sqrt{\mathcal{G}M/R^3}$ of Jupiter and the Sun. The modulation due to the planetary core and the different cutoff frequencies make

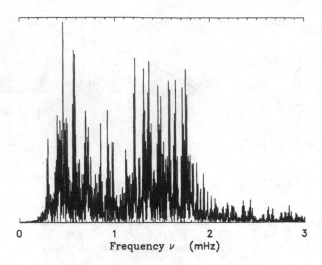

Figure 5.   Fourier spectrum of the data recorded at OHP in 1987. The most evident
signature is due to non-continuous observations. Many doublets separated by two times
the rotational frequency (28 μHz) can be detected.

the global aspect of the theoretical spectra very different.

4. Observations

4.1. RESONANCE SEISMOMETRY

Full disk observations of Jupiter with a sodium cell resonance were performed at the Observatoire de Haute-Provence, in november 1987, during 6
nights close to the planetary opposition (Schmider *et al.* 1991). The Doppler
shifts of the solar sodium line reflected by the planet are analyzed by the local sodium reference. The relative variations are related to a velocity change
by the factor $2.10^{-5}/(\text{m.s}^{-1})$. The detection of the signature of the rotational removal of degeneracy acts in favor of the detection of the pressure
mode signal (Mosser *et al.* 1991). Various tests showed that this signature was unambiguous, and not related to photometric changes, despite a
non-favorable duty cycle.

4.2. FOURIER TRANSFORM SEISMOMETRY

The principle of Fourier Transform seismometry (Mosser *et al.* 1993, Maillard 1996) consists of searching for the Doppler signal in the interferogram
of the planetary spectrum. Working in the Fourier space makes possible to
benefit of the multiple advantage. The Jovian spectrum exhibits at 1.1 μm

a complex rotational-vibrational pattern, due to the $3\nu_3$ methane band. More than 40 lines appear around $\sigma_0 = 9100$ cm^{-1}, and induce strong signatures in the interferogram at path differences near 1 cm. The pattern at a selected path difference δ is sensitive to the velocity v and develops schematically as:

$$I(\delta) \propto \cos 2\pi\sigma\delta = \cos 2\pi\sigma_0\delta \left[1 + \frac{v}{c}\right]$$

Stability of the method is provided by the metrologic He-Ne laser of the Fourier Transform Spectrometer based at the CFH 3.6-m telescope. The method was employed first on Jupiter. This mono-site observation provided results in agreement with the one of Schmider et al. (1991). Intensity changes at a zero crossing of the interferogram were registered (Mosser et al. 1993).

$$dI = I_{\max} \, 2\pi\sigma_0\delta \, \frac{dv}{c}$$

The field of view being limited to 12" (1/4 of the Jovian diameter at planetary opposition), it was therefore necessary to cancel out the guiding errors. Due to the fast planetary rotation, 1" drift on the planet is equivalent a velocity shift about 0.5 km.s^{-1}. A rapid modulation parallel to the Jovian axis of rotation permitted this. Furthermore, this modulation cancelled out the contribution of telluric water. However, this method appeared to be very sensitive to any modification of the working point.

In order to increase the performance of the detection, the principle of the observation was changed into a phase measurement. New InGaAs detectors permitted to increase the quantum efficiency of the detection and to obtain a 24" field of view. This new method was used in July 1996. A sample of the phase determination is given in Fig. 6.

$$d\varphi = 2\pi\sigma_0\delta \, \frac{dv}{c}$$

This second method presents great advantages compared to the first one. First it is not contaminated by photometric changes, since the determination of the phase and the amplitude are independent. Second the sensitivity is no longer affected by unavoidable low frequency drifts of the working point. Furthermore, it makes possible an absolute calibration of the signal. The calibration factor is simply $c/2\pi\sigma_0\delta$, $\simeq 5$ km.s^{-1}

4.3. IR PHOTOMETRY

The previous methods are sensitive to the velocity field of the oscillation. IR monitoring of the Jovian disk allows the measurement of temperature

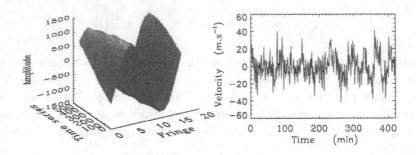

Figure 6.   Registration of the phase of the scanned portion of the interferogram, and traduction in terms of Doppler velocity

variations associated to the adiabatic wave (Deming *et al.* 1989). The temperature perturbation of an adiabatic wave is related to its velocity by

$$\frac{\mathrm{d}T}{T} = (1 - \gamma)\,\frac{\mathrm{d}v}{c_\mathrm{S}}$$

where c_S is the local sound speed, about 0.8 km.s^{-1} in the Jovian troposphere, and γ the adiabatic exponent (about 1.4). When sounding the Jovian troposphere in the 10-μm region, sensitive to the tropospheric levels around $T_0 \simeq 135$ K, the relative flux variations are related to the thermal variations by:

$$\frac{\mathrm{d}\Phi}{\Phi_0} \simeq 12\,\frac{\mathrm{d}T}{T_0}$$

Finally, the relative flux variations are related to a Doppler term by the ratio 5.10^{-3} / m.s^{-1}. This method was first investigated by Deming *et al.* (1989), but unsuccessfully. The detector was a 20-element linear array, limiting the sensitivity of the observations to only zonal high-degree modes ($|m| = \ell \geq 10$). Other attempts conducted with a 20×64 pixels camera (Fisher 1994) were also unsuccessful.

Observations made in July 1994, during and after the collision of comet Shoemaker Levy 9 with Jupiter, used this technique. Daytime observations are possible, what was necessary for the recording of the SL9 events, that occurred in the daytime when Jupiter was far from opposition. Spatial

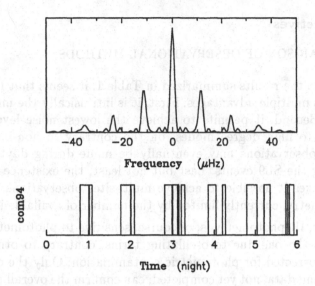

Figure 7. Window function of the 3 runs conducted with IR cameras in July 1994 after the SL9 impacts. The duty cycle is about 34%.

resolution is related to the number p of pixels along the Jovian diameter. It is theoretically equal to $p/2$. Multi-site observations were conducted, from the Canary Islands, Chile and Hawaii (Fig. 7). On the contrary to Doppler measurements, this method offers no internal stable reference. Each visible feature on the Jovian thermal map has a non-negligible contribution to the signal. The flux variation between two neighbour pixels is about 1/100 the mean flux, as high as the typical rms noise level.

TABLE 1. Comparison of observational techniques

	Na cell seismometry OHP 1987	FT seismometry CFH 1996	IR photometry ESO 1994
stability	atomic line	He-Ne laser	×
sensitivity $(1/\text{m.s}^{-1})$	2.10^{-5}	3.10^{-5}	5.10^{-3}
noise level (1σ, 1 min integration)	25 m.s^{-1}	15 m.s^{-1}	3 m.s^{-1}
corrected for photometry	yes	yes	×
multi-site (1996)	yes	no	yes
high ℓ modes	no	no	yes

5. Perspectives

5.1. COMPARISON OF OBSERVATIONAL METHODS

According to the results summarized in Table 1, it seems that IR photometry presents multiple advantages. First, it is intrinsically the most sensitive
technique. Second, it permits to achieve the lowest noise level. Third, it
is sensitive to high degree modes, on the contrary to non-imaging techniques. IR observations may eventually be made during daytime, as was
done during the SL9 events. Last but not least, the existence of many IR
cameras makes it possible to achieve multi-sites observations, contrary to
FT seismometry, currently limited by the number of available instruments.

However, IR photometry is, of course, sensitive to photometric changes,
not only to the one due to oscillating terms, contrary to other methods
which are corrected for photometric contamination. Only the careful analysis of existing data, not yet completed, can confirm the overall performance
of this technique. The complexity due to the inhomogeneous thermal field
has not yet been completely solved. Further IR observations should be conducted with a filter selecting atmospheric regions where the Jovian thermal
map is the most uniform.

In fact both photometric and spectrometric observations have advantages. IR imaging techniques are necessary for the possible detection of
the plasma phase transition of hydrogen, whereas full disk observations, as
made with current spectrometric techniques, are of prime interest for the
determination of the core structure (Mosser 1994). In both cases, the duty
cycle of the observations must be as long as possible.

5.2. OBSERVATIONAL PROJECTS

The observational results are not satisfying. Even if the detection of the
signature of the rotational splitting implies the detection of waves in the
frequency range [1, 2 mHz], no new constraint has been carried out by seismology. On the other hand, the seismic interest concerning Jupiter, with
all the potential constraints it implies, has induced many other works. A
new generation of Jovian interior models (Guillot et al. 1994, 1995) was
constructed, including up to date equations of state, based on rigorous
physical assumptions, and numerically more robust than the previous generation. Some techniques proposed for the monitoring of the Jovian oscillations proved to be not convenient (Mosser 1992); on the other hand,
Fourier-transform seismometry is directly applicable for asteroseismology
(Maillard 1996).

Finally, it is possible to define guidelines for future observations:

- At least 4 nights observation with multi-sites observations providing a duty cycle better than 50% are necessary to infer physical results from the oscillation pattern.
- The field of view of either imaging techniques or spectrometric observations must accept the entire planetary image.
- Spectrometric measurements must include the correction for photometry.
- "Exotic" technique of detection must not be excluded, such as proposed by Marley & Porco (1993) for Saturn, considering the possible resonance between fundamental modes and gravitational perturbations in the planetary rings.
- Observations from space provide two major requirements for seismology: excellent duty cycle, stable photometry. Jupiter is a secondary objective of the spatial project Corot (Catala et al. 1995).

Acknowledgments

I wish to thank Philippe Delache and Daniel Gautier, who introduced me to giant planets seismology, and all the colleagues involved in the progress of this new topic.

This work has been supported by the Programme National de Planétologie from the Institut National des Sciences de l'Univers (INSU) and by the Groupement de Recherches 'Structure interne des étoiles et des planètes géantes' from the Centre National de la Recherche Scientifique (CNRS).

References

Baglin A. (1991) Adv. Space Res., 11, 4, (4) 133
Bercovici D., G. Schubert (1987) Icarus 69, 557
Cacciani A. et al. (1995) GRL 22, 17, 2437
Campbell J. K., S. P. Synnot (1985) Astron. J. 90, 364
Catala C. et al. (1995) COROT: a space project devoted to the study of convection and rotation in the stars. in 4th SOHO Workshop: Helioseismology, eds J.T. Hoeksema, V. Domingo, B. Fleck &B. Battrick, ESA SP 376 vol 2, p 549
Chabrier G., D. Saumon, W.B. Hubbard, J.I. Lunine (1992) Saturn. Astrophys. J 391, 817
Christensen-Dalsgaard (1988) in Advances in helio- and asteroseismology, eds. J. Christensen-Dalsgaard and S. Frandsen, p. 295
Deming D., M. J. Mumma, F. Espenak, D. E. Jennings, Th. Kostiuk, G. Wiedemann, R. Loewenstein, and J. Piscitelli (1989) Astrophys. J 343, 456
Deming D. (1994) Geophys. Res. Let. 20, 1095
Fisher B.M. (1994) High time-resolution infrared imaging observations of Jupiter. PhD thesis, University of California, San Diego.
Galdemard Ph., B. Mosser, P.O. Lagage, E. Pantin. The seismic response of Jupiter to the SL9 impacts: A 3-D analysis of the Camiras infrared images. Submitted to PSS.
Gough D.O. (1994) Mon. Not. R. Astron. Soc. 269, L17

Gough D.O. (1986) In *Hydrodynamics and MHD Problems in the Sun and Stars* (Y.Osaki, Ed.), 117 *University of Tokyo Press*.

Gudkova T., Mosser B., Provost J., Chabrier G., Gautier D., Guillot T. (1995) *Astron. Astrophys.* **303**, 594

Guillot T., Morel P. (1995) *Astron. Astrophys. Suppl. Ser.* **109**, 109

Guillot T., Chabrier G., Morel P., Gautier D. (1994) *Astron. Astrophys.* **303**, 594

Hunten D.M., Hoffman W.F., Spague A.L. (1994) *Geophys. Res. Let.* **20**, 1091

Kanamori H. (1993) *Geophys. Res. Let.* **20**, 2921

Lederer S. M., M. S. Marley, B. Mosser , J-P. Maillard, N. J. Chanover, R. Beebe (1995) *Icarus* **114**, 269

Lee U. (1993) *Astrophys. J.* **405**, 359

Lee U., Van Horn H.M. (1994) *Astroph. J.* **405**, 359

Lognonné Ph., B. Mosser (1993) *Survey in Geophysics* **14**, 239

Lognonné Ph., Mosser B., Dahlen F. (1994) *Icarus* **110**, 180

Maillard J.P. (1996) *Applied Optics* **35, 16**, 2734

Marley M.S. (1991) *Icarus* **94**, 420

Marley M.S., C.C. Porco (1993) *Icarus* **106**, 508

Marley M.S. (1994) *Astrophys. J.* **427**, L63

Mosser B., D. Gautier, Ph. Delache (1988) In *Seismology of the Sun and Sun like-stars*. Tenerife, Spain, Sept. 1988. *Proc. ESA*, **SP-286**, 593

Mosser B. (1990) *Icarus* **87**, 198

Mosser B., F.-X. Schmider, Ph. Delache, D. Gautier (1991) *Astrophys. J.* **251**, 356

Mosser B., D. Gautier, Th. Kostiuk (1992) *Icarus* **96**, 15

Mosser B. (1992) Etude des oscillations globales de Jupiter et des planètes géantes. *PhD thesis*, Université Paris-XI, Orsay.

Mosser B., D. Mékarnia, J.-P. Maillard, J. Gay, D. Gautier, Ph. Delache (1993) *Astron. Astrophys.* **267**, 604

Mosser B. (1994) Jovian seismology. in *The equation of state in astrophysics*, IAU colloquium 147, St-Malo, France Eds. G. Chabrier and E. Schatzman, Cambridge University Press, p. 481

Mosser B., Gudkova T., Guillot T. (1994) *Astron. Astrophys.* **291**, 1019

Mosser B. (1995) *Astron. Astrophys.* **293**, 586

Mosser B., Ph. Galdemard, P.O. Lagage, E. Pantin, M. Sauvage, Ph. Lognonné, D.Gautier, F. Billebaud, T. Livengood, H.U. Käufl (1996) *Icarus* **121**, 331

Provost J., B. Mosser, G. Berthomieu (1993) *Astron. Astrophys.* **274**, 595

Schmider F.-X., B. Mosser, E. Fossat (1991) *Astron. Astrophys.* **248**, 281

Stevenson D.J. (1985) *Icarus* **62**, 4

Tassoul M. (1980) *Astrophys. J. Suppl* **43**, 469

Unno W. ,Y. Osaki, H. Ando, H. Shibashi (1979) *Nonradial oscillation of stars*, (W. Unno, Ed.), University of Tokyo press. p. 149

Vorontsov S.V., V.N. Zharkov, V.M. Lubimov (1976) *Icarus* **27**, 109

Vorontsov S.V. , V.N. Zharkov (1981) *Astron. Zh.* **58**, 1101, [*Sov. Astron.* **25**, 627, 1982].

Vorontsov S.V. (1981) *Astron. Zh.* **58**, 1275, [*Sov. Astron.* **25**, 724, 1982].

Vorontsov S.V. (1984a) *Astron. Zh.* **61**, 700, [*Sov. Astron.* **28**, 410, 1984].

Vorontsov S.V. (1984b) *Astron. Zh.* **61**, 854, [*Sov. Astron.* **28**, 500, 1984].

Vorontsov S.V., T.V. Gudkova, V.N. Zharkov (1989) *Pis. Astron. Zur.* **15**, 646

von Zahn U., D.M. Hunten (1996) *Science* **272**, 849

Walter C.M., M.S. Marley, D.M. Hunten, A.L. Sprague, W.K. Wells, A. Dayal, W.F. Hoffman, M.V. Sykes, U.K. Deutsch, G.G. Fazio, J.L. Hora (1996) *Icarus* **121**, 341

APPARENT SOLAR DIAMETER AND VARIABILITY

A. VIGOUROUX

Laboratoire G.D. Cassini, U.M.R. CNRS 6529
Observatoire de la Côte d'Azur
B.P. 4229, 06304 Nice Cedex 04, France

1. Introduction

Philippe Delache brought me to solar physics in late June 1992. The first things he began with was the diameter of the Sun as it is measured at the "plateau de Calern", and my first work was to analyze this time series in order to separate the "noise" from the pertinent information (section 2). Wet then applied the technique developed for this purpose to the study of the 11-year solar cycle as seen in the Wolf sunspot number (section 3) and I continued in collaboration with Judit Pap with the study of total irradiance and some other solar indicators which could help to understand its variations (section 4).

2. Study of the diameter

2.1. WHY DO WE STUDY THE DIAMETER OF THE SUN ?

The diameter is a by-product of astrometric measurements: the best data are obtained from transit times through the meridian or through an elevation circle. The most homogeneous time series obtained from an astrolabe is due to F. Laclare from Observatoire de la Côte d'Azur, France (Laclare *et al.* 1996) but other measurements are available from astrolabes established in Belgrade (Ribes *et al.* 1988) or Brazil (Leister and Benevides-Soares 1990).

The diameter can also be a by-product of measurements of solar oblateness or more generally of the limb figure (obtained for example from solar images from space).

The ground-based measurements provide the necessary long-term time series whereas the satellite measurements may provide a way to separate

265

J. Provost and F.-X. Schmider (eds.), Sounding Solar and Stellar Interiors, 265-273.
© 1997 *IAU. Printed in the Netherlands.*

possible variations induced by terrestrial atmosphere from real solar variations.

In 1983, F. Laclare found that the diameter might vary in time (Laclare 1983). From this, Philippe Delache made many different studies in order to determine and understand the origin of the observed variations.
Measurements of the diameter and its possible variations are of prime importance since the present theory predicts no observable changes (Spruit 1994) whereas measurements do show variations (up to 0.1").
We can ask a few questions about the sources of such variations:

- what is the influence of the terrestrial atmosphere ?
- is there any variations in the limb profile ?
- is there any variation in the energy production ?
- what are the magnetic fields and/or convection effects ?

2.2. THE TIME SERIES OF THE CERGA ASTROLABE

Reliable and coherent results are obtained when observations are homogeneous (a single observer) and available over a long enough time span (at least a solar cycle): the CERGA astrolabe provides such time series from 1975 up to now.

The instrument is a Danjon-type astrolabe which has been adapted for solar observations. The single equilateral prism has been replaced by a set of 11 reflector prisms which allow observations at various zenith distances. Two images of the Sun are observed: a direct one and an other reflected off a level mercury surface. The procedure consists in timing the point at which the upper edge of the direct image is in contact with the inner edge of the reflected image. The second timing is proceeded when the reverse occurs. The radius is half the difference between the zenith distances computed at the times of the two crossings (Laclare *et al.* 1996) and is represented on figure 1a.

The differential refraction should not affect the measurements of the vertical diameter since the opposite sides of the Sun are measured at equal altitudes. However, the thickness of the atmosphere changes with the zenith distance. Laclare *et al.* (1996) have indeed underlined a relation between measurements of the diameter and altitude: the diameter decreases when the zenith distance increases whereas, as could be guessed, the standard deviation of the measurements increases with altitude. This had been also noticed by Philippe Delache who had used this relation to work out the error bars. Distinguishing East and West measurements, he has, for each year and for each zenith distance, calculated the deviation from the corresponding mean of each data set. Then, a mean deviation over all years is calculated for each zenith distance and for East and West observations. Then, he

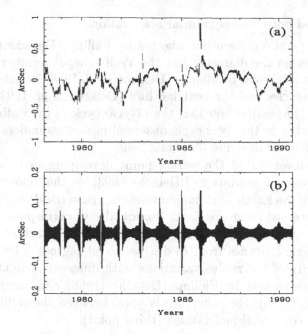

Figure 1. *(a):* Solar data observations: 1-σ error bars.
(b): Time series of errors bars centered around zero. Notice the large variety of σ_i's amplitudes and the signature of seasons, the large errors bars corresponding to winters, especially in early years.

could assess an error bar to each individual measurement, given its zenith distance and direction of observation. Figure 1a shows 3-month average of the radius time series (only one point over nine is plotted) together with the corresponding error bars which are calculated as the reverse of the square root of the sum of the individual error bars within 3 month. Those mean error bars are represented on figure 1b (being centering around 0): some error bars are larger, for example at the beginning of the time series where the number of individual measurements was not as high as it became later. They are also larger during winter, when observations are scarce, but also when the measurements are mostly obtained at larger zenith distance (low over the horizon) which increases the individual error bars of the data.

2.3. HOW DO WE ASSESS THE REALITY OF OBSERVED VARIATIONS ?

We can have at least three possibilities to find out whether the variations are solar in their origin or not:

- correlation between diameter and other solar time series
- sorting signal from noise, and assessing significance of model parameters such as periodicities or amplitudes

– compared polar and equatorial observations

The first point has been investigated by Philippe Delache through correlations between the diameter and the Wolf sunspot number (Delache *et
al.* 1985), the total solar irradiance (Delache *et al.* 1988, Delache, 1988), the
p-mode frequencies and the neutrino flux (Delache *et al.* 1993, Gavryusev
et al. 1994). The results show that the 11-year cycle of the radius variations
is anti-correlated to the solar cycle observed in various indices but may be
correlated with the neutrino flux variations.

We have investigated the second point developing methods using the
wavelet analysis (Vigouroux and Delache 1993, see the following section).

Because of the Earth's orbital parameters, the vertical (horizontal) solar
diameters observed from the Earth do not always correspond to the polar
(equatorial) solar diameter. With its 11 prisms, F. Laclare can observe
the heliographic latitudes from 15 degrees to 80 degrees. This has enabled
a careful study of the radius variations with different inclinations. It has
been first investigated by Philippe Delache (1988) and more recently by
Laclare *et al.* (1996): the coherence is good between the radius variations
at polar and equatorial inclinations (third point).

2.4. ANALYSIS OF THE CERGA DIAMETER TIME SERIES

Philippe Delache wanted to assess the reality of the observed variations of
the Sun's radius. We therefore had to sort out the signal from the noise.
The simplest analysis consists in making the Fourier transform of the data
and cutting-off some of the high frequencies, and reconstructing the cleaned
signal from the retained frequencies. This raises two problems:

– firstly: how can we determine in the Fourier plane the noise level ? In
  other words, are we sure that there is no signal in what we call noise ?
– secondly: when we retain frequencies considered significant, are we sure
  that the corresponding standing oscillations (or waves) are present all
  along the time series ? With the Fourier analysis, we cannot see if a
  frequency occurs here and there in the time series.

In order to answer the first question, we have studied in detail the nature (or origin) of uncertainties which constitute the so-called "error bars".
We may then simulate the noise issued by the error bars, so that we are
able to assess some significant cut-off level in the frequency domain. Simulating noisy data permits then to retain only the significant parts of actual
signals.

To answer the second question, Philippe Delache proposed to make wavelet
transform analysis. This can be viewed as a time and scale transform: it
allows to detect structures at different scales (or within several frequency
bands) and to locate them in time. Since the classical Fourier transform

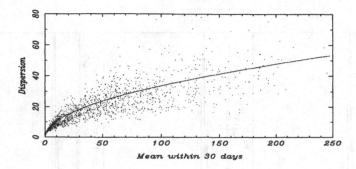

Figure 2.   Dispersion of the daily value calculated within 30 days as a function of the corresponding mean. The parabolic fit is representative of a Poisson distribution of the daily Wolf number within 30 days.

gives only informations on frequency, the wavelets seem to be more appropriate for removing noise in a time series. Indeed, the noise may vary in time (as showed on figure 1b: the error bars are larger in winter than summer) and the Fourier techniques does not take into account its variations contrary to the wavelets.

We performed Fourier and wavelet analysis of the radius time series, assuming that the noise distribution within an error bar is Gaussian. The results obtained through both techniques are in favor of the wavelet analysis. Firstly because the wavelet reconstruction leaves out large error bars. Secondly the number of independent parameters needed for a reconstruction of the same quality is smaller in the wavelet analysis case: this latter representation is then simpler.

3. Study of the Wolf number

We then proposed to apply similar techniques to historical solar activity as measured by sunspot numbers. We still wanted to assess the degree of significance of the elements of Fourier or wavelet transforms of the data, taking care of their statistical properties. Here again, inverting the transform yields to "models" for the variation of initial data which retain only these significant elements.

The Wolf sunspot number contains no uncertainties, contrary to the radius time series. However, some of the day-to-day variations may not be completely deterministic. We used the monthly sunspot number as our time series to analyze and we proposed to associate to every monthly data the dispersion of the daily sunspot numbers that it includes, considering the dispersion as the relevant "uncertainty". The latter is however more solar than instrumental in its origin. Moreover, it appeared that the dispersion strongly depends upon the mean activity (figure 2). Actually a parabolic

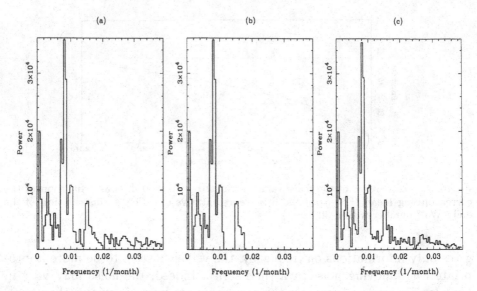

Figure 3. Fourier transform of the Wolf sunspot number *(a)*, of its Fourier *(b)* and wavelet *(c)* reconstruction.

fit adequately represents the cloud of points. This has confirmed the model of sunspot occurrence which has been proposed by Morfill *et al.* (1991). Indeed, on observational and physical grounds, those authors have suggested that a Poisson distribution can mimic the behavior of the solar activity over times of the order of the solar rotation, which corresponds reasonably well to the month over which are averaged the daily sunspot numbers. We were thus entitled to use this parabolic fit in order to assess a likely "uncertainty" to any actual value of the monthly sunspot number according to the fitting relation.

We then proceed to the sunspot number analysis in a similar way as our previous work on the solar radius. The results show that the number of parameters retained for the wavelet and Fourier reconstruction is closer, for a reconstruction of the same quality (Vigouroux and Delache 1994). However, Fourier analysis needs a more severe criterion to reach this required quality.
Examination of the spectra of the original and reconstructed time series (figure 3) shows that all spectral estimates corresponding to frequencies below $(0.018 \text{ month})^{-1}$ are preserved as significant. Both wavelet and Fourier analysis tell us that a large frequency band conveys significant informations. A purely modal description of the data as interferences between a few standing oscillations is therefore a dangerous biased selection of the actual information contained in the data.

4. Study of the total solar irradiance

In collaboration with Judit Pap (from JPL and UCLA) we have made a detailed study of the total solar irradiance data set. We have investigated the real origins of "error bars" and their relation to the solar cycle (Pap *et al.* 1996). Plotting the irradiance error bars as a function of their corresponding value provides an interesting tool to separate the random fluctuations related to instrumental effects from real solar variability. Putting all the daily variability within 30-day into error bars calculated as dispersion allows again to study the cycle dependence of those dispersions. We showed that the dispersion values plots as function of their 30-day corresponding mean values may help in determining the length and precise dates for the solar cycle minimum. This study applied to other solar indices related to total irradiance (full disk integrated magnetic flux, Mg core-to-wing ratio, Photometric Sunspot Index which is an indicator of the irradiance deficit due to sunspots) has underlined some differences related to the minimum length between those time series.

Using several techniques (wavelet analysis to retrieve the solar cycle and then cross-correlation between scales of the wavelet transform of the analysis residual time series), we then found that the magnetic field and Mg c/w ratio (or total irradiance corrected from the sunspots darkening) have the same behavior but for time scales shorter than 8 months. It should be pointed out that those 8 months are related to the complex of activity evolution across the solar disk and that all of those three indices are related to them. We then found a correlation for 10 months delay between the total irradiance corrected from the sunspots darkening and the Mg c/w ratio when the latter is leading the former (Vigouroux and Pap 1995, Vigouroux *et al.* 1997). The observed time delays between various data sets representing photospheric and chromospheric conditions indicate that the response of the chromospheric layers to the magnetic field variations is quite different from that of the photosphere. The observations of some instruments on the SOlar Helioseismology Observatory (SOHO) satellite will dramatically improve our knowledge and capability in interpreting those results. Furthermore, analysis and interpretation of the time-delays found between the magnetic field and solar radiation emitted from different layers of the solar atmosphere will lead to a better understanding firstly of the dynamics taking place below, in, and above the photosphere and secondly of the basic mechanism governing the solar variability.

5. Conclusion

I am indebted to Philippe Delache for his help and for all he taught me. Although he was the director of the Observatoire de la Côte d'Azur at the

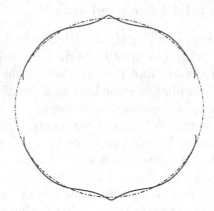

Figure 4.   Perfect spherical Sun (*dashed line*) compared to the shape of the Sun as observed by F. Laclare (*solid line*).

time I begun my thesis, he nevertheless saved time for discussions and advice. He also always encouraged me to meet other scientists, to participate to colloquium and summer school, to present the work I did under his direction, instead of being sticked to the computer looking for a better solution to some problem. He introduced me into the international community.

I cannot end without speaking again of the radius. In section 2, I mentioned that Francis Laclare spans a large range of heliographic latitudes thanks to his 11 prisms. It has appeared that the radius has not the same value whether it is measured near poles, near equator or near 45° of heliographic latitude. Figure 4 shows a perfect circle representative of a perfect Sun and some distorted circle which is the actual shape of the Sun as measured by F. Laclare (the deviations from the perfect circle have been multiplied by 500 in order to make them clear): I believe Philippe Delache would have loved to see the Sun as a lemon.

References

Delache, Ph., Laclare, F. and Sadsaoud, H. (1985) Long period oscillations in solar diameter measurements, *Nature*, **317**, 416

Delache, Ph. (1988) Variability of the solar diameter, *Adv. Space Res.*, **8**, n° **7**, 119

Delache, Ph., Laclare, F. and Sadsaoud, H. (1988) Long periods in diameter, irradiance and activity of the Sun, in *Advances in helio-asteroseismology*, eds. J. Christensen-Dalsgaard et S. Frandsen, pp. 223

Delache, Ph., Gavryusev, V., Gavryuseva, E., Laclare, F., Régulo, C. and Roca Cortés, T. (1993) Time correlation between solar structural parameters: *p*-mode frequencies, radius, and neutrino flux, *Astrophys. J*, **407**, 801

Gavryusev, V., Gavryuseva, E., Delache, Ph. and Laclare, F. (1994) Periodicities in solar radius measurements, *Astron. Astrophys.*, **286**, 305

Laclare, F. (1983) Mesures du diamètre solaire à l'astrolabe, *Astron. Astrophys.*, **125**, 200

Laclare, F., Delmas, C., Coin, J.P. and Irbah, A. (1996) Measurements and variations of the solar diameter, *Solar Physics*, **166**, 211

Leister, N.V. and Benevides-Soares, P. (1990) Variations du diamètre solaire, *C.R. Acad. Sci. Paris*, t. **311**, **Série II**, 399

Morfill, G. E., Scheingraber, H., Voges, W. and Sonett, C. P. (1991) Sunspot number variations: stochastic or chaotic, in *The Sun in time*, eds. C.P. Sonett, M.S. Giampapa et M.S. Matthews, University of Arizona Press, Tucson, pp. 30

Pap, J. M., Vigouroux, A. and Delache, Ph. (1996) Study of the distribution of daily fluctuations in observed solar irradiances and other full disk indices of solar activity, *Solar Physics*, **167**, 125

Ribes, E., Ribes, J. C., Vince, I. et Merlin, Ph. (1988) A survey of historical and recent solar diameter observations, *Adv. Space Res.*, **8**, 129

Spruit, H.C. (1994) Theories of radius and luminosity variations, in *The solar engine and its influence on terrestrial atmosphere and climate*, ed. E. Nesme-Ribes, NATO A.S.I. Series **25**, Springer-Verlag Berlin Heidelberg, pp. 107

Vigouroux, A. and Delache, Ph. (1993) Fourier versus wavelet analysis of solar diameter variability, *Astron. Astrophys.*, **278**, 607

Vigouroux, A. and Delache, Ph. (1994) Sunspot numbers uncertainties and parametric representations of solar activity variations, *Solar Physics*, **152**, 267

Vigouroux, A. and Pap, J. M. (1995) Studying solar irradiance variability with wavelet technique, in *Solar drivers of interplanetary and terrestrial disturbances*, A.S.P. Conference Series **95**, eds. K.S. Balasubramaniam, S.L. Keil et R.N. Smartt, pp. 586

Vigouroux, A., Pap, J. M. and Delache, Ph. (1997) Estimating long-term solar irradiance variability: a new approach, *Solar Physics*, in press

INPUTS FROM HELIOSEISMOLOGY
TO SOLAR PHYSICS

SOLAR ACTIVITY AND SOLAR OSCILLATIONS

Y. ELSWORTH

University of Birmingham
Edgbaston, Birmingham B15 2TT, U.K.

1. Introduction

Helioseismology provides us with the tools to probe solar activity. So that we can consider how the solar oscillations are influenced by that activity, we first consider the phenomena that we associate with the active Sun. The surface of the Sun is not quiet but shows evidence of convection on a wide range of scales from a few hundred kilometres through to several tens-of-thousands of kilometres. The surface temperature shows signs of the convection structures with the temperature in the bright granules being some 100 K to 200 K hotter than the surrounding dark lanes. Sunspots, which are regions of high magnetic field that suppress convective flows, are clearly visible to even quite crude observations. They are several tens-of-thousands of kilometres in diameter and about 2000 K cooler than their surroundings. Ultraviolet and X-ray pictures from satellites show that the higher layers of the solar atmosphere are very non-uniform with bright regions of high activity. Contemporaneous magnetograms show that these regions are associated with sunspots. Flares – regions of magnetic reconnections – are seen at all wavelengths from X-ray through the visible to radio. They are the non-thermal component of the radio emission of the Sun. There are many other indicators of activity on the Sun.

Precise correlations between individual regions of solar activity and the observations of solar oscillations are still scarce and in this paper we consider principally the correlations between activity and the average behaviour of the Sun. We are concerned with how solar oscillations probe solar activity and we seek correlations between these signatures of activity and some characteristics of the oscillations. The oscillations themselves can be described by the "degree" and frequency of the modes. In general, instruments which are good at making measurements on the very low-degree modes are not sensitive to the higher-degree modes, while those instruments

J. Provost and F.-X. Schmider (eds.), Sounding Solar and Stellar Interiors, 277-285.
© 1997 IAU. Printed in the Netherlands.

that have spatial resolution to measure the higher degrees do not produce very precise measurements of the low-degree modes. Modes of different degrees probe different volumes in the Sun. According to asymptotic theory, the inner turning point is determined by $\nu/[\ell(\ell+1)]$, where ν is the frequency of the mode and ℓ is its angular degree. The upper turning points are also a function of the mode under consideration, but in this case, up to moderate ℓ, the modes are essentially vertical at the surface and their behaviour is a function of frequency alone and not angular degree. The lower-frequency modes have upper turning points deeper in the Sun than do higher-frequency modes and so, for a phenomenon localised very near to the surface of the Sun, we can expect that its effect will depend principally on the frequency of the mode.

2. Solar Cycle

Activity levels on the Sun are well known to vary on many different timescales. In order to characterise the activity we use various proxies – the most famous and longest measured of which is the sunspot number. There are other several other measures that can be used: for example the 10.7-cm radio flux, which comes primarily from the higher levels in the solar atmosphere; the MPSI magnetic plage strength index from the Mount Wilson magnetograms; KPMI, the magnetic index from Kitt Peak; and UVI, a measurement of the solar UV variability as determined by the Mg II 280 nm core-to-wing ratio. The historical sunspot record shows that the activity levels on the Sun vary with a periodicity of about 11 years. The investigation of the magnetic field associated with the sunspots shows that the underlying magnetic cycle has a period of twice that indicated by the sunspot number. As a matter of interest, there was a long interval in the second half of the 17th century when no sunspots were recorded and which coincided with a period of extreme cold in Europe. We wish to probe this magnetic activity using observations of solar p-mode oscillations.

The frequency of a given p mode depends, among other parameters, on the cavity within which the mode is "constrained". It is plausible to suggest that the cavity will in some way be altered during the solar cycle and hence one might expect that the frequency of a mode would change with solar cycle. Moreover, as different modes see different cavities, the size and even the sign of the change will vary from mode to mode. Note that the magnetic effects are limited to a band of active latitudes and are not spread over all the Sun. According to Spörer's law, the latitude at which sunspots first appear is itself a function of the phase within the solar cycle. At the beginning of a new solar cycle, sunspots appear at latitudes of about $\pm 40°$, and at lower latitudes of $\sim \pm 5°$ at the minimum. Thus we might also expect some latitude dependence in oscillations variation. The earliest observations of the p-modes were of modes of medium degree using instruments which resolved the Sun. However, these were short data sets not well suited to the study of the variability of the modes.

Observations of the low-degree p modes started in 1975 with the Birmingham measurements made from Tenerife. Those measurements have been continued and augmented until up to present day. There are also measurements from the late 1970's by Fossat (Gelly *et al.* 1988). It is not only the oscillation observed in velocity which we have to consider. The total solar irradiance in the ecliptic plane varies by about 0.1 per cent during the solar cycle with the luminosity being highest at solar maximum. (Wilson & Hudson 1988; Foukal & Jean 1990). Solar Irradiance oscillations were seen by ACRIM on SMM from mid February 1980 until the loss of the spacecraft fine pointing in December 1980 (Woodard & Noyes 1985). Medium and high-ℓ observations have been made by several teams in the years since 1979 (e.g., Duvall *et al.* 1988; Rhodes, Cacciani & Korzennik 1991; Libbrecht & Woodard 1991, Bachmann & Brown 1993). Details of the observations – including details of the precise epoch for which data are available from each instrument – are given in the review by Rhodes *et al.*.

There were trends visible in the observations of the frequencies which indicated that the frequencies were decreasing with time, but there was some confusion in the reports. However, in 1989 at the Hakone meeting Libbrecht (see Libbrecht & Woodard 1990) presented strong evidence for a solar cycle variation in frequencies based on BBSO observations in 1986 and 1988 of moderate degree data ($5 \leq \ell \leq 60$). The scale of the variation had, in fact, been predicted by Kuhn (1989), using limb-photometry data and an analysis of temporal changes in the even-Legendre frequency splittings. Closely following the moderate degree analysis came the publication of the Birmingham data (Palle *et al.* 1989, Elsworth *et al.* 1990) from the BiSON network in a data set which extended back to before the solar maximum in 1981. The observed shift was in these low degree modes was of the order of 3 nHz per Rz (i.e., per sunspot number). Taking into account the rather weak ℓ-dependence seen, this value was not incompatible with the higher-ℓ shifts.

The question now is how do we interpret the observations in terms of an understanding of how the Sun is changing during the solar cycle? The main features of the BBSO results are as follows. The year 1986 – which is near the low point in solar activity – is taken as the reference year with which all other years are compared. The frequencies increase progressively as the the solar activity increases: the shift between 1986 and 1989 is greater than that between 1986 and 1988. With high-quality data, it is possible to determine that the observed shift is a function of frequency as well as solar activity, with the shift effectively disappearing below about 1.5 mHz. The shifts themselves are small: the observed shift at about 4 mHz in 1989 is 800 nHz, i.e., an effect at the level of 2 parts in 10,000.

An important step in the understanding of what it is in the Sun that is changing comes from the fact that it can be shown that the observed frequency shifts up to a certain frequency appear to be inversely proportional to the mode inertia. The mode inertia itself is strongly dependent on the conditions which hold in the outer layers where the mode is evanescent, and not on the structure of the deep interior. This is because the normalisation of the eigenfunction is carried out at the surface of the star where the mode is observed and not in the interior where most of the energy is. Very near to the surface of the Sun, all p modes of low and moderate degree have essentially the same spatial structure and so those modes sample the near-surface layers of the Sun in the same way. So the mode-mass dependence suggests that the structural changes are located in the outermost layers of the Sun. Theoretical analyses by various authors showed and that both magnetic and thermal influences are required to explain the observations. The authors have considered either the influence of the chromospheric magnetic cavity (Campbell & Roberts 1989) or sub-photospheric flux tubes (Goldreich et al. 1991) or changes in the mixing length (Gough & Thompson 1988). For an up-to-date discussion of the theory see also the review by Roberts (1996).

Above about 4 mHz, where the shifts get smaller, leakage into a chromospheric cavity is invoked. Interestingly, this places the seat of the variation away from the region where one might expect the solar dynamo to be sited. An evolving magnetic field at the base of the convection zone could, in principle, influence the frequencies – especially those of the low-order modes. However, unless the fields are of the order of a mega-Gauss, the implied frequency change is small. At the base of the convection zone, a field of 10^6 G would contribute a magnetic pressure of only 7×10^{-4} of the gas pressure. The balance is different high in the atmosphere where the gas pressure is much lower.

Activity on the Sun does not change in a completely smooth way – there are fluctuations of the timescale of days and months as well as years. As an indication that there are short-term fluctuations, consider the scatter in the estimate of a frequency with the formal errors derived from the fit to the data. Formal errors are underestimate the observed scatter on the frequency measurements (Chaplin et al. 1996). This is, in part, due to the random nature of the excitation and the consequent variation in the line shape. Many people have considered direct correlations between shift and activity indices. Woodard et al. (1991) show very clear correlations between magnetic activity as measured by the mean magnetic field, smoothed to remove effects of the solar rotation. Palle (1996) has shown that the shifts observed in the rising and falling phases of a given cycle are not always the same, even for the same level of magnetic activity. However the magnetic activity in the two phases are also different – strictly speaking, it is therefore not enough just to correlate with an activity index.

The frequencies of the oscillations are a measure of the conditions inside the Sun. Inversion techniques are used to infer the interior structure. To get the best quality data over a wide range of frequencies and degrees, it is often convenient to combine data sets. However, given that the frequencies are actually changing with activity levels on the Sun and hence with time, it is very important to use contemporaneous data sets if the inferred structure of the interior is not to be corrupted by activity effects (Basu *et al.* 1996). The problem does not stop there. Because the activity is concentrated at certain latitudes, one must also take account of the degree dependence of the frequency shift (Dziembowski & Goode 1996).

There are many more observations to be made. The existing groups can build on their historical records and there are the observations to be made from the SOHO satellite. As many of the posters at this meeting, and in particular those in the session, testify we can expect to improve our understanding in the near future. Perhaps we will even see a signature of the 22-year magnetic cycle of the Sun.

3. Sunspots

Leighton (1962), in the paper which announced the observation of local solar oscillations, commented that the oscillation amplitude appeared to be suppressed in active regions. Later Braun (1987) showed strong evidence for the absorption in sunspots for modes of degree 200 and higher. It seemed likely that the absorption was a property of strong magnetic fields. The f mode was most strongly absorbed. Much more recent work by Braun suggests that the 5-minute power is from forced oscillations of the surrounding photosphere, whereas the 3-minute power is due to actual resonance in sunspots. These effects are strongest in chromospheric lines (Thomas *et al.* 1984). It is hoped that the work will lead to an observation of the magnetic field strength as a function of depth in thin flux tubes and sunspots. However, there needs to be a proper understanding of the mechanism as right now one cannot disentangle temperature and magnetic field. There is clearly a strong interaction between sunspot magnetic fields and p-modes. (See Spruit 1996; Braun 1987; 1988; 1994).

In 1909, at the Kodaikanal Observatory in India, Evershed studied the horizontal outflows in sunspots. These flows are observed in the penumbral photosphere but stop abruptly at the boundary to the undisturbed photosphere. There are flows in the dark lanes which constitute the background for the bright penumbral grains. The flow variable, tends to increase outwards, and may reach velocities of several kilometres per second. The observations have been interpreted as a siphon flow along magnetic flux tubes. Thus there is an imbalance in the system. The Evershed flows take material away from the centre of the spot and magnetic pressure also pushes outwards, therefore there must be something to stabilise the system. This could be provided by a downward flow and studies of sunspot have searched for such a flow.

Time travel maps of sound waves propagating in local regions of the convection zone provide for such an observation. Measurements are made by estimating the cross correlation functions of fluctuations of brightness on the solar surface caused by solar oscillations. Because of the high noise level, the data must be averaged over large surface areas. Typically mean travel times are obtained between a central point and surrounding annuli. A sort of tomography is then used to interpret the data. In the data observed by Duvall and co-workers (Duvall 1995,1996), the four annuli chosen trace rays which propagate through the upper 30 per cent of the convection zone. The data can be interpreted to show strong downflows of order of 1km/s around the sunspot studied. This is a very important piece of evidence in our understanding of why sunspots can exist.

4. Solar noise

The limit to the solar oscillation modes which can be detected is set by the background solar noise in velocity or intensity. The Harvey model attempts to quantify the components of the noise which are important to helioseismology. The velocity field is considered to be determined by granulation, mesogranulation, supergranulation and active regions. Convection cells have magnetic fields on the boundaries and the flow is restricted to the edges. Auto-correlation analyses suggest that it is appropriate to represent each of the these phenomena by a characteristic decay time and a strength. It was proposed that each component of the solar noise be represented by a function of the form

$$P(\nu) = \frac{4\sigma^2}{1 + (2\pi\nu\tau)^2}.$$

The combined effect of all these using the original estimates for the parameters is shown as the *Harvey model* in Fig. 1.

Observations from the BiSON instrument in the Canary Islands and analysed by the Tenerife group confirmed the general shape of the prediction (Regulo *et al.* 1994). There is the requirement that, for a mode to be observable, it must be detectable, against the noise background, within one

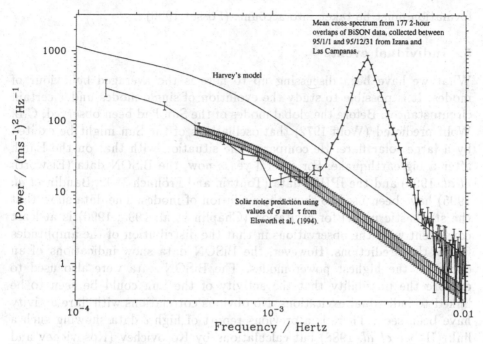

Figure 1. Solar noise power as measured by BiSON and compared with the original Harvey prediction.

lifetime of the mode. Longer observations improve on the signal to noise. If this condition is not met, then the signal to noise increases only as the fourth power of the observing time. The amplitude of the *p* modes is such that they are not difficult to observe. For *g* modes, the strength of the signal is weaker and the solar noise level is higher than in the *p*-mode region. The *g* modes are expected to be very coherent and so long observing times can be used. BiSON observations (Elsworth *et al.* 1993) have already shown that the solar noise levels are below those estimated by Harvey (see Fig 1). At 1 mHz, the value observed by cross-correlation of data from well-separated stations is about a factor of 5 below the original estimate. Recent BiSON data taken near solar minimum show that the observations have changed very little with solar cycle (Chaplin *et al.* 1996)

The GOLF and Virgo instruments on SOHO will shed further light on the signal levels in velocity and intensity. It is important to remember that the observed strength of the oscillation signal is a function both of the height in the solar atmosphere at which the oscillations are formed, and the frequency of the oscillation. To make comparisons for example between the BiSON value which is a measurement in the potassium line and the GOLF value which is a measurement in the sodium line, the different heights of formation must be taken into account. (Fossat 1996)

5. Individual events

What we have been discussing up to here is the averaged behaviour of modes. It is possible to study the evolution of single modes under certain circumstances. Before the global modes of the Sun had been observed, C.L. Wolff predicted (Wolff 1972) that oscillations of the Sun might be excited by a large solar flare. He compared the situation with that on the Earth after a big earthquake. For several years now, the BiSON data (Elsworth *et al.* 1995) and the IPHIR data (Toutain and Fröhlich 1992, Baudin *et al.* 1995) have been used to show the evolution of modes. The data show that the stochastic model for excitation (Chaplin et al. 1995; 1996) is at least consistent with the observations in that the distribution of the amplitudes follow the predictions. However, the BiSON data show indications of an excess of the highest power modes. The BiSON data were also used to explore the possibility that the activity of the Sun could be seen to be linked to individual excitations. No obvious correlations with flare activity have been seen. There is a previous report of high-ℓ data showing such a link (Haber *et al.* 1988) but calculations by Kosovichev (Kosovichev and Zharkova 1995) suggest that there is not enough energy in flares for them to significantly influence the excitation process. However, more energetic events such as coronal mass ejections or the demise of a sun-grazing comet (Isaak 1978) might provide enough energy.

6. Summary and Conclusions

Activity is clearly varying periodically. The future holds great potential for studying that variation However, to date, only the ground-based networks have long databases which go back beyond the beginning of the previous solar cycle in the case of the BiSON data. We have seen that the frequency of the solar oscillations varies with that activity. The amplitude of the modes also varies in a systematic way (Anguera Gubau *et al.* 1992, Elsworth *et al.* 1993). The modes are strongest when the activity is at its minimum For the behaviour of other aspects of the oscillations – there is little clear evidence for systematic trends in the observed values. For example, according to that data there is currently no evidence for changes in the widths (Chaplin *et al.* 1996) of the low-degree modes. Equally the solar noise background and the overall solar rotation rate do not appear to vary with the solar cycle (Fig. 1).

References

Anguera Gubau, M., Pallé, P. L., Perez Hernandez, F., Regulo, C. & Roca Cortes T.

(1992) A&A, **255**, 363

Bachmann K. T. & Brown T. M. (1993) ApJ, **411**, L45

Basu S. et al. (1997) these proceedings

Baudin F., Gabriel A., Gibert D., Palle P. & Regulo C. (1995) in Fourth SOHO Workshop: Helioseismology, eds. Hoeksema J. T., Domingo V., Fleck B. & Battrick B., ESA SP-376, vol 2, p 323

Braun D. C. (1987) ApJ **319**, L27

Braun D. C. (1988) ApJ **335**, 1015

Campbell, W. R., and Roberts, B. (1989) ApJ **338**, 538

Chaplin W. J., Elsworth Y., Howe R., Isaak G. R., McLeod C. P., Miller B. A. & New R. (1995) in Fourth SOHO Workshop: Helioseismology, eds. Hoeksema J. T., Domingo V., Fleck B. & Batrick B., ESA SP-376 vol 2, p 335

Chaplin W. J., Elsworth Y., Howe R., Isaak G. R., McLeod C. P., Miller B. A. & New R. (1996) MNRAS, submitted

Chaplin W. J., Elsworth Y., Isaak G. R., McLeod C. P., Miller . B. A, New R. & Underhill. C. J (1997) these proceedings

Chaplin et al. (1996), in preparation

Duvall T. et al. (1988) ApJ **324**, 1158

Duvall T. L. Jr., Jefferies S. M. & Harvey J.W. (1995) Bull. Am. Ast. Soc. **25**, 950

Duvall T. L. Jr., D'Silva S., Jefferies S. M., Harvey J.W. & Schou J. (1996) Nature **379**, 235

Dziembowski W. A. & Goode P. R. (1996) A&A **317**, 919

Elsworth Y., Howe R., Isaak G. R., McLeod C. P. & New R. (1990) Nature **345**, 322

Elsworth Y., Howe R., Isaak G. R., McLeod C. P., Miller B. A., New R., Wheeler S. J. (1993) MNRAS **265**, 888

Elsworth Y., Howe R., Isaak G. R., McLeod C. P., Miller B. A., Wheeler S. J., New R. (1995) in GONG '94 ,Helio- and Astero-Seismology from Space, eds. Ulrich, R., Rhodes, E., & Däppen, W., p. 318

Fossat E. (1996), private communication

Foukal P. & Lean J. (1990) Science **247**, 556

Gelly, B.,Fossat., E., & Grec, G. (1988) in Seismology of the Sun and sun-like stars, eds. ESA SP-286, p. 275.

Goldreich P., et al. (1991) ApJ **370**, 752

Gough D. O. & Thompson M. J. (1988) in Advances in Helio- & Asteroseismology, Symp. IAU 123, eds. Christensen-Dalsgaard J. & Frandsen S., 175

Haber D. A., Toomre J., Hill F., &Gough D. O. (1988) in Proc. Symp. Seismology of the Sun and sun-like stars, Ed. Rolfe E. J., ESA SP-286, p. 301

Isaak G. R. (1978) Physics Bulletin **27**, 127

Kosovichev A. G. & Zharkova V. V. (1995) Fourth SOHO Workshop: Helioseismology, eds. Hoeksema J. T., Domingo V., Fleck B. & Battrick B., ESA SP-376, p. 341

Kuhn J. R. (1989) ApJ **339**, L45

Leighton, R. B., Noyes, R. W., & Simon, G., W. (1962) ApJ **135**, 474

Libbrecht K. G. & Woodard M. F. (1990) Nature **345**, 779

Palle P. L., Regulo C. & Roca Cortes (1989) A&A bf 169, 313

Palle P. L. et al. (1996) ApJ, submitted

Rhodes E. J., Cacciani A. & Korzennik S. C. (1991) Adv. Space. Res. **11** (4), 17

Roberts B (1996) Bull. Astr. Soc. India **24**, 199

Spruit, H. C. (1996) Bull. Astr. Soc. India **24**, 211

Toutain, Th., & Fröhlich, C. (1992) A&A **257**, 287

Wilson R. C. & Hudson H. S. (1988) Nature **319**, 654

Wolff C. L. (1972) ApJ **177**, L87

Woodard M. F., & Noyes R. W. (1985) Nature **318**, 449

Woodard, M. F., Libbrecht, K. G., Kuhn, J. R., & Murray, N. (1991) ApJ, **373**, L81

EXCITATION OF SOLAR ACOUSTIC OSCILLATIONS

PAWAN KUMAR[1]
Institute for Advanced Study, Princeton, NJ 08540

Abstract. The stochastic excitation of solar oscillations due to turbulent convection is reviewed. A number of different observational results that provide test for solar p-mode excitation theories are described. I discuss how well the stochastic excitation theory does in explaining these observations. The location and properties of sources that excite solar p-modes are also described. Finally, I discuss why solar g-modes should be linearly stable, and estimate the surface velocity amplitudes of low degree g-modes assuming that they are stochastically excited by the turbulent convection in the sun.

1. Introduction

It was realized about 25 years ago that the Sun, our nearest star, is a variable star. Millions of acoustic normal modes (p-modes) of the sun are seen to be excited with a typical surface velocity amplitude of only a few cm s^{-1}, whereas other pulsating stars have a few modes excited to large amplitudes. Considering this dramatic difference between the pulsation property of the sun and other variable stars, it should not be surprising that the solar oscillations are excited by a mechanism that is different from the overstability mechanism believed to be responsible for the pulsation of other stars (overstability can arise for instance when the radiative flux is converted to mechanical energy of pulsation due to an increase of opacity with temperature). A number of early papers in the field proposed that the solar p-modes are excited by some overstability mechanism (Ulrich 1970, Leibacher and Stein 1971, Wolf 1972, Ando and Osaki 1975). However, the margin of instability for solar p-modes is found to be small and different ways of handling radiative transfer and/or the interaction of convection with oscillation

---

[1] Alfred P. Sloan Fellow & NSF Young Investigator

J. Provost and F.-X. Schmider (eds.), Sounding Solar and Stellar Interiors, 287-305.
© 1997 IAU. Printed in the Netherlands.

seems to change the sign of stability e.g. Goldreich and Keeley 1977a, Antia et al. 1982 and 1988, Christensen-Dalsgaard and Frandsen 1983, Balmforth and Gough 1990, Balmforth 1992 (the last two papers used a sophisticated version of the mixing length theory of Gough, 1977); see Cox et al. 1991 for a more complete list of references. If we assume that the solar p-modes are overstable then their amplitudes grow exponentially until some nonlinear mechanism limits their growth. By considering all possible 3-mode nonlinear couplings amongst overstable and stable p-modes in the sun, which is the most efficient process for saturating the amplitudes of overstable modes, Kumar and Goldreich (1989) and Kumar, Goldreich and Kerswell (1991) showed that the amplitudes of overstable modes saturate at a value that is several orders of magnitude larger than the observed value. This suggests that solar p-modes are linearly stable.

In this article we will assume that solar p-modes are stable, and describe how they can be excited by acoustic waves generated by turbulent convection. The basic idea is that the broad band acoustic noise generated by the turbulent flow in the convection zone is selectively amplified at frequencies corresponding to the normal mode frequencies of the sun. The process of wave generation by homogeneous turbulence was first studied systematically and in some detail by Lighthill (1952). Stein (1967) and Kulsrud (1955) applied it to the heating of the solar chromosphere/corona by acoustic and MHD waves respectively. Goldreich and Keeley (1977b) carried out a careful calculation of the stochastic excitation of solar normal modes by turbulent convection (for an excellent general review of wave generation due to turbulent fluid please see Crighton 1975). We describe the stochastic excitation process for the simple case of a homogeneous sphere below and later discuss its generalization to the Sun (§2). In §3 we describe various observations that any theory for excitation of solar p-modes must be able to explain and discuss how well the stochastic excitation theory performs when confronted with these observations. The estimate for the surface velocity amplitude of low degree g-modes, assuming that they are stochastically excited, is given in §4.

2. Stochastic Excitation

Let us consider a homogeneous gas sphere with a surface that reflects acoustic waves. Some fraction of the fluid inside this sphere is assumed to be in the state of turbulence which acts as a source of sound waves. Following Lighthill (1952) we write the perturbed mass and momentum equations as

$$\rho_1 + \nabla \cdot (\rho \boldsymbol{\xi}) = 0, \tag{1}$$

and

$$\frac{\partial^2 \rho \xi_i}{\partial t^2} + c^2 \frac{\partial \rho_1}{\partial x_i} = -\frac{\partial T_{ij}}{\partial x_j},$$ (2)

where c and ρ are unperturbed mean sound speed and density of the medium, ξ is fluid displacement, and

$$T_{ij} \equiv \rho v_i v_j + p\delta_{ij} - \rho c^2 \delta_{ij}.$$ (3)

These equations can be combined to yield the following inhomogeneous wave equation

$$\frac{\partial^2 \rho \xi_i}{\partial t^2} - c^2 \nabla^2 (\rho \xi_i) = -\frac{\partial T_{ij}}{\partial x_j},$$ (4)

Expanding ξ in the terms of normal modes of the system

$$\xi = \frac{1}{\sqrt{2}} \sum_q A_q \xi_q \exp(-i\omega t) + c.c.,$$ (5)

where ξ_q is displacement eigenfunction of mode q which is normalized to unit energy i.e.

$$\omega_q^2 \int d^3 x \, \rho \, \xi_q \cdot \xi_q^* = 1,$$ (6)

and substituting this expansion into equation (4) we find the following equation for the mode amplitude A_q

$$\frac{dA_q}{dt} \approx -\frac{i\omega_q}{\sqrt{2}} \exp(i\omega_q t) \int d^3 x \, \xi_{qi} \frac{\partial T_{ij}}{\partial x_j} = \frac{i\omega_q}{\sqrt{2}} \exp(i\omega_q t) \int d^3 x \, \frac{\partial \xi_{qi}}{\partial x_j} T_{ij}.$$ (7)

Turbulent flow is crudely described as consisting of critically damped eddies. The velocity v_h of an eddy of size h is related to the largest or the energy bearing eddy (size H and velocity v_H) by the Kolmogorov scaling i.e.

$$v_h = v_H \left(\frac{h}{H}\right)^{1/3}.$$ (8)

Moreover, following Lighthill (1952), we take $T_{ij} \approx \rho v^2 \delta_{ij}$. Since the displacement eigenfunction, for low ℓ modes, near the surface of the sphere is in the radial direction, therefore equation (7) reduces to

$$\frac{dA_q}{dt} \approx \frac{i\omega_q}{\sqrt{2}} \exp(i\omega_q t) \int d^3 x \, \rho v^2 \frac{\partial \xi_{qr}}{\partial r}.$$ (9)

where ξ_{qr} is the radial displacement eigenfunction of mode q. The mean energy input rate in mode q can be obtained from the above equation and is given by

$$\frac{dE_q}{dt} \equiv \frac{d\langle|A_q|^2\rangle}{dt} \approx 2\pi\omega_q^2 \int dr\ r^2\ \rho^2 v_\omega^3 h_\omega^4 \left[\frac{\partial \xi_{qr}}{\partial r}\right]^2, \tag{10}$$

where v_ω and h_ω are the velocity and size of the eddies which have characteristic time, $\tau_h \equiv h_\omega/v_\omega$, approximately equal to the mode period. This equation is valid not only for the homogeneous gas sphere considered here but also for more general systems including the excitation of solar p-modes by the Reynolds stress as discussed below. Of course, we must use the eigenfunction ξ_q and turbulent velocity appropriate for the system being considered.

It can be easily shown that the solution of the homogeneous wave equation (eq. [4] with right side set equal to zero), in the limit of large n (mode order) is

$$\xi_{qr} \approx B\ j_\ell\left(\omega_q r/c\right) \approx B\frac{\sin\left(r\omega_q/c - \pi\ell/2\right)}{r\omega_q}, \tag{11}$$

where j_ℓ is spherical Bessel function, and B is a constant factor independent of mode frequency for properly normalized mode eigenfunction (condition expressed by eq. [6]). Substituting this into equation (10) we find

$$\dot{E}_q \equiv \frac{dE_q}{dt} \propto \omega_q^2 h_\omega^4 v_\omega^4. \tag{12}$$

Let us assume that the turbulent velocity field in the sphere is concentrated in a thin layer of thickness H located near the surface of the sphere. We shall take the size of the largest eddies to be H and their rms speed to be v_H. The p-modes of period greater than $\tau_H = H/v_H$ are predominantly excited by the largest size eddies and the resultant energy input rate in these modes is proportional to ω_q^2, as can be seen immediately from the above equation. Modes of higher frequency ($\omega_q \gtrsim \tau_H^{-1}$) couple best to inertial range eddies which have characteristic time of order the wave period. Making use of the Kolmogorov scaling (eq. [8]) and equation (12) we see that $\dot{E}_q$ scales as $\omega_q^{-5.5}$. Thus the energy input rate into p-modes of this homogeneous system shows a break at frequency $1/\tau_H$ where the power law index changes by 7.5.

The generalization of above equations to describe the excitation of solar modes is not difficult. Equation (2) is replaced by the linearized momentum equation valid for a stratified medium i.e.

$$\rho\frac{\partial^2 \xi}{\partial t^2} + \nabla p_1 - \rho_1 g = -\nabla\cdot(\rho v v) \equiv F, \tag{13}$$

and the linearized equation of state is

$$p_1 = \frac{\partial p}{\partial \rho}\rho_1 + \frac{\partial p}{\partial s}s_1, \tag{14}$$

where

$$s_1 = \tilde{s} - (\boldsymbol{\xi}\cdot\boldsymbol{\nabla})s. \tag{15}$$

Here $\boldsymbol{\nabla}s$ denotes the background entropy gradient, and $\tilde{s}$ is the entropy fluctuation associated with turbulent convection. Equation (15) is the Eulerian version of the statement that the Lagrangian entropy perturbation is due entirely to turbulent convection. In other words, we approximate the waves as adiabatic (these equations are adopted directly from Goldreich et al. 1994). Combining equations (1) and (13)-(15) we obtain the following inhomogeneous wave equation, which is the generalization of equation (4) and describes the stochastic excitation of solar oscillations:

$$\rho\frac{\partial^2 \boldsymbol{\xi}}{\partial t^2} - \boldsymbol{\nabla}\left[c^2\boldsymbol{\nabla}\cdot(\rho\boldsymbol{\xi}) + \rho\boldsymbol{\xi}\cdot g - c^2\rho\boldsymbol{\xi}\cdot\boldsymbol{\nabla}\ln\rho\right] + g\boldsymbol{\nabla}\cdot(\rho\boldsymbol{\xi})$$

$$= -\boldsymbol{\nabla}\left(\frac{\partial p}{\partial s}\tilde{s}\right) - \boldsymbol{\nabla}\cdot(\rho vv). \tag{16}$$

This equation describes wave generation due to Reynolds stress as well as entropy fluctuation. As the entropy of a fluid element fluctuates, so does its volume. The fluctuating volume is a monopole source for acoustic waves. In a stratified medium the fluctuating buoyancy force adds a dipole source. By transferring momentum among neighboring fluid elements, the Reynolds stress acts as a quadrupole source.[2] The anisotropy of a stratified medium blurs the distinction between monopole, dipole, and quadrupole sources. It allows for destructive interference between the monopole and dipole amplitudes. Although the monopole and dipole amplitudes are individually larger than the quadrupole amplitude, their sum is of comparable size to that of the quadrupole. That this applies to energy bearing eddies follows directly from equations (16) and the relation between entropy and velocity fluctuation for convective eddies. The justification for inertial range eddies requires a subtle argument (cf. Goldreich and Kumar, 1990).

The new energy equation (which replaces eq. [10]) is given below

$$\dot{E}_\alpha \sim 2\pi\omega_\alpha^2 \int dr\, r^2 \rho^2 \left|\frac{\partial \xi_\alpha^r}{\partial r}\right|^2 v_\omega^3 h_\omega^4 \left(\mathcal{C}_\alpha^2\mathcal{R}^2 + 1\right)\mathcal{S}^2, \tag{17}$$

---

[2] We classify acoustic sources as monopole, dipole, or quadrupole according to whether they produce a change in volume, add net momentum, or merely redistribute momentum.

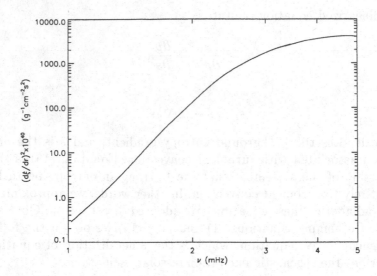

Figure 1. The plot of $(d\xi_r/dr)^2$, at the top of the solar convection zone, as a function of p-mode frequency (the mode degree is zero). The solar model used here is due to J. Christensen-Dalsgaard.

where $\mathcal{C}_\alpha$ is wave compressibility defined by

$$\nabla\cdot\xi_\alpha = \mathcal{C}_\alpha \frac{\partial \xi_\alpha^r}{\partial r}, \tag{18}$$

the shape parameter $\mathcal{S}$ describes the ratio of the horizontal to vertical correlation lengths of turbulent eddies, and $\mathcal{R}$ is given by

$$\mathcal{R} \equiv \frac{4H}{\Lambda} \left(\frac{\partial \ln p}{\partial \ln \rho} \right)_s, \tag{19}$$

with Λ the mixing length and H the pressure scale height. The factor $\mathcal{C}_\alpha^2 \mathcal{R}^2$ measures the ratio of the excitation by entropy fluctuations to that due to fluctuating Reynolds stress. Note that the frequency spectra of waves excited due to entropy fluctuation and that due to Reynolds stress are identical except of course for an over all normalization factor. The compressibility, $\mathcal{C}_\alpha$, for p-modes near the top of the convection zone, where the excitation takes place, is close to 1, and the value of $\mathcal{R}^2$ in this region is about 10 (see Goldreich et al. 1994). Therefore, the excitation of p-modes is dominated by entropy fluctuations (Stein and Nordlund 1991, arrived at the same conclusion using their numerical simulations of solar convection). On the other hand f-modes are nearly incompressible ($\mathcal{C}_\alpha \approx 0$) and so they are not excited by entropy fluctuation. This is perhaps why the power in f-modes is observed to be smaller than a p-mode of similar frequency.

The observed rate of energy input into solar p-modes can now be readily understood. One of the differences between the homogeneous gas sphere system and the sun is in the shape of the eigenfunction especially near the surface where the wave excitation takes place. It can be shown that the radial derivative of the normalized radial displacement eigenfunction for p-modes just below the photosphere scales as $\nu_q^{3.8}$ for $\nu_q \lesssim 3.0$mHz and as $\nu_q^{1.1}$ for $\nu_q \gtrsim 3.5$mHz (see Figure 1). Substituting this scaling in equation (17), or equation (10), we find that the energy input rate in the p-modes, at a fixed degree, scales as ν_q^7 for $\nu_q \lesssim 3.0$mHz and as $\nu_q^{-4.4}$ for $\nu_q \gtrsim 3.5$mHz which is in good agreement with the observations (Libbrecht and Woodard 1991); please see Goldreich et al. (1994) for a more detailed analysis and comparison with the observed energy input rate.

3. Observational constraints for the excitation theory of solar p-modes

A valid theory for the excitation of solar p-modes should be able to explain the observed rate of energy input in different modes. In addition, there are four other observational results that the theory must be able to explain and provide a fit to the data. These observations are: mode linewidth, the deviation of p-mode line profiles from symmetric Lorentzian shape, the statistics for the fluctuation of mode energy, and the presence of peaks in the power spectrum above the acoustic cutoff frequency ($\nu \gtrsim 5.3$mHz).

The agreement between the observed and the theoretically calculated energy input rate in solar p-modes due to stochastic excitation was described in the last section. We describe below the other four observations and compare them with the results of the stochastic excitation theory.

3.1. MODE LINEWIDTH

A number of groups have measured p-mode linewidth as a function of mode frequency (cf. Duvall et al. 1988, Libbrecht 1988, Elsworth et al. 1990). It is found that the mode linewidth at a fixed degree increases monotonically with mode frequency. At 2 mHz the linewidth is about 0.5 μHz, or mode lifetime is 20 days ($Q \approx 4 \times 10^3$) and at 4 mHz the linewidth is 10 μHz. The observed linewidth for $2\text{mHz} \lesssim \nu \lesssim 3\text{mHz}$ increases as $\nu^{4.2}$ whereas numerical calculations show that the mode linewidth due to radiative and turbulent damping increases as ν^8 for frequencies below $\sim$4mHz (cf. Christensen-Dalsgaard and Frandsen 1983, Balmforth 1992; Goldreich and Kumar 1991, give a simple analytical derivation of these results). This suggests that the mode damping at frequencies greater than about 2 mHz is due to some process other than the radiative and turbulent viscosity. A number of alternate mechanisms have been suggested to account for

the observed mode linewidth. These include modulation of convective flux
(Christensen-Dalsgaard et al. 1989), scattering of low degree p-modes by
turbulent convection to high degree modes (Goldreich and Murray 1994),
scattering of p-modes by magnetic fields (Bogdan et al. 1996). Goldreich
and Murray (1994) have carried out a detailed calculation of the scattering
process and find that an almost elastic scattering of p-modes by convective eddies is an important contributor to the mode linewidth at frequencies
$\nu \gtrsim 2\text{mHz}$, and the computed linewidth has the same frequency dependence
as the observed width (see also Murray, 1993).

Recent observational results indicate that the linewidth scales as ν^7 for
$\nu \lesssim 2$ mHz (Chaplin et al. 1996, and Tomczyk 1996). Perhaps below 2
mHz there are few modes available for p-modes to scatter into, and thus
the linewidth falls off more rapidly with decreasing frequency. According to
Jefferies (personal communication) the observed mode linewidth peaks at
a frequency of about 5mHz, followed by a slight decline, and then remains
constant at higher frequencies. This is a puzzling result for which as far as
I know no explanation has been offered.

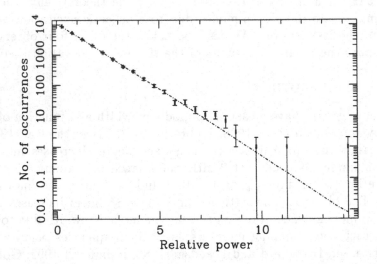

Figure 2.   The statistics of power fluctuation in low degree p-modes. The straight line
is the exponential distribution, which is the theoretically expected distribution if modes
are stochastically excited due to turbulent convection. The data is kindly provided by
BiSON (please see Chaplin et al. 1995 for details).

3.2. ENERGY STATISTICS

Modes excited by their interaction with a Gaussian random field (turbulent convection) have fluctuating amplitudes that follows the Gaussian distribution. The correlation time for mode amplitude, which is of order the mode lifetime, is typically much larger than the mode period (see §3.1) or the characteristic time of resonant eddies. This is because a mode interacts with a large number of eddies each of which contribute only a small fraction of the total energy in the mode. A good analogy is a pendulum placed in contact with a thermal heat bath of molecules. The mean energy in the pendulum is one third the mean kinetic energy of molecules, however it takes a large number of collisions (of order the ratio of the pendulum mass to molecular mass) in order for the amplitude of the pendulum to change. The statistics of energy fluctuation in a solar p-mode, if stochastically excited, like the energy of the pendulum placed in a heat bath, follows Boltzmann distribution (see Kumar et al. 1988 for a rigorous derivation of this result). At least two different groups (Toutain and Frohlich 1992; and Elsworth et al. 1995) have looked for the statistics of energy fluctuation in the solar p-modes and find it to be in good agreement with the theoretical expectation for stochastic excitation i.e. Boltzmann or exponential distribution (see fig. 2).

3.3. PEAKS AT HIGH FREQUENCIES

Acoustic waves of frequency less than about 5 mHz (the acoustic cutoff frequency at the temperature minimum) are reflected at the solar photosphere and thus trapped inside the sun. The reflectivity however drops off rapidly at higher frequencies; at 6mHz about 2% of the incident wave energy is reflected at the photosphere whereas at 7mHz the reflectivity drops to less than 0.3%. A number of observations indicate that high frequency acoustic waves (waves of frequency greater than about 5mHz) suffer little reflection at the chromosphere/corona as well (Duvall et al. 1993, Kumar et al. 1994, Jefferies 1996). If high frequency acoustic waves were significantly reflected at the chromosphere/corona boundary then the frequency spacing between modes of adjacent order would fluctuate with mode frequency (see figure 3); this is because of the interference of waves partially trapped between two cavities above and below the temperature minimum. The observations, however, show, no evidence for such a behavior (fig. 3) and thus provide an upper limit of about 10% to the reflection at the chromosphere/corona boundary (Kumar et al. 1991; Jefferies, personal communication).

In the absence of wave reflection at the solar surface these high frequency acoustic waves are not trapped in the sun, and thus it was expected that the power spectrum above the acoustic cutoff frequency should be featureless

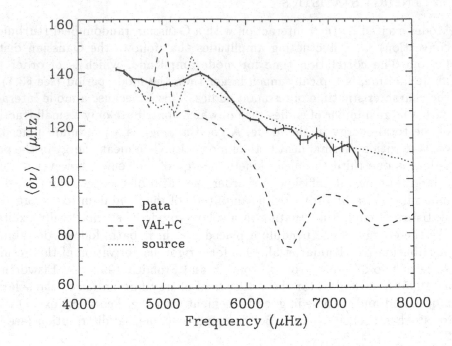

Figure 3. Average frequency spacing between adjacent peaks in the power spectrum, $\langle \delta\nu \rangle$, as a function of frequency. The averaging over frequency bins of width 100 μHz, and ℓ range of 80 and 150 has been carried out after subtracting a linear term in ℓ (0.6981 ℓ μHz) from ($\nu_{n+1,\ell} - \nu_{n,\ell}$). The observational data (thick solid curve) was obtained by Duvall et al. (1993) at the geographical South Pole in 1988. The dotted curve, labeled 'source' in the legend, is the result of calculation of peak frequencies in the theoretically computed power spectra for Christensen-Dalsgaard's solar model with sources lying about 140 km below the photosphere. The dashed curve labled 'VAL+C' is the frequency spacing calculated for JC-D solar model that includes the "mean quiet sun" chromospheric structure of Vernazza et al. (1981) as well as an isothermal corona at a temperature of 10^6 K.

i.e. devoid of peaks. However, the observed spectra contain very regular peaks that are seen upto the Nyquist frequency of observations. One of the best recent data set obtained at the South Pole in 1994 shows peaks extending to almost 11 mHz which makes the length of the spectrum above the acoustic cutoff frequency larger than the observed spectrum below the cutoff frequency!

The existence of these high frequency peaks provides one of the strongest evidence that solar acoustic oscillations above 5 mHz are not excited by some overstability mechanism[3], and since power spectrum varies smoothly

---

[3]Considering the poor reflectivity of high frequency waves at the chromosphere/corona, the energy flux in the solar atmosphere associated with them represents a net loss of their energy. So if these waves are to be excited due to an overstability

from below the acoustic cutoff frequency to above the cutoff frequency we infer that the trapped p-modes in the sun are also not overstable (Kumar et al. 1989).

The peaks at high frequencies can be understood as arising quite naturally if waves are stochastically excited. These peaks form because of the constructive interference between waves propagating from the source (located in the convection zone) upward to the photosphere and waves traveling downward from the source that is refracted back up due to increasing sound speed and thus end up at the photosphere (Kumar and Lu 1991). Therefore the frequencies of peaks above the acoustic cutoff ($\sim$ 5 mHz for the sun) depend on the difference between these two paths or in other words on the depth of acoustic sources. A good fit to the high frequency power spectrum is obtained by placing sources (assumed to be quadrupole) approximately 140 km below the photosphere (Kumar 1994); please see figure 4. It should be emphasized that unlike the lower frequency trapped p-modes (frequency less than about 5 mHz) the frequencies of peaks at high frequencies is not a property **of** the equilibrium model of the sun alone but depends in a sensitive way on the location of sources that excite these oscillations.

As discussed in Kumar (1994) if the acoustic sources are assumed to be dipolar instead, then no matter where these sources are placed in the solar convection zone they do not provide a fit to the observed power spectrum. This suggests that the acoustic sources, at least for the high frequency waves, are not dipole but quadrupole, which is consistent with the work of Goldreich et al. (1994).

High frequency acoustic waves also provide information about the power spectrum of turbulent convection in the sun (Kumar 1994); we can constrain the spectrum of turbulent convection in the region where acoustic emission is significant. The theoretical power spectrum shown in figure 4 was computed using the Kolmogorov power spectrum of turbulence, *i.e.*, $P(k) \propto k^{-5/3}$. Evidently, this provides a good fit to the observed spectrum between 5.5 and 10 mHz. In order to determine the power law index α of solar turbulence, $P(k) \propto k^{-\alpha}$, from the high frequency interference peaks, we relate the fluctuating velocity, v_h, of sub-energy bearing eddies to the velocity v_H of scale-height size eddies as follows:

$$v_h \approx v_H \left(\frac{h}{H} \right)^{(\alpha-1)/2},$$

mechanism, their e-folding time must be less than about an hour, and thus these waves can at best be amplified by a factor of $\sim e$ as they make one passage through the solar interior. Thus we need a mechanism that provides a large seed amplitude, within a factor of e of the observed value, and clearly in this case it seems most natural that the same mechanism generates the full observed amplitude.

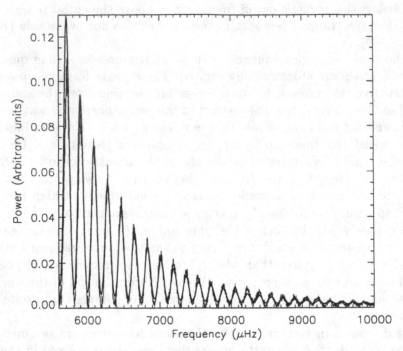

Figure 4.   Observed power spectrum (thin solid line) from the 1994 South Pole observation (courtesy of S. Jefferies) for ℓ=90, and theoretically calculated power spectrum for sources lying 140km below (thick solid line) the photosphere. The Nyquist frequency of the data is 11.9 mHz. Both the ℓ–leakage and the Nyquist folding has been included in the theoretically computed spectrum; to model the ℓ–leakage the theoretically calculated power spectra for ℓ=88 to 92 were added together with weighting factors of 0.147, 0.68, 1.0, 0.61, and 0.10 respectively which corresponds to the 1994 South Pole observations (Jefferies, personal communication). The radial extent of the sources is taken to be 50km, and the spectrum of turbulent convection is Kolmogorov. A frequency dependent background has been subtracted from the observed spectrum.

where h is the size of the eddy. This equation is a generalization of equation (8). Using this relation, we find that the frequency dependence of the source function is $(\omega\tau_H)^{-(3\alpha+5)/(3-\alpha)}$ (the derivation is similar to the one leading to eq. [12]). Therefore, a change in the spectral index of turbulence from 5/3 to 1.4 decreases the dependence of the acoustic power spectrum on frequency by $\omega^{1.75}$, which results in a poor fit to the observed spectrum. We find that the observed high frequency power spectra suggest that the power law index for the solar turbulence lies between 1.5 and 1.7.

We note that the energy input rate for p–modes in the frequency range between 3.5 and 5mHz is proportional to $\omega^{-4.4}$, which is understood most naturally if the spectrum of turbulence near the top of the solar convection zone is taken to be Kolmogorov (Goldreich et al. 1993). The result described above extends this range to 10mHz.

3.4. ASYMMETRIC LINE PROFILES OF P-MODES

The last observational result I would like to describe is the asymmetry of low frequency p-mode line profiles. The spectrum of individual p-modes is fitted very well by a Lorentzian profile. However, Duvall et al. (1993) discovered that the line profiles do not have perfect Lorentzian shape and in particular the power spectrum falls off more rapidly on one side of the peak than the other i.e. the lines are asymmetrical. The data from GONG and SOHO clearly show that line-profiles for low frequency modes are asymmetrical. Duvall et al. had also proposed in their original paper an explanation for why the lines are asymmetrical which is found to be basically correct by a number of independent investigations (Gabriel 1992, 1995; Abrams and Kumar 1996). The line profile for a p-mode of frequency 1.9 mHz and $\ell = 1$, calculated using JC-D solar model, is shown in figure 5.

The physical explanation for line asymmetry is simplest when sources lie in the region of the sun where acoustic waves can propagate (this case

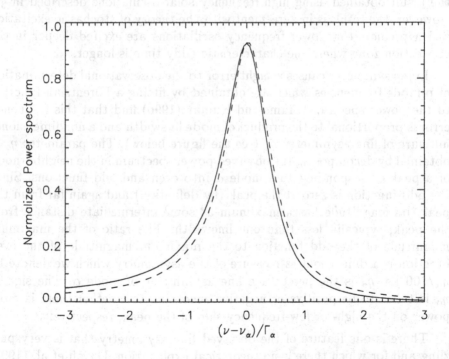

Figure 5. Line profiles of a p-mode of frequency 1.9 mHz and $\ell = 1$ of a solar model due to J. Christensen-Dalsgaard. The power spectrum plotted using a continuous line corresponds to sources placed at the upper turning point of the mode. The other power spectrum, dashed curve, arises when sources are placed 300 km below the upper turning point. The radial extent of sources in both cases was taken to be 100 km.

does not seem to apply to solar oscillations however which are excited by sources that lie in the evanescent region). Consider a source lying close to the node of a p-mode. As the frequency of acoustic waves is varied in the neighborhood of this p-mode frequency the position of the node changes with respect to the source position. Thus waves of frequencies lying symmetrically on the opposite side of the p-mode frequency gets excited to different amplitudes making the resultant power spectrum asymmetrical. It is clear from this rather simple example that not all p-mode line profiles are expected to be equally asymmetrical (as is observed) and also that the degree of asymmetry depends on the location of sources. In fact Duvall et al (1993) had recognized this in their original paper and used this to determine the depth of sources that are exciting p-modes. The recent paper of Abrams and Kumar (1996) uses a realistic solar model due to Christensen-Dalsgaard to calculate the p-mode power spectrum and finds that in order to reproduce the magnitude of asymmetry observed by Duvall et al. (1993) the sources responsible for exciting low frequency p-modes should lie about 250 km below the photosphere. This might appear to be in conflict with the result obtained using high frequency solar oscillations described in §3. However, the result is in agreement with the theory of stochastic excitation which predicts that lower frequency oscillations are excited deeper in the convection zone where the characteristic eddy time is longer.

Line asymmetry causes a slight error to the observational determination of p-mode frequencies which are obtained by fitting a Lorentzian function to the power spectra. Abrams and Kumar (1996) find that this frequency error is proportional to the product of mode linewidth and a nondimensional measure of line asymmetry η_α (see the figure below). The parameter η_α is obtained by decomposing the observed power spectrum in the neighborhood of a peak corresponding to a mode α into even and odd functions. Since the odd function is zero at the peak (by definition) and again far from the peak, its magnitude has a maximum at some intermediate distance from the peak, typically less than one linewidth. The ratio of the maximum magnitude of the odd function to the maximum magnitude of the even function is a dimensionless measure of the asymmetry which we denote by $\eta_\alpha/100$ i.e. η_α is the percentage line asymmetry of mode α. The sign of η_α is taken to be positive or negative according to whether there is more power on the high- or low-frequency side of the peak, respectively.

There is one feature of the observed line asymmetry that is very puzzling and for which there is no theoretical explanation. Duvall et al. (1993) reported that the sense of asymmetry reverses in the velocity and the intensity power spectra for the p-mode. This behavior has been confirmed by the most recent GONG data. The difference between the velocity and the intensity power spectra can arise as a result of line formation in the

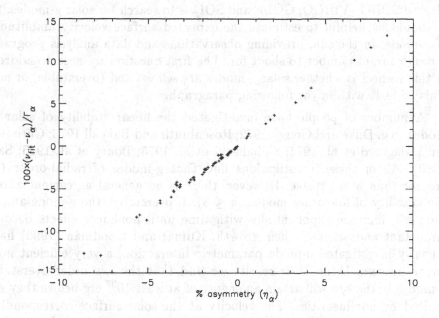

Figure 6.   Error in the measurement of mode frequency (expressed as percentages of corresponding linewidths) as a result of asymmetry of lines in the power spectrum; $_{fit}$ is the frequency obtained by fitting a Lorentzian function to the power spectrum, ν_α is the mode eigenfrequency, and Γ_α is mode linewidth. The frequency error is shown as a function of a dimensionless measure of line asymmetry (η_α) defined in the text. The slope of the line is approximately 1.5. The power spectra were calculated by solving an inhomogeneous wave equation which included radiative damping of waves (see Abrams and Kumar 1996, for details). The solar model used in this calculation was kindly provided by J. Christensen-Dalsgaard.

presence of oscillations. However, it is not clear what process can cause a reversal of the sign of η_α in the two spectra; the process has to be extremely frequency sensitive so that it can modify the spectrum in an interval of only a few μHz.

4. Can we detect gravity modes in the sun?

Gravity mode oscillations of the sun are primarily confined to its radiative interior and their observation would thus provide a wealth of information about the energy generating region which is poorly probed by the p-modes. In the past 20 years a number of different groups have claimed to detect g-modes in the sun (e.g. Brookes et al. 1976; Brown et al. 1978; Delache and Scherrer 1983; Scherrer et al. 1979; Severny et al. 1976, Thomson et al. 1995; for a detailed review of the observations please see the article by Pallé, 1991, and references therein), but thusfar there is no consensus that g-modes have in fact been observed. One of the objectives of the instruments aboard SOHO (VIRGO, GOLF and SOI) is to search for solar g-modes. So it should be helpful to estimate the expected surface velocity amplitudes of g-modes in the sun, providing observations and data analysis programs a rough target number to shoot for. The first question we need to address in this respect is whether solar g-modes are self-excited (overstable) or not. This is dealt with in the following paragraph.

A number of people have investigated the linear stability of solar g-modes (e.g. Dilke and Gough 1972; Rosenbluth and Bahcall 1973; Christen-sen-Dalsgaard et al. 1974; Shibahashi et al. 1975; Boury et al. 1975; Saio 1980). All of these investigations find that g-modes of radial-order (n) greater than 3 are stable. However, there is no general agreement about the stability of low order modes ($n \leq 3$). If overstable, the g-mode amplitude will increase exponentially with time until nonlinear effects become important and saturate their growth. Kumar and Goodman (1995) have recently investigated 3-mode parametric interaction, a very efficient nonlinear process. Using their results we find that the low order overstable g-modes in the sun will attain an energy of at least 10^{37} erg before they are limited by nonlinearities. The velocity at the solar surface corresponding to this energy is $\sim 10^2$ cm s^{-1}, which is an order of magnitude larger than the observational limit of Pallé (1991). Thus even low order g-modes of frequency greater than about 150 μHz are unlikely to be overstable.

However, g-modes can be stochastically excited. A number of people have estimated g-mode amplitude assuming that they are linearly stable and stochastically excited. Keeley (1980) applied his theory, developed with Goldreich in 1977, of the excitation of solar modes to estimate the amplitude of the 160 minute oscillation, and found the theoretical amplitude to be much smaller than claimed by the observations; much more sensitive observational searches since then have not detected this oscillation (cf. Pallé 1991). Gough (1985) carried out an application of the energy partition result of Goldreich and Keeely (1977) to solar g-modes and estimated the surface velocity amplitude of low n and ℓ g-modes ($n \leq 3$, $\ell \leq 2$) to be about 1-2 mm s^{-1}. Kumar et al. (1996) estimated the g-mode amplitude using the recent theoretical work of Goldreich et al. (1994) on stochastic excitation of waves, which reproduces the observed energy input rate into solar p-modes of all frequencies (see §2), and taking into account the radiative and viscous turbulent dampings. They find the surface velocity amplitude of low order g-modes to be about 0.4 mm s^{-1} (see figure 7). Recently Anderson has carried out numerical simulation of g-mode excitation as a result of turbulent flow associated with penetrative convection. He finds that the transverse surface velocity amplitudes of g-modes of degree about 6 is ~ 0.2 mm s^{-1} in the case when he assumes that 10^3 modes are excited by this process (Anderson, 1996). Thus several different calculations suggest that the am-

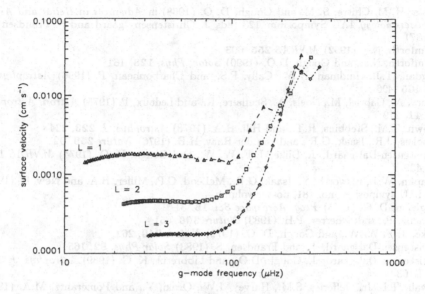

Figure 7. Magnitude of the surface velocity amplitude as a function of frequency for low degree solar g-modes excited by coupling with turbulent convection. The surface velocity amplitude falls off rapidly with increasing ℓ, thus only low degree g-modes are expected to be observable.

plitudes for low degree g-modes are of order 0.5 mm s^{-1}. The uncertainty in this estimate is at least a factor of a few. If the nature turns out to be cooperative and the actual amplitudes of solar g-modes are a factor of a few larger than these estimates, then instruments aboard SOHO have a good chance of detecting g-modes and opening up a new window in the study of the solar core.

Acknowledgment

I am grateful to Peter Goldreich for collaboration over a number of years and for his insights on the topic discussed in this review. I thank Joergen Christensen-Dalsgaard for the use of his solar model, and for pointing out several references. I am indebted to Tom Duvall, Jack Harvey and Stuart Jefferies for sharing with me their excellent data from the South Pole. This work was supported by a NASA grant NAGW-3936.

References

Abrams, D., and Kumar, P. (1996) *Astrophys. J.* **472**, 882
Anderson, B.N. (1996) *Astron. Astrophys.* **312**, 610
Ando, H. and Osaki, Y. (1975) *Publ. Ast. Soc. Japan*, **27**, 581
Antia, H. M., Chitre, S. M. and Narasimha, D. (1982) *Solar Physics*, **77**, 303

Antia, H. M. Chitre, S. M. and Gough, D. O. (1988) in *Advances in Helio- and Aster-oseismology*, IAU Symposium 123 (eds J. Christensen- gaard and S. Frandsen) p. 371

Balmforth, N.J. (1992) *MNRAS* **255**, 603

Balmforth, N.J., and Gough, D.O. (1990) *Solar. Phys.* **128**, 161

Bogdan, T.J., Hindman, B.W., Cally, P.S., and Charbonneau, P. (1996) *Astrophys. J.* **465**, 406

Boury, A., Gabriel, M., Noels, A., Scuflaire, R., and Ledoux, P. (1975) *Astron. Astrophys.* **41**, 279

Brown, T.M., Stebbins, R.T., and Hill, H.A. (1978) *Astrophys. J.* **223**, 324

Brookes, J.R., Isaak, G.R., and van der Raay, H.B. (1976) *Nature* **259**, 92

Christensen-Dalsgaard, J., Dilke, J. F. W. W., and Gough, D. O. (1974) *MNRAS* **169**, 429

Chaplin, W.J., Elsworth, Y., Isaak, G.R., McLeod, C.P., Miller, B.A. and New, R. (1996) IAU symposium no. 181, poster volume

Crighton, D. G. (1975) *Prog. Aerospace Sci.* **16**. 31

Delache, P., and Scherrer, P.H. (1983) *Nature* **306**, 651

Dilke, J. F. W. W., and Gough, D. O. (1972) *Nature* **240**, 262

Christensen-Dalsgaard, J., and Frandsen, S. (1983) *Solar Phys.* **82**, 165

Christensen-Dalsgaard, J., Gough, D.O., and Libbrecht, K. G. (1989) *Astrophys. J.* **341**, L103

Duvall, T.L. Jr., Jefferies, S.M., Harvey, J.W., Osaki, Y., and Pomerantz, M.A. (1993) *Astrophys. J.* **410**, 829

Duvall, T.L., Jefferies, S.M., Harvey, J.W., and Pomerantz, M.A. (1993) *Nature* **362**, 430

Duvall, T. L., Harvey, J. W., and Pomerantz, M. A. (1988) in *Advances in Helio- and As-teroseismology*, IAU Symposium 123, eds. J. Christensen- Dalsgaard and S. Frandsen, p. 37

Elsworth, Y., Isaak, G. R., Jefferies, S. M., McLeod, C. P., New, R., Pallé, P. L., Regulo, C. Roca Cortés, T. (1990) *MNRAS* **242**, 135

Elsworth, Y, *et al.* (1995) in *Fourth SOHO Workshop in Helioseismology*, ed. J.T. Hoek-sema, V. Domingo, B. Fleck and B. Battrick

Gabriel, M. (1992) *Astron. Astrophys.* **265**, 771

Gabriel, M. (1995) *Astron. Astrophys.* **299**, 245

Goldreich, P. and Keeley, D. A. (1977a) *Astrophys. J.* **211**, 934

Goldreich, P. and Keeley, D.K. (1977b) *Astrophys. J.* **212**, 243

Goldreich, P. and Kumar, P. (1990) *Astrophys. J.* **363**, 694

Goldreich, P. and Kumar, P. (1991) *Astrophys. J.* **374**, 366

Goldreich, P., and Murray, N. (1994) *Astrophys. J.* **424**, 480

Goldreich, P., Murray, N., and Kumar, P. (1994) *Astrophys. J.* **424**, 466

Gough, D.O. (1977) *Astrophys. J.* **214**, 196

Gough, D.O. (1985) in *Future missions in solar, heliospheric and space plasma physics*, eds. E.J. Rolfe and B. Battrick, ESA SP-235, 183

Jefferies, S.M. (1996) personal communication

Kulsrud, R. M. (1955) *Astrophys. J.* 121, 461

Leibacher, J. and Stein, R.F. (1971) *Astrophysics Lett.* **7**, 191

Libbrecht, K.G. (1988) *Astrophys. J.* **334**, 510

Libbrecht, K.G., and Woodard, M.F. (1991) *Science* **253**, 152

Lighthill, M.J. (1952) *Proc. Roy. Soc. A* **211**, 564

Keeley, D.A. (1980) in *Nonradial and Nonlinear Stellar Pulsation*, (eds H. A. Hill and W. A. Dziembowski), Springer, Berlin p. 245

Kumar, P., and Lu, E. (1991) *Astrophys. J.* **375**, L35

Kumar, P. (1994) *Astrophys. J.* **428**, 827

Kumar, P, Fardal, M.A., Jefferies, S.M., Duvall, T.L. Jr., Harvey J.W., and Pomerantz, M.A. (1994) *Astrophys. J.* **422**, L29

Kumar, P., Franklin, J., and Goldreich, P.(1988) *Astrophys. J.* **328**, 879

Kumar, P. and Goldreich, P. (1989) *Astrophys. J.* **342**, 558

Kumar, P. and Goldreich, P., and Kerswell, R. (1994) *Astrophys. J.* **427**, 483

Kumar, P., and Goodman, J. (1996) *Astrophys. J.* **466**, 946

Kumar, P., Quataert, E. J., and Bahcall, J.N. (1996) *Astrophys. J.* **458**, L83

Murray, N. (1993) in *Seismic Investigation of the Sun and Stars*, ASP conference series, vol. 42, ed. T. Brown, p. 3

Pallé, P. L. (1991) *Adv. Space Res.* **11**, 4, 29

Rosenbluth, M., and Bahcall, J. N. (1973) *Astrophys. J.* **184**, 9

Saio, H. (1980) *Astrophys. J.* **240**, 685

Scherrer, P.H., Wilcox, J.M., Kotov, V.A., Severny, A.B., and Tsap, T.T. (1979) *Nature* **277**, 635

Severny, A.B., Kotov, V.A., and Tsap, T.T. (1976) *Nature* **259**, 87

Shibahashi, H., Osaki, Y., and Unno, W. (1975) *Publ. Astron. Soc. Japan* **27**, 401

Stein, R.F. (1967) *Solar Phys.* **2**, 385

Stein, R.F., and Nordlund, A. (1991) in *Challenges to Theories of the Structure of Moderate-Mass Stars*, ed. D. Gough and J. Toomre (Berlin: Springer-Verlag), 195

Thomson, David J., Maclennan, Carol G., Lanzerotti, Louis J. (1995) *Nature* **376**, 139

Tomczyk, S., 1996, personal communication

Toutain, T., and Fröhlich, C. (1992) *Astron. Astrophys.* **257**, 287

Ulrich R. K. (1970) *Astrophys. J.* **162**, 993

Vernazza, J. E., Avrett, E. H., and Loeser, R. (1981) *Astrophys. J.S* **45**, 635

Wolf, C.L. (1972) *Astrophys. J.* **177**, L87

AN OBSERVATIONAL STUDY OF THE INTERACTION OF IMPULSIVE AND PERIODIC PERTURBATIONS IN THE SOLAR ATMOSPHERE

F.-L. DEUBNER AND R. KLEINEISEL

Astronomisches Institut

Am Hubland, D-97074 Würzburg

Abstract. Time series of simultaneous high resolution filtergram observations in white light and in CaK_2 have been employed to study the dynamical interdependence of the 'low' and the 'high' layers of the solar atmosphere, the 3-D propagation of perturbations in the stratification, and the excitation of p-mode and other oscillations. Impulsive and oscillatory perturbations have been isolated in the wavenumber and frequency regime, and their mutual interaction was also studied.

Several of our findings challenge interpretation and deserve further study.

• The occurrence of a granule can be clearly recognized in the upper chromosphere.

• A granule induces *low* brightness and *low* oscillation amplitudes at the CaK_2 level, whereas the opposite happens at the intergranular border.

• It is not always evident that the 'source' of the perturbation is located at the lower, i.e. the photospheric layer.

• It is far from being clear which components of the various perturbations contribute to a net outward mechanical energy flux.

1. Introduction

Stochastic excitation of p-modes by turbulent convection is generally considered a viable and satisfactory model of the origin of the 5-min oscillations of the sun (cf. e.g. Kumar *et al.*, 1990). This process incorporates the explanation of the pseudo p-modes that occur at frequencies, where resonant eigenmodes of the global p-mode cavity do not exist (Kumar, 1993).

How the energy is transferred to the wave field from individual convective eddies (granules, downdrafts etc.), however, is not being considered

J. Provost and F.-X. Schmider (eds.), Sounding Solar and Stellar Interiors, 307-314.

in detail in this model. Also, observations that could help treating this problem quantitatively are scarce, and are scattered irregularly over more than 30 years that elapsed since the first detailed spatio-temporal studies of solar p-modes were performed. Even taking the difficulty of obtaining observations from the ground with sufficient spatial resolution (preferably $\lesssim 1$ arc sec) during extended periods of time ($\gtrsim 20$ min) into account, the material available makes a rather casual impression, and progress towards a more comprehensive treatment of the subject appears to be lacking.

Two paradigms have been drawn up in the course of this long interval since Evans and Michard (1962) first discussed the vertical progression of a photospheric perturbation (notably a granule) into the solar atmosphere.

Paradigm #1: *The granular piston*

According to the description of this event given by Evans and Michard, the initial brightening of a granule is followed first by upward motion, and then by regular oscillations, at successively higher layers in the atmosphere (cf. also Fig. 2 in Noyes, 1967, for illustration). Since the period of oscillation in the upper part of the atmosphere is shorter than below, the initial phase lag of the upper layers is quickly compensated, and the whole atmosphere swings in phase before the motion decays. A rough estimate of the speed of propagation of the initial perturbation yields the value of the local sound velocity.

The hydrodynamic problem of describing the motions excited by a piston driven from below into an isothermal atmosphere overlaid by a hot corona was solved analytically in 2-D by Meyer and Schmidt (1967). The good qualitative agreement with the observations lent strong support to the long held belief that in general the 5-min oscillations were primarily a local phenomenon. Much later, Goode *et al.* (1992) presented a numerical calculation of a rather similar process in a realistic solar atmosphere in 1-D. Evidently, the evolution of the perturbation with height in the atmosphere (cf. Fig. 1 of this paper) has a high degree of likeness with the Figure in Noyes (1962). Unfortunately, the horizontal dimension is missing in this calculation.

As a first observational approach of the two-dimensional problem, Deubner (1975) has selected from a 32 min time series of the evolution of a 320 arc sec wide region at the center of the disk, observed simultaneously in four spectral lines, the one event characterized by the highest amplitude of vertical oscillatory motions in the 3 mHz range. The analysis of this particular event did not only corroborate the granular piston paradigm in displaying p-mode oscillations with decreasing period at increasing height on top of a prominent granule. It also shows long period g-modes excited at the flanks of the granule as predicted by Meyer and Schmidt, as a consequence of the lateral propagation of the perturbation at higher layers.

With regard to spectral intensity, the picture in this observation is much less clear; but that was not a subject of the Meyer and Schmidt paper either.

Paradigm #2: *The intergranular downdraft*

The source of oscillations in this paradigm is thought to be turbulent motion created within the downdraft funnel by high horizontal velocity gradients giving rise to high frequency sound waves that appear in the photosphere.

This idea was checked observationally by Deubner and Laufer (1983) who used a similar data set as in the 1975 paper, selecting this time 132 events with increased power in the 11 - 45 mHz range. The averaged temporal sequence clearly shows at the center the signature of a downdraft with reduced brightness at the photospheric level about 2 min before the high frequency event. Also in the low chromosphere (NaD), downward motion is associated with this scenario.

Observational evidence was further elaborated by Rimmele *et al.* (1995) who found in a 1-D analysis of spatial high resolution data that the temporally averaged high frequency acoustic flux is inversely correlated with the normalized intensity in the Fe 5434 line wing at +20 mÅ from the line center. Moreover it was confirmed that the flux reaches a maximum about 2 min after minimum brightness in the continuum. Espagnet *et al.* (1996) documented the association of intergranular darkening and prominent oscillations in the 5-min range by applying to one and the same set of 2-D data two different k-ω filters that isolate the oscillatory from the convective regime.

Looking at this observational material, it has been our impression so far, that the two paradigms are not really contradictory; rather they seem to emphasize different aspects of the interaction between granulation and oscillations. Obviously the goal of future observations had to be to fully deploy the 2-D aspect of the local phenomena as well as to provide insight into the height dependence of the perturbation. Another important aspect in view of the omnipresence of the high amplitude global p-modes is the necessity to acquire a sufficient volume of statistically independent data in order to sift the local response from that intrusive oscillatory background.

2. Observations

The new observational material discussed in the following section consists of a 2.5 hours time series of filtergrams taken simultaneously in the continuum at 5000 Å, and with a 0.6 Å passband centered on CaK, with the German VTT of the Teide Observatory on Tenerife. The original field of view in the quiet center of the disk is 180 x 180 arc sec, corresponding to 1024 x 1024 pixels of the CCD arrays. The exposure time was 1 sec, at a 12 sec cadence. The seeing conditions varied from excellent to good.

To separate the 'convective' from the 'oscillatory' components of the observed brightness fluctuations we have applied two different spatio-temporal filters similar to those described in Espagnet et $al.$ (1996). A 'subsonic' filter transmits all fluctuations with a ratio $\nu/k_H \leq 6.5$ km s^{-1} corresponding to the velocity of horizontally propagating sound waves; it transmits brightness fluctuations due to gravity waves as well as convective perturbations. The other filter is a band pass transmitting frequencies from 2 mHz to 10 mHz with a limiting ratio of $\nu/k_H \geq 9$ km s^{-1}; there we find evanescent and propagating p-modes in the observed range of frequencies. In addition, a time dependent r.m.s. amplitude of the oscillatory component was defined for each picture element with the help of the band pass filtered data set, using a 144 sec running mean.

After filtering the data we have selected four data cubes with a volume of 12.7 Mm x 12.7 Mm x 20 min each, to search for suitable reference locations, or 'starting points', for the subsequent spatio-temporal analysis, taking the distribution of several scattered fragments of the chromospheric network into account. We have defined a total of 256 starting points by finding the 128 brightest and the 128 darkest pixels that are both spatially and temporally unrelated, within the 'subsonic' continuum cube.

Centered around these starting points we have finally averaged the evolution of the brightness distribution both azimuthally and over all 128 events, covering a radial range of 7.6 Mm, and a time interval of 600 sec before and 600 sec after maximum (granule) or minimum (intergranule) brightness in the continuum. Figures 1 and 2 display the average evolution of brightness and r.m.s. amplitude in continuum light and in the light of the CaK line.

3. Results and discussion

Inspecting Figure 1, we shall discuss what appears relevant for an improved understanding of the convection - oscillation interaction.

Testing Paradigm #1 (granule events): In white light, i.e. in the photosphere, we observe that the occurrence of a granule is quickly followed by a substantial decrease of oscillatory power (10% to 15% below average) lasting about 600 sec. The extent of the region with reduced power remains essentially confined within a constant area during this interval, and does not spread beyond the size of a granule. We note that the granular event is preceded by a maximum of oscillatory power about 300 sec before the 'starting' time. In the light of CaK we observe the arrival of the

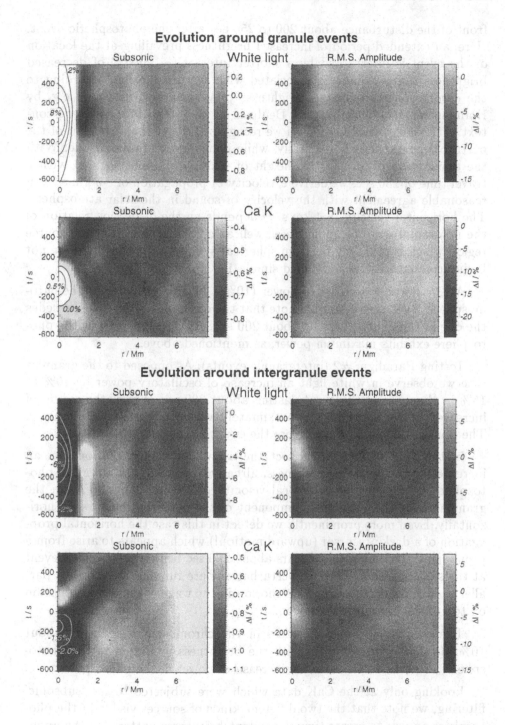

Figure 1. Subsonic intensity and oscillation r.m.s. in photosphere and chromosphere

front of the disturbance about 200 to 250 sec after the photospheric event. Here, an extended period of increased brightness prevailing at the location of the event is interrupted by a shorter interval ($\sim$100 sec) of decreased brightness that we know is associated with upward motion, according to the chromospheric velocity - brightness phase relations discussed e.g. by Hofmann *et al.* (1996), and by Deubner *et al.* (1996) in a different context. At this chromospheric level we recognize further that the front of the perturbation expands horizontally, while it also grows in amplitude. With the assumption of an effective height of CaK$_{232}$ of 1500 km, and with a travel time of 250 sec, we derive a velocity of propagation of $\sim$6 km s^{-1}, in reasonable agreement with the velocity of sound in the solar atmosphere. The horizontal velocity ($\sim$4 km s^{-1}) depends on the dispersion relation of the non-radial wave components as well as on the geometry of the source region which is probably far from spherical symmetry; it can therefore not be interpreted without a detailed simulation.

A considerable decrease of power (10% - 15%) is observed at the chromospheric level as well, and we note that the onset of this decrease precedes the one in the photosphere by about 200 sec! At the same instant the photosphere exhibits maximum power, as mentioned above.

Testing Paradigm #2 (intergranule events): As opposed to the granular case we observe in white light an increase of oscillatory power by 10% to 15% in the wake of the event, lasting about 300 sec. Again, the region of increased power is confined approximately within one granular diameter. The power is below average before the event at the source position.

In the light of CaK we detect the arrival of a perturbation, i.e. an increase by about 1% in brightness, already about 150 sec after the photospheric event, yielding a vertical velocity of about 8 km s^{-1}. As in the granular case, the 'subsonic' component of the perturbation spreads horizontally. Even more prominently we detect in this case the horizontal propagation of a dark precursor (upward motion!) which appears to arise from a pronounced darkening that occurs about 200 sec before the specified event at the same location. The two perturbations are running more or less parallel to each other, the downward motion in the wake of the upward motion on top of the intergranular region.

The oscillatory power increases in the chromosphere again by about 10%. Its maximum coincides with the brightness minimum in the photosphere; however, it precedes the increase of power in the photosphere.

Looking only at the CaK data which were subjected to the 'subsonic' filtering, we note that the two different kinds of sources visible in the photosphere appear to create time dependent brightness patterns in the upper layers that are very nearly negative copies of each other.

4. Conclusions

We have attempted to work out the 2-D response of the solar atmosphere, as observed at two different levels, to a local perturbation induced by impulsive convective motions. We have evaluated the evolution of the two cases, bright photospheric sources ('granules') and dark sources ('intergranular events'), separately, but with the same procedure. Averaging of a large number of events (128) was necessary to isolate the local response from the strong uncorrelated background signal contributed by the global p-mode field.

In either case we recognize the propagation of the photospheric disturbance upon its arrival at the chromospheric level by identifying a distinct local darkening (brightening) with the upward (downward) motion of the arriving pulse. The velocity of propagation in the vertical direction agrees with the sound velocity. It appears to be marginally higher in the second case (intergranule). In either case we find a distinct trace of the horizontal propagation of the disturbance that shows a maximum velocity of about 4 km s^{-1}, and originates from the central up(down)flow according to the above identification with the brightness variations. This horizontal wave attains a radial distance of up to 1.5 Mm in our data, and appears most prominent at a slightly smaller distance. The horizontal velocity is consistent with the velocity of propagation of internal gravity waves, and indeed our observations corroborate the theoretical results of Meyer and Schmidt (1967), and the observation of Deubner (1975), who have associated the excitation of gravity waves with the occurrence of strong granules.

Up to this point our findings are symmetric with regard to the polarity of the source (up- or downward pulse). The new observations confront us with puzzling results as this symmetry appears violated: The granular type source causes a decrease of p-mode power in the atmospheric column on top of it (but not next to it), and the flow connected to a strong darkening in the intergranular network causes an increase of similar magnitude. Would wavefront retardation due to quasi-stationary up(down)flow in conjunction with an increase (decrease) of the local sound speed at the source location be sufficient to disperse (concentrate) the ambient p-mode power by a substantial amount, as suggested by Deubner and Laufer (1983)?

Even more alarming appears the observation that the onset, and the maximum of the response of the p-mode power to the perturbation occurs distinctly earlier (200 sec) in CaK (that is in the upper layer!) than in the continuum, i.e. even earlier (100 - 200 sec) than the photospheric manifestation of what we have called the source. This finding is compelling evidence that we need to investigate the local response of convective instability in the superadiabatic transition region with regard to the ambient p-mode wave field.

Regrettably, radial velocity data, which are fundamental for studies of solar atmospheric dynamics, have not been obtained in this investigation. The inferences based on earlier work on 2-D velocity-brightness (V-I) phase differences are helpful, but certainly not satisfactory. With advanced spectroscopic systems that permit either rapid spatial or rapid spectral scanning, the requisite kind of data is gradually becoming available.

Further, the exchange of energy between convective eddies and the atmospheric wave field can be fully understood only on the basis of a quantitative description of the observed phase relations. It is therefore planned to extend our study by including wavelet analyses, that allow the characterization of waveforms together with the determination of the temporal delays that occur between the observed fluctuations.

References

Deubner. F.-L. (1975) *Solar Phys.* **40**, 333

Deubner, F.-L., Laufer,J. (1983) *Solar Phys.* **82**, 151

Deubner, F.-L., Waldschik, Th., Steffens, S. (1996) *Astron. Astrophys.* **307**, 936

Espagnet, O., Muller, R., Roudier, Th., Mein, P., Mein, N., Malherbe, J.-M. (1996) *Astron. Astrophys.* **313**, 297

Evans, J. W., Michard, R. (1962) *ApJ* **136**, 493

Goode, P. R., Gough, D., Kosovichev, A. (1992) *ApJ* **387**, 707

Hofmann, J., Steffens, S., Deubner, F.-L. (1996) *Astron. Astrophys.* **308**, 192

Kumar, P. (1993) in *Seismic Investigation of the Sun and Stars* T.Brown ed., ASP Conf. Ser. 42, p. 15

Kumar, P., Duvall, T. L., Harvey, J. W., Jefferies, S. M., Pomerantz, M. A., Thompson, M. J. (1990) in *Proc. Oji International Seminar* Y. Osaki and H. Shibahashi eds., Springer Berlin, p. 87

Meyer, F., Schmidt, H. U. (1967) *Z. Astrophys.* **65**, 274

Noyes, R. (1967) in *Aerodynamic Phenomena in Stellar Atmospheres* R.N.Thomas ed., Acad. Press London, p. 293

Rimmele, T. R., Goode, P. R., Strous, L. H., Stebbins, R. T. (1995) in *Proc. of Fourth SOHO Workshop: Helioseismology* J.T.Hoeksema, V.Domingo, B.Fleck, B.Battrick eds., ESA SP-376.2, p. 329

ASTEROSEISMOLOGY:
THEORY AND METHODS

SOUNDING STELLAR INTERIORS

W. A. DZIEMBOWSKI

Copernicus Astronomical Center
ul.Bartycka 18, 00716 Warszawa, Poland

1. Introduction

Multimode pulsations have been discovered in a large variety of main sequence stars and white dwarfs. There are abundant and accurate data on p- and g- mode frequencies in these objects. So far, however, use of these data for sounding interiors has been very limited. The main obstacle is the difficulty in identification of detected modes that is in assignment of spherical harmonic degree, l, azimuthal order, m, and radial order, n, to the modes detected in power spectra of variable stars. This has never been a problem in helioseismology. Even in the case of whole disc measurements a simple structure of power spectra enables an unambiguous mode identification. Such a situation is very rare in asteroseismology. Perhaps oscillating white dwarfs PG 1159-035 and GD 358 are the only examples. Methods of determining the l and m from observational data on the excited modes are available. However, their implementation to – typically low amplitude – multimode pulsators is difficult. In most cases mode identification cannot be separated from determination of stellar parameters.

The number of modes detected in individual stars, and in fact in all stars put together, is by orders of magnitude smaller than that in the Sun. Again the two white dwarfs each with over 100 excited modes are on top of the list of multimode pulsating stars. Not surprisingly, astrophysics made best use of asteroseismic data on these objects. I will return to these two stars in section 3.

Most of this review, in particular sections 4 and 5, is devoted to δ Scuti and β Cephei stars. This reflects primarily my personal interest. The truth is that for no object of these two types credible seismic sounding is available. Here the maximum number of excited modes is 20, which is much less than the number of unstable modes with $l \leq 2$ found in corresponding models. The possibility of identification with higher degree modes is usually

317

J. Provost and F.-X. Schmider (eds.), Sounding Solar and Stellar Interiors, 317-330.
© 1997 IAU. Printed in the Netherlands.

neglected. This simplifies mode identification but, unfortunately, is not fully justified.

The ultimate aim of asteroseismology is to construct stellar seismic models i.e. stellar models with adjusted parameters to fit observed frequencies within the measurement errors. We are not at this stage yet. In section 7 I outline the methodology which, I hope, will be soon applied.

The true asset of stellar seismology are g-modes which are definitely detected in quite a variety of pulsating stars. These modes are indeed much better design to probe interiors than p-modes – still the only modes available for helioseismic sounding. In many stars, like in the Sun, the g-modes have large amplitudes in deep interior. Furthermore, their frequencies are very sensitive to the mean molecular weight gradient. The possibility of probing a chemically inhomogeneous interior is crucial for testing stellar evolution theory.

2. White Dwarf and Main Sequence Pulsators

Naturally, asteroseismologists are interested in stars with many excited modes and such stars are found almost exclusively among white dwarfs and stars in the main sequence band. The exceptions are some of δ Scuti stars which are in the phase of hydrogen burning in a thick shell surrounding a helium-rich core. Basic data on types of pulsating stars of interest are summarized in Table 1 and in Fig. 2, showing positions of the objects in the H-R diagram.

Oscillating white dwarfs are spread over a wide range of effective temperatures. Objects denoted as DAV and DAB have, respectively, pure H and He atmospheres. The theory predicts excitation of p-modes in these stars but the presence of such modes has not been confirmed. All oscillating white dwarfs are g-mode pulsators. Abundant data on these stars have been collected in campaigns of the Whole World Telescope (Nather *et al.*, 1990).

A large diversity of pulsation is found among the upper main sequence stars. There are two types of high-order g-mode pulsators: Slowly Pulsating B stars (SPB) (Waelkens, 1991) and γ Doradus stars (Balona *et al.*, 1994). These long-period variable stars are only of potential interest for asteroseismic sounding. Years of observations are required to gather sufficient amount of data. Use of automatic telescopes seems the best option. Data on β Cep and δ Sct stars collected for decades are far more abundant. Two networks – STEPHI (Belmonte *et al.*, 1993) and DSN (e.g. Breger *et al.*, 1994)– are devoted to observations of the latter type. Their operation led to the discovery of truly multimode objects. High-order p-mode pulsators in the upper main sequence are represented by roAp stars. The objects share

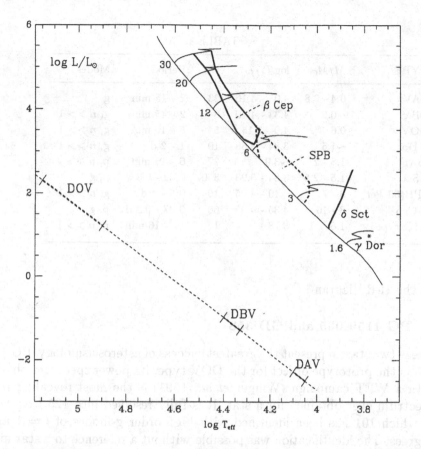

Figure 1. The position of the pulsating stars in the H-R diagram. Evolutionary tracks for Population I stars with indicated masses are shown. For β Cep and SPB stars theoretical instability domains are shown. For δ Sct only the high temperature boundary is given. The asterisk marks the η Boo position.

their H-R position with δ Sct stars but are chemically peculiar and have strong magnetic fields. Prospects for the roAp star seismology are discussed in these proceedings by J.Matthews.

In all these objects the most likely (and in many cases certain) cause of mode excitation is the opacity mechanism. In cooler stars we expect excitation of solar-like oscillations. Christensen-Dalsgaard and Frandsen (1983) provided a crude prediction of the amplitudes and period ranges. Unfortunately, none of the announced detections has been confirmed by independent teams of observers. The most convincing case is that of η Boo (Kjeldsen *et al.*, 1995).

In Table 1 I provide some data on white dwarf and main sequence pulsators (N* denotes the number of objects). Figure 1 shows the objects

TABLE 1.

TYPE	$M/M_\odot$	$\log T_{eff}$	N*	Periods	Modes
DAV	0.4 - 0.8	4.05 - 4.10	23	2 - 15 min	g
DBV	~0.6	4.33 - 4.40	8	2 - 15 min	g, $n \gg 1$
DOV	0.6	4.9 - 5.15	5	5 - 16 min	g, $n \gg 1$
γ Dor	~1.5	3.84 - 3.88	10	1 - 2 d	g, $n \gg 1$
roAp	1.8 - 2	~3.9	27	6 - 15 min	p, $n \gg 1$
δ Sct	1.5 - 2.5	3.84 - 3.93	350	0.02 - 0.3 d	p, g
SPB(53 Per)	3 - 7	4.10 - 4.25	10	0.5 - 4 d	g, $n \gg 1$
β Cep	8 - 16	4.35 - 4.45	60	0.07 - 0.3 d	p, g
solar type	1.6	3.78	1	12 - 16 min	p, $n \gg 1$

on the H-R diagram.

3. PG 1159-035 and GD 358

These two stars represent the greatest success of asteroseismology. PG 1159-035 is the prototype object for the DOV type. Its power spectrum obtained with a WET campaign (Winget *et al.*, 1991) is the most revealing power spectrum ever obtained for a star. It is fully resolved into 125 frequencies of which 101 has been identified with high-order g-modes of $l = 1$ and 2 degrees. The identification was possible without a reference to a star model because the frequencies obey approximately the two following asymptotic relations valid for high-order g-modes:

$$P_{l,n+1,0} - P_{l,n,0} \approx \frac{2\pi^2}{\sqrt{l(l+1)} \int_0^R N d\ln r}$$

and

$$\nu_{l,n,m} - \nu_{l,n,0} \approx m \frac{\Omega}{2\pi}(1 - \frac{1}{l(l+1)}),$$

where P is the period and N is the Brunt-Väisälä frequency.

The same relations were earlier used by Delache and Scherrer (1983) in their search for g-modes in solar oscillation spectra. However, in the g-mode frequency range the solar power spectrum is far less clean than that of PG 1159-035. The latter is more similar to that of solar p-modes obtained with the whole disc measurements. This analogy goes further. In both cases the main patterns of the power spectra are modeled with just two numbers. The approximate validity of asymptotics allows to separate mode identification from model fitting.

The preliminary analysis of the power spectrum yields very important information about star rotation: the value of period, $P_{rot} = 1.35$ d, and the evidence that it is nearly uniform. Let us recall that we had no useful information about solar rotation from the early whole-disc data and even now there are discrepant inferences. The measured mean period separation enabled an estimate of the mass: $M/M_\odot \approx 0.59$.

If the data had satisfied Eq.(1) exactly then that would have been the end of seismic analysis and no seismic probing of the internal structure would have been possible. Fortunately, this is not the case. There is a departure from the asymptotic value, which plotted as function of P exhibits oscillatory behavior which is caused by a nearly discontinuous transition between regions of different chemical composition. Detailed models reproducing the observed behavior were constructed by Kawaler & Bradley (1994).

GD 358 is a DBV type object. Its power spectrum obtained with a WET campaign (Winget et al., 1994) is the richest ever obtained for a distant star. It is more complicated than that of PG 1159-035 and a smaller fraction of 180 peaks has been connected to oscillation mode frequencies. A sequence of hypothetical $l = 1$ triplets was identified. However, such an identification requires nonuniform rotation and the presence of a strong magnetic field. If confirmed, these findings would have interesting ramifications.

4. Unstable Modes in Upper Main Sequence Stars

Figure 2 shows how the instability to low degree modes appears in β Cep star models (12 $M_\odot$ sequence) and in δ Sct star models (1.8 $M_\odot$ sequence). In the former case the driving effect arises in the metal opacity bump at $T \approx 3 \times 10^5$ K and in the latter in the HeII ionization zone at $T \approx 5 \times 10^4$ K.

Typically, a large number of modes is simultaneously unstable. The instability is essentially independent of the azimuthal order. Thus, each dot actually represents $2l + 1$ modes. Unstable modes with $l > 2$ occur in the same frequency ranges as those for $l \leq 2$. Even if, invoking visibility arguments, we disregard high degree modes we still find in cooler δ Sct stars over 80 unstable modes – more than four times as many as detected in the richest periodograms. For β Cep stars the number of unstable modes is significantly smaller than in δ Sct stars. However, it is still much larger than the maximum number of detected modes which is six. In both types, the frequency range of the detected modes agrees with the theoretical prediction. What we do not understand is why we do not see most of the unstable modes. We do not know whether they are not excited or have amplitudes below the detection level.

Only at $l = 0$ do we see an almost equidistant frequency separation

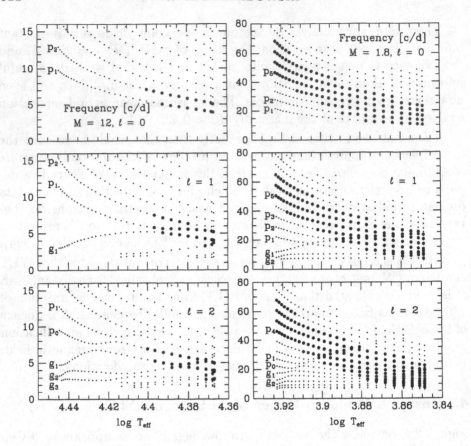

Figure 2. Frequencies (in c/d units, 1 c/d = 11.57μHz) of low-order p- and g-modes of low degree for models of 12 $M_\odot$ and 1.8 $M_\odot$ stars in the main sequence evolutionary phase. Big dots denote unstable modes.

between consecutive modes, which is a property of p-modes. At $l > 0$ the picture is complicated by the occurrence of unstable g-modes, whose frequencies increase during the initial phases of the evolution. Except close to ZAMS all the unstable nonradial modes have a mixed g- and p-mode character. Instability of g-modes is a good news for prospects of seismic probing but it complicates mode identification procedure. Another complication is caused by rotation. At the rates typical for these stars the calculated frequency splitting departs significantly from the simple equidistant pattern seen in the PG 1159-035 power spectra.

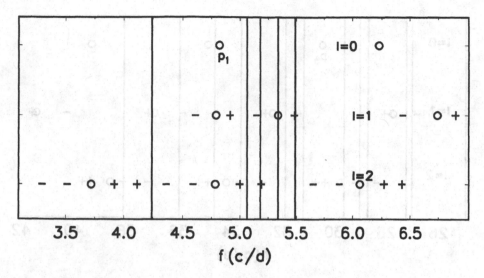

Figure 3. Oscillation frequencies in DD Lac (shown as vertical lines) compared with those of unstable modes in an approximate model ($M = 12M_\odot$, $\log T_{eff} = 4.378$, $V_{rot} = 92$ km/s, $\log g = 3.74$). Frequencies of calculated modes with $m < 0$, $m = 0$, and $m > 0$ are shown with -, o, and +, respectively.

5. Selected Power Spectra

Here I compare periodograms for one β Cep star and two δ Sct stars with model predictions. Chosen models have parameters within the range allowed by the photometric and spectroscopic data for the objects but in no case effort was made to fit frequencies except of one mode in β Cep type star. Thus, these are not even approximate seismic models in the sense explained in section 7.

In this comparison I consider only modes with $l \leq 2$. Modes of higher degrees are also unstable but they are less likely to reach detectable amplitudes. For such modes the integration over the stellar disk includes comparable terms having opposite signs and this leads to a considerable reduction of observable amplitudes. As clearly seen in the whole disc solar data, a large increase in the amplitude reduction takes place between $l = 2$ and 3 degrees. Assuming $l \leq 2$, identification is a useful working hypothesis but it lacks a solid justification.

5.1. DD LAC

This β Cep star is a subject of an ongoing analysis done in collaboration with Mike Jerzykiewicz. The figure and the discussion were presented in a poster at this meeting. The equidistant triplets suggest the $l = 1$ interpre-

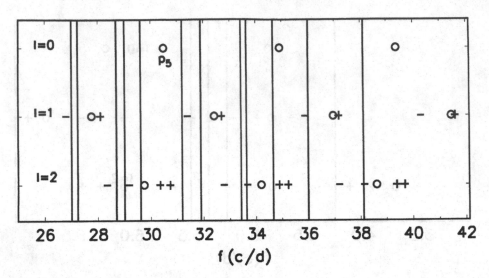

Figure 4. Oscillation frequencies in CD-24 7599 (shown as vertical lines) compared with those of unstable modes in an approximate model ($M = 1.95 M_\odot$, $\log T_{eff} = 3.910$, $V_{rot} = 70$ km/s, $\log g = 4.06$). Frequencies of calculated modes with $m < 0$, $m = 0$, and $m > 0$ are shown with -, o, and +, respectively.

tation, which is also consistent with the photometric data. A sequence of models was constructed that fits the central peak of the triplet to $l = 1$, $m = 1$ modes. The model shown in Figure 3 is an example. A common problem for all these models is that the quadratic effect of rotation causes departure from constant separation far greater than in the data. One possible explanation suggested is the phase locking resulting from resonant interaction between the three modes (Buchler *et al.*, 1995).

In all models of DD Lac consistent with the data two close $l = 1$ triplets occur in the frequency range of the observed peaks, because all these models are close to avoided crossing between g_1 and p_1 modes. The occurrence of this effect may be seen in Figure 2. Such modes are of special interest for asteroseismology because their frequencies are very sensitive to the extent of convective overshooting (Dziembowski & Pamyatnykh, 1991; Audard *et al.*, 1995).

5.2. CD-24 7599

This recently discovered δ Scuti star in a joint campaign of WET and DSN networks (Handler *et al.*, 1996) was selected by us (a collaboration including Pamyatnykh, Handler, Goode and Pikall) to practice the methodology of constructing seismic stellar models. The object seems a good choice. It has many detected modes and is located near ZAMS. One may see in Fig. 2

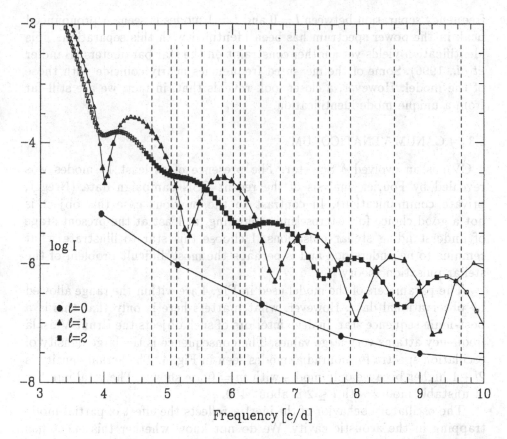

Figure 5. Oscillation frequencies in 4 Canum Venaticorum (shown as vertical lines) compared with those in an approximate model ($M = 2.3M_\odot$, $\log T_{eff} = 3.84$, $V_{rot} = 88$ km/s, $\log g = 3.39$). The ordinate gives moment of inertia (I) which is evaluated assuming the same radial displacement at the surface. Modes partially trapped in the envelope are characterized by low I. Only the $m = 0$ modes are shown. Empty symbols denote stable modes.

how the complexity of the oscillation spectra increases with the progress of the evolution. The project is not finished. Some preliminary results are included in a paper which is now in press (Handler *et al.*, 1996). In Fig. 4 I use one of the approximate models used in that project.

The star is located near the high-T_{eff} boundary of the δ Sct domain in the H-R diagram. In such a case only higher order modes are unstable beginning with p_4 or p_5, at $l = 0$. Stability calculations yield in this case useful constraints on stellar parameters, which come within the ranges allowed by the data.

The departure from a constant rotational splitting in the model spectrum is larger than in the case of DD Lac. However, a nearly constant frequency separation between $l = 0$ and $l = 1$ modes is seen. A pronounced peak in the power spectrum has been identified with this separation. This identification yields yet another constraint on the star parameters (Handler *et al.*, 1996). Some of the measured frequencies nearly coincide with those of the model. However, a closer look reveals that, in fact, we are still far from a unique mode identification.

5.3. 4 CANUM VENATICORUM

4 CVn is an evolved δ Sct star. The presence of at least 17 modes was revealed by Fourier analysis of the recent DSN campaign data (Breger, private communication). In contrast to the previous case this object is not a good choice for asteroseismic sounding, at least at the present stage of understanding stellar pulsations. I choose this star to illustrate what remains to be understood and is perhaps the most difficult problem of the stellar pulsation theory.

The parameters of the model used in Fig. 4 are within the range allowed by observational data. However, what matters here is only that this is a post-main sequence star. In deep interiors of such objects the Brunt-Väisälä frequency attains very large values. The consequence is the large density of oscillation spectra for nonradial modes seen in Fig. 4. The actual density is $2l + 1$ higher because only modes with $m = 0$ are shown. The total number of unstable modes with $l \leq 2$ is about 450.

The oscillatory behavior of the inertia reflects the effect of partial mode trapping in the acoustic cavity. We do not know whether this effect has anything to do with mode selection. It is possible that the amplitudes are random quantities and that there is no rule of mode selection. Without such a rule we will never be able to make use of nonradial mode frequencies for seismic probing.

6. Mode Identification Using Additional Observables

The way to a seismic model of an oscillating star is much easier if we know the spherical harmonics for any of the excited modes. This information may be, in principle, inferred from observational data.

Stamford and Watson (1977) first noted that ratios of intensity amplitudes measured at various wavelengths and ratios involving the amplitude of radial velocity are sensitive to the l-value. Balona & Stobie (1979) showed that the color to intensity amplitude ratio *versus* corresponding phase difference diagrams yields a particularly revealing diagnostic. This method of l determination has been developed in a number of subsequent papers and applied to various objects. Recently, Cugier *et al.* (1994) published an extensive survey of diagnostic diagrams for β Cep stars based on accurate calculations of stellar oscillations. In several types of the diagrams the loci of $l = 0$, 1, and 2 modes are well separated. Hence, it is possible to discriminate between the three identifications without knowledge of stellar parameters. For $l = 0$, having some information about the parameters, one may use the diagrams to discriminate between the $n = 1$ and $n = 2$ modes.

The amplitude ratios and the phase differences are independent of the mode azimuthal order, m, and of the inclination angle of the rotation axis, i. Both quantities, which are very useful in constructing seismic models, may be inferred from line profile changes. A systematic method of determination of m, l, and i from line profile data was developed by Balona (1986) and by Aerts (1996), who applied it with a moderate success to the β Cep star 12 Lac. Unfortunately, in the case of δ Sct stars with rich oscillation spectra we do not have a single mode with reliably determined l and m values. The situation is somewhat better in the case of β Cep but, in fact, none of the suggested identifications could be regarded as fully reliable.

7. Seismic Stellar Models

Even in best cases the number of modes detected in individual stars is far too small to determine the radial structure directly from measured frequencies, like it is done in helioseismology. The best we may hope for is using observed frequencies to determine global parameters characterizing the star and certain parameters of the theory. Basic equations for the seismic model may be written in the following form

$$\nu_{j,obs} = \nu_{j,cal}(l_j, m_j, n_j, \vec{P}_S, \vec{P}_T),$$

where j labels measured frequencies, $\vec{P}_S$ gives the set of parameters characterizing the model, and $\vec{P}_T$ the set of free parameters of the theory.

The parameters given by $\vec{P}_S$ are those characterizing the evolutionary sequence like initial mass, M, chemical composition, X_0 and Z_0, and the angular momentum and a single parameter identifying the model, for which age is always a good choice. Alternatively, one may use the initial equatorial velocity $V_{rot,0}$ instead of the angular momentum and, in application to objects in the expansion phase of the Main Sequence evolution, $\log T_{eff}$ instead of the age. In any case there are five parameters in $\vec{P}_S$. The quantities one may include in $\vec{P}_T$ are mixing length parameter α, overshooting distance d, as well as parameters describing angular momentum evolution and mass loss. In applications to Upper Main Sequence objects α is rather unimportant. Adopted defaults in our projects are $\alpha = 1$, $d = 0$, uniform rotation, and a global angular momentum conservation.

The model must be consistent with photometric and spectroscopic data on the star. However, the accuracy of such data is usually much poorer than that of the frequency measurement and they allow several different identifications for the modes. In our project a preliminary fit of the frequencies is obtained by adjusting values of M, $V_{rot,0}$, $\log T_{eff}$, using the tabulation of model frequencies for all modes with $l \leq 2$ in the specified frequency range. Identification of the excited modes is based on the fit quality, measured by

$$\chi^2 = \frac{1}{J}\sum_{j=1}^{J}\left(\frac{\nu_{obs} - \nu_{cal}}{\sigma_{obs}}\right)_j^2,$$

where J is the number of modes in the data set. In the example shown in Fig. 4 we see that a unique assignment of modes to all measured frequencies is not possible. At such point one may either continue the search in the parameter space or consider an identification with higher degree modes. A formal determination of all parameters by a least square method is the goal, which we hope to achieve at some point. At present, we still do not know whether this procedure will lead to a unique mode identification.

For seismic models of the Sun (Dziembowski *et al.* 1995, Basu & Thompson 1996) we have $\chi \sim 1$. We are very far from such good fits in asteroseismology. Therefore, all published models may only be regarded as approximate seismic models. Perhaps the most advanced are the models of the white dwarfs PG1159-035 (Kawaler & Bradley, 1994) and GD 358 (Bradley & Winget, 1994). Models of the δ Scuti stars: GX Peg (Goupil *et al.*, 1993) and FG Vir (Guzik & Bradley, 1995) as well as that of the β Cep star EN Lac (Dziembowski & Jerzykiewicz, 1996) are crude and based on inadequate treatment of the effects of rotation on oscillation frequencies. Typical equatorial velocities of rotation in δ Sct and β Cep stars are in the range 50-200 km/s. At such velocities the effect of rotation on oscillation is not reduced to a simple Zeeman-like frequency splitting. Accurate treatment of rotation is essential for the construction of seismic models of these stars.

8. What We May Expect from Space Asteroseismology

The amplitude resolution in observations from space is between 1 and 10 μmag which means between two and three orders of magnitude better than in ground-based photometry. If everything goes as planned the EVRIS instrument installed on board of MARS96 will soon start providing us with data. It is likely that the more ambitious, devoted mainly to asteroseismology, COROT mission will follow (Baglin in these proceedings). Observations from space will open a new epoch in this field. We will certainly have much more frequency data. It is, however, less clear whether we will be able to make good use of them.

Almost certainly solar-like oscillations in a number of stars will be detected and two parameters known as large, $\Delta\nu_l = \nu_{l,n+1} - \nu_{l,n}$, and small,

$D_0 = 0.5(\nu_{0,n} - 0.5(\nu_{1,n+1} + \nu_{1,n}))$, frequency separations will be determined. These measurements will allow us to locate the objects in the Christensen-Dalsgaard (1988) seismic H-R diagram and constrain their global parameters (e.g. Gough, 1995). It is less certain whether oscillatory features in the $\Delta\nu_l(\nu)$ dependence will be detectable. These features as Gough (1990) showed probe the lower boundary of the convective zone.

Another safe prediction is that many new modes will be detected in δ Scuti and in other types of known pulsating stars. The obvious candidates are nonradial modes with $l > 2$. Goupil et al. (1996) argued that with data from space a detection of modes up to $l = 7$ is likely. Further, they considered a conjectured set of unstable modes with $l = 1$ to 7 in δ Scuti star models and showed that if we had measurements of the rotational splitting for a fraction of these modes we could quite precisely determine the rotation rate behavior in the interior, including the chemically inhomogeneous zone surrounding the convective core. The latter possibility follows from the presence of a number of g-modes in the conjectured set. The most optimistic aspect of that exercise is that we will be able to identify the modes. We should have no illusion that this will be an easy task.

Acknowledgements

I thank the Local Organizing Committee for the financial help. This review contains some unpublished results of work done in collaboration with Gerald Handler, Phil Goode, Mike Jerzykiewicz, Alosha Pamyatnykh, and Holger Pikall. My special thanks to Alosha Pamyatnykh who made figures 1, 2 and 5. He and Pawel Moskalik read the preliminary version of this paper and suggested improvements. This work was supported by KBN-2P304-013-07 grant.

References

Aerts, C. (1996) Mode identification with the moment method, A & A, **314**, pp. 115–122.
Audard, N., Provost, J., and Christensen-Dalsgaard, J. (1995) Seismological effects of convective-core overshooting in stars of intermediate mass, A&A, **297**, pp. 427–440.
Balona L.A. (1986) Mode identification from line profile variations, MNRAS, **219**, pp. 111–129.
Balona, L.A., Krisciunas, K., and Cousins, A.W.J. (1994) γ Doradus: evidence for a new class of pulsating stars, MNRAS, **270**, pp. 905–913.
Balona L.A. and Stobie, R.S. (1979) The effect of radial and non-radial stellar oscillations on the light, colour and velocity variations, MNRAS, **189**, pp. 649–658.
Basu, S. and Thompson, M.J. (1996) On constructing seismic models of the Sun, A & A, **305**, pp. 631–624.
Belmonte, J.A. et al. (1993) STEPHI: a new approach to δ Scuti asteroseismology, in

W.W. Weiss and A. Baglin eds. *Inside the Stars, IAU Coll. 137*, Astron.Soc.Pac. Conf.Ser., **40**, pp. 739–741.

Bradley, P.A. and Winget, D.E. (1994) Asteroseismological determination of the structure of the DBV white dwarf GD 358, *ApJ*, **430**, pp. 850–857.

Breger, M. *et al.* (1994) EP Cancri: a nonradially pulsating δ Scuti star, *A&A*, **281**, pp. 90–94.

Buchler, J.R., Goupil, M.-J., and Serre, T. (1995) Dynamic pulsation amplitudes for rotationally split nonradial mode: the $l = 1$ case, *A&A*, **296**, pp. 405–417.

Christensen-Dalsgaard, J. (1988) ,A Hertzsprung-Russel diagram for Stellar Oscillations, in J. Christensen-Dalsgaard and S. Frandsen *Advances in Helio- and Asteroseismology*, (Reidel) pp. 295–298.

Christensen-Dalsgaard, J. and Frandsen, S. (1983) Stellar 5 min Oscillations, *Solar Phys.*, **83**, pp. 169–486.

Cugier, H., Dziembowski, W.A., and Pamyatnykh, A.A. (1994) Nonadiabatic observables in β Cephei models, *A&A*, **291**, pp. 143–154.

Delache, P. and Scherrer, P.H. (1983) Detection of solar gravity mode oscillations, *Nature*, **306**, pp. 651–653.

Dziembowski, W.A., Goode, P.R., Pamyatnykh, A.A., and Sienkiewicz, R. (1995) Updated Seismic Solar Model *Astrophys.J.*, **445**, pp. 509–510.

Dziembowski, W.A. and Jerzykiewicz, M. (1996) Asteroseismology of the β Cephei stars I. 16 (EN) Lacertae, 1996,*A&A*, **306**, pp. 436–442.

Dziembowski, W.A. and Pamyatnykh, A.A. (1991) A potential asteroseismological test for convective overshooting theories, *A&A*,**248**, pp. L11–14.

Gough, D.O. (1990) Comments on Helioseismic inference, in Y. Osaki and H. Shibahashi *Progress of Seismology of the Sun and Stars*, Lecture Note in Physics, **367**, pp. 283–318.

Gough, D.O. (1995) Prospects for asteroseismic inference, in R.K. Ulrich, E.J. Rhodes, Jr. and W. Däppen eds.,*GONG'94: Helio- and Astero-Seismology from the Earth and Space*, Astron.Soc.Pac. Conf.Ser., **76**, pp. 551–567.

Goupil, M.-J., Dziembowski, W.A., Goode, P.R., and Michel, E. (1996) Can we measure the rotation rate inside stars?, *A&A*, **305**, pp. 487–497.

Goupil, M.-J., Michel, E., Libretto, Y., and Baglin, A. (1993) Seismology of δ Scuti stars–GX Pegasi, *A&A*, **268**, pp. 546–560.

Guzik, J.A. and Bradley P.A. (1995) Models and Mode identifications of the δ Scuti Star FG Vir, *Baltic Astronomy*, **4**, pp. 442–452.

Handler, G., Breger., M., Sullivan, D.J., *et al.* (1996) Nonradial pulsation of the unevolved hot δ Scuti star CD-24 7599 discovered with the Whole Earth Telescope, *A&A*, **307**. pp. 529-538.

Handler,G., Pikall, H., O'Donoghue, D., *et al.* (1996) New Whole Earth Telescope observations of CD-24 7599: steps towards Delta Scuti star seismology, *MNRAS*, in press.

Kawaler, S.D. and Bradley, P.A. (1994) Asteroseismology of PG 1159 Stars, *ApJ*, **427**, pp. 415–428.

Kjeldsen, H., Bedding, T.R., Viskum, M., and Frandsen, S. (1995) Solarlike oscillations in η Boo, *AJ*, **109**, pp. 1313–1319.

Nather, R.E., Winget, D.E., Clemens, J.C., Hansen, C.J., and Hine, B.P., (1990) The Whole Earth Telescope: A New Astronomical Instrument, *ApJ*, **361**, pp. 309–317.

Stamford, P.A. and Watson, R.D. (1977) Observational characteristics of simple radial and non-radial β Cephei models, *MNRAS*, **180**, pp. 551–565.

Waelkens, C. (1991) Slowly pulsating B stars, *A&A*, **246**, pp. 453–468.

Winget, D.E., Nather, R.E., Clemens, J.C., *et al.* (1991) Asteroseismology of the DOV Star PG 1159–035 with the Whole Earth Telescope, *ApJ*, **378**, pp. 326–346.

Winget, D.E., Nather, R.E., Clemens, J.C., *et al.* (1994) Whole Earth Telescope Observations of the DBV White Dwarf GD 358, *ApJ*, **430**, pp. 839–844.

ASTEROSEISMOLOGY: GROUND BASED OBSERVATIONAL METHODS

S. FRANDSEN
Institute for Physics and Astronomy
Århus University
Bygning 520
DK 8000 Århus C, Denmark

Abstract. The improvement in observational techniques to study pulsations in solarlike stars is discussed. For higher mass stars the problem is to allocate specific eigenmodes to a set of observed frequencies. Additional observables can be used to sort out the present ambiguities.

1. Introduction: the photometry-astrometry connection

The study of the interior of stars by seismic techniques imply that observations are made of periodic changes in one or more parameters. The variations looked for are as low as a few ppm (parts per million). Observers therefore strive to increase the sensitivity of the measurements as much as possible.

To reach high precision in photometry with a CCD, it is necessary to have fairly large stellar images with soft edges (Badiali *et al.*, 1996). First, this averages out the combined effect of pointing variations and pixel calibration errors. It is impossible in practice to calibrate the response of the CCD as well as one would like, especially at the subpixel level. Thus, when the image moves around, an error signal is introduced. Secondly, sharp images will saturate the detector faster than large (defocused) images. The duty cycle decreases and precious photons are lost. Confusion with other stars puts a limit to the size of images.

To some it might be a surprise that the principles for doing precise relative photometry carry over to astrometry. The best solution for measuring precise relative positions within one CCD frame is to have large stellar images. Once more the finite size of the detector elements and related cali-

331

J. Provost and F.-X. Schmider (eds.), Sounding Solar and Stellar Interiors, 331-344.

bration problems mean that spreading the image over an area increases the accuracy.

A measuring scheme optimised for precise photometry turns out to be very close to the ideal scheme for doing precise astrometry.

If we turn to spectroscopy (one dimensional data), one can think of photometry as the measurements of equivalent widths. Astrometry corresponds to measurements of line positions, i.e. measurements of the radial velocity of the object.

Using the same arguments as in the two dimensional case, high resolution may not be the best choice for radial velocity measurements. High resolution in spectroscopy is equivalent to sharp images, and we know that this is not recommended for high precision photometry or astrometry. This is illustrated by a small diagram:

The High Precision Connections

Imaging (Two Dimensional Case)
Photometry $\Longleftrightarrow$ Astrometry

$\Updownarrow$

Spectroscopy (One Dimensional Case)
Equivalent Width $\Longleftrightarrow$ Radial Velocity

The principles for reaching high precision should be kept in mind in the discussions that follow.

2. Solar type oscillations: the detection

For solar type stars with oscillations driven by stochastic excitation the main problem is to detect modes. The amplitudes are tiny for stars similar to the Sun, and even though they increase in amplitude in more luminous, hotter stars, so do the periods, and longer observing runs are needed to resolve the modes.

This section is split in three parts: first the signal to be observed is described; second, the different techniques to search for the signal are discussed and finally, the current status is given of the attempts to measure oscillations.

2.1. WHAT DOES THE SIGNAL LOOK LIKE?

For stars we can only make full disk measurements. Apart from a scaling of frequencies and amplitudes, we expect to find oscillation spectra similar to the solar full disk spectrum.

The effect of the p modes on the spectral distribution of light is a complicated story. The p modes affect the spectral lines by the Doppler shift due to the sound velocity and by the temperature (and density) fluctuations associated with the waves. Looking at a spectral line, the intensity is affected in opposite direction on each side of the center by the velocity and in the same direction by the temperature perturbation.

The detailed behaviour of a single spectral line was studied by Andersen (1984) and Frandsen (1984). The feature to be noticed was the increase in intensity amplitude measured going towards the center of the spectral line. A much larger spectral range was measured by Ronan et $al.$ (1991), and the results show a similar behaviour of many lines in the solar spectrum.

Looking at the results from Ronan et $al.$ (1991) it is evident that the oscillation signal appears mainly in spectral lines. How one might combine everything into one observable (a seismic signal) is obviously not simple, especially when instrumental efficiency and atmospheric degradation of the signal have to be taken into account.

2.2. MEASURING TECHNIQUES

A number of different attempts have been made to try to detect oscillations in solar type stars. Doppler measurements were made first, but followed by different observations, when the required precision was obtained with Doppler measurements.

2.2.1. *Photometry*

The atmosphere is the major source of problems, when ppm level photometry is at issue. Scintillation dominates all other sources of noise for bright stars all the way down to magnitudes fainter than 10. The noise scales with telescope diameter as $D^{-2/3}$. To beat down the noise by using large telescopes, a campaign was organised to try to detect oscillations in a few F stars in M67 (Gilliland et $al.$, 1993).

There was no major problems in reaching the scintillation limit, so technically the campaign was a success. Unfortunately, the noise level achieved when all data were combined (6 ppm) was barely enough to make an unambiguous detection of modes in the F stars.

2.2.2. *Doppler Techniques*

The difficulty here lies in the extreme precision required that make the measurement vulnerable to instrumental instabilities of all types. The amplitude of the Doppler shift expected in a solarlike star typically corresponds to a shift of the spectrum of less than 10^{-4} pixel.

The procedures applied are described in detail in the context of the search for planets around stars by Butler *et al.* (1996). A similar precision has been achieved with the AFOE instrument (Brown *et al.*, 1994).

2.2.3. *Equivalent Width (EW) Techniques*

The equivalent width can be considered as an approximate indicator of the temperature. Variations are somewhat larger than flux or luminosity variations (Bedding *et al.*, 1996). Inspired by the line of argumentation pursued in the introduction, combined with efforts to avoid the influence of the atmosphere, new observing strategies have been devised by Kjeldsen *et al.* (1995) and Noyes *et al.* (1995). Kjeldsen *et al.* (1995) use a type of Hydrogen line index similar to the photometric β index. The idea is to work in a relatively narrow spectral range to reduce the influence of scintillation, and to measure something in a smooth way minimising the influence of instrumental instability and detector non-uniformities. Noyes *et al.* (1995) find a spectral mask by looking at the difference between spectra from two models with slightly different temperature. This looks like a very *efficient* method, because all the information in a wide spectral range is used. The method has to be modified to make it less sensitive to instrumental stability - the present noise levels are five times the photon noise limit. To detect oscillation in stars like α Cen and other solar 'twins', high efficiency is extremely important.

2.3. OBSERVATIONAL STATUS

A careful study of the observational status late 1993 exists (Kjeldsen & Bedding (KB), 1995) and there is no reason to repeat what is already told. This status report thus only adds to the KB discussion.

New *Doppler studies* have been carried out by several groups mentioned in Table 1, which is an update of Table 4 in KB. As in Table 4 the quality of the entries should be judged by reading the references. Strictly speaking, nobody has come up with a convincing claim for a detection of modes. But in several cases the performance achieved is very close to a level, where oscillations are expected to show up either as power excess or as underlying structure in the oscillation power spectrum. Several observers find such signatures.

Photometry has come equally close, and when better measurements can be done (from space with EVRIS), it would not come as a big surprise if some of the frequencies or separation of frequencies that have been quoted are confirmed.

The *equivalent width techniques* have led to two claims of detection of modes, none of them confirmed by other groups. The first star with p mode

TABLE 1. Measurements to detect solar oscillations, 1994-1996

Reference	v_{osc} (cm s^{-1})	EQW (ppm)	$\Delta\nu_0$ (μHz)	$\delta\nu_0$ (μHz)
• α Cen A				
Edmonds & Cram (1995)	< 70	–	110.6	–
Kjeldsen et al. (1996)	–	7	105.5 ± 0.1	6 ± 1
• η Boo				
Kjeldsen et al. (1995)	–	35	40.3	–
Noyes et al. (1995)	< 150			
Harvey et al. (1996)		test observations		
• Procyon				
Bedford et al. (1995)	< 100	–	70 ± 4	–
Brown et al. (1996)		no data published		
Harvey et al. (1996)		test observations		
Kjeldsen et al. (1996)	–	20–25	52.6	–
Noyes et al. (1996)		test observations		
• β Hyi				
Edmonds & Cram (1995)	< 150–200	–	–	–
• α Tri				
Harvey et al. (1996)		test observations		

data was η Boo (Kjeldsen *et al.*, 1995) and the second one α Cen (Kjeldsen *et al.*, 1996).

To summarise I will give the status for four different objects. Other stars have been observed, but the main efforts have been devoted to these targets.

1. **The Sun.** The Sun has served as a reference object for testing the equivalent width techniques. There has been discussion about whether any signal appeared in the Hydrogen lines at all. The measurements by Ronan *et al.* (1991) indicated that the Balmer lines vary only weakly. A signal with an amplitude of the order 6 ppm (as predicted) was seen in daytime observations of the blue sky, both in the case of the η Boo observations and the α Cen observations. The latest solar data obtained during the α Cen campaign gives the large separation of 135 μHz within a tenth of a μHz. Harvey *et al.* (1996) observed the Hα line in the Sun at Kitt Peak and were able to confirm the presence of a signal with expected amplitude. Thus, there seems to be no doubt, that the signal shows up in the Balmer lines at the expected amplitude.

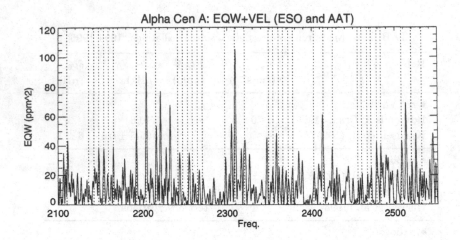

Figure 1. The power spectrum from the Hα line measurements combining an equivalent width and a velocity. The dashed line indicates the expected position of peaks with the large and small separation that have been selected (see text)

2. **Procyon.** Nobody has been able to observe single modes, but there seems to be excess power in the right frequency band around 1.5mHz (see references in KB). New observations confirm the excess (Kjeldsen *et al.*, 1996). One can still discuss the significance of the data (as KB do), and a new, *large* observing run is needed to find out, whether the power seen is instrumental or not. A campaign is planned early 1997 under the auspices of the SONG (see later) initiative.

3. **η Boo.** The detection of oscillations in this star by Kjeldsen *et al.* (1995) still awaits confirmation. An attempt by Noyes *et al.* (1995) to reobserve the modes did not show any of the formerly detected modes. The discrepancy is still being disputed.

4. **α Cen A.** This binary component is a key object for stellar modelling and seismology, due to its similarity to the Sun. A large operation involving the ESO 3.6m and the AAT 3.9m telescopes for 6 consecutive nights was launched to try to detect modes with EW amplitudes expected at a level of 7ppm (Kjeldsen *et al.*, 1996). There is evidently a tendency for peaks to show up at equidistant intervals in the power spectrum (Fig. 1). The peaks are mostly in the range 2-4 times the noise, and none of them are unambiguous detections in their own right. This regularity of peaks around 2.3mHz permits a determination of the large separation $\Delta\nu = 105.48 \pm 0.1\mu$Hz. The small separation has been determined as well, $\delta\nu = 6 \pm 1\mu$Hz. The combination of this value and the two site alias peaks leads to some problems in selecting the correct large separation.

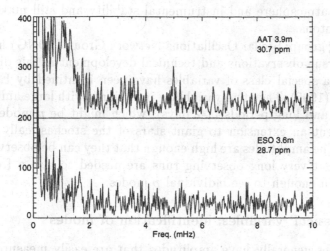

Figure 2. The amplitude spectrum of a continuum index formed by comparing two narrow (0.5Å) bands 1Å apart. The spectra hae been generated from 10 hours of observations at each site. The photon noise is 10-12 ppm. The power is seen to be larger at the AAT site than at ESO and to have a stronger 1/f component. Vertical axis is in ppm. White noise levels are given for each site. The difference in noise levels corresponds to the difference in airmass at the two sites.

The photon noise limit is not reached for reasons that begin to be understood. One of the limitations of the equivalent width measurements is the differential effect of scintillation. Even for a small piece of the spectrum the continuum is seen to wobble like a rubber band. A comparison of an index of the variation in the continuum shows a stronger variation at the Australian site, which is situated at lower altitudes than La Silla and more sensitive to atmospheric conditions. This can be seen in Fig. 2.

This has as a consequence, that the ESO data turn out slightly better, even though a more stable spectrograph was used on a larger telescope at the AAT .

Recent years have moved the observational limit to the ppm levels if large telescopes are used. It is not a question whether we are able to detect solar type oscillations unambiguously. The question is how soon? *Large telescopes are not necessary*, if the photons are used in an efficient way. The latest α Cen results prove that we can reach noise levels that allow us to measure oscillations in stars like Procyon and η Boo.

It is also clear that, when using the Balmer lines to look for oscillations at equivalent width amplitudes below 10ppm, the atmosphere has a disturbing influence. There are still many experiments to be performed before an optimal technique has been developed, which circumvent the detrimental effects of the atmosphere and instrumental stability and still makes good use of the photons.

A working group: Stellar Oscillations Network Group (SONG) has been formed to pursue observations and technical development in this direction.

Recently, a special class of variables have been identified by Edmonds and Gilliland (1996). They find variability in K giants with large amplitudes (5-15 mmag) and long periods (1-2 days), which might be p modes. They would represent an extension to giant stars of the stochastically excited oscillations. The amplitudes are high enough that they can be observed from the ground, but very long observing runs are needed to give a frequency resolution high enough to see individual p modes.

3. The 'classical' variables: identification of modes

These variables generally have amplitudes that are easily measured, even with one channel photometers on a few nights. When more efficient and extensive observing programmes are invoked, many more low amplitude modes appear. A multitude of modes makes asteroseismology much more interesting, and the efforts involved in using more advanced measurements are fully compensated by the wealth of information one recovers.

I will mainly discuss δ Scuti stars, but much of the discussion and the results apply to β Cephei, roAP stars and other short period ($P < 4h$) variables.

3.1. WHAT DO WE NEED TO CLASSIFY?

The δ Scuti stars often rotate with a considerable velocity, typically $V \sin i = 100$ kms^{-1}. Evolved stars have complex oscillation frequency spectra with avoided crossings and modes that are a mix of g and p modes. Even when a large set of modes (> 20) have been observed (Breger et al., 1995), these have to be matched to a set of theoretical modes, which is 3–4 times larger. If only the observed frequencies are known, it is not an easy task.

If we could determine, not only the frequency, but also the character of the mode in terms of the angular quantum numbers ℓ and m, the match of observed and theoretical frequencies would be much easier.

A number of techniques have been suggested to determine ℓ and possibly m, which I will discuss one by one. I will also present a few examples of results obtained recently.

3.2. PHOTOMETRY

Observations of the variables must resolve the oscillation spectra. Modes are typically found with separations down to one or two μHz or less. New observations confirm what has been expected for a long time: more modes appear when the detection level decreases. Observations from one site consequently must have an extraordinary S/N to make it possible to handle the alias problems created by the daily side lobes. The photometric data needed for asteroseismic studies can in practice only be obtained by extensive multisite campaigns. Excellent results have been obtained by the DSN, STEPHI and the WET groups, the best results when the two networks (DSN+WET) work together (Breger *et al.*, 1995).

As more modes show up every time the detection level is lowered, efficient measuring techniques are important. In comparison to photoelectric photometry, relative CCD photometry gives an important increase in precision per minute of observation time. It comes at the cost of more complicated operations and more data reduction. It depends on the availability of reference stars within the field of view, but it is less of a problem now than a few years ago with the advent of large format detectors.

There is a big gain in efficiency, when many stars can be observed within a single CCD frame. A few open clusters that permit simultaneous measurements are known, but with a small set of variables (2-6). Praesepe (Belmonte *et al.*, 1996) has a rich population of δ Scuti stars (> 10 multiperiodic pulsators), but at most telescopes at most two stars can be observed simultaneously. Needless to say, the membership of an open cluster provide you with some very useful additional information, when modelling the evolution of the cluster.

3.3. MODE CLASSIFICATION

The techniques for identifying the quantum numbers characterising a mode depends on the effect on the observable of the spherical function belonging to an eigenmode. The difference in the center to limb behaviour of the observable combined with the usual limb darkening gives a different amplitude and phase for each spherical wave function. Some degeneracy is present when the star is a slow rotator. For moderate to fast rotators the line profiles can be analysed using Doppler imaging techniques. This works well for moderate to high degree modes ($\ell \gg 1, |m| \approx \ell$).

Observations of phase differences between colours and magnitudes were among the first methods applied. Recent extensions of the method to include amplitude ratios is the reason for having the discussion of this technique at the end of the section.

3.3.1. *Moment Methods*

The influence of oscillations on the spectral line profiles was studied early by Balona (1986a), and a mode identification technique developed in a series of papers (1986b;1987). Later improvements are due to Aerts (1992;1996). Recent examples of the use of the moment method are Mathias & Aerts (1996) and Mantegazza & Poretti (1996).

The advantages of the technique is that the moments are easy to calculate directly from the observed profiles. The moments describe a decreasing series in spatial scale size on the stellar surface. The theoretical interpretation is not straight forward. It involves the computation of the change of spectral line profiles caused by the oscillations, and the subsequent generation of the moments for combinations of the independent parameters in the problem. Not only the quantum numbers enter, but also the rotation velocity, the inclination of the rotation axis and a few more.

Often the result is a small selection of possible solutions that all fit equally well.

The method demands that the star does not rotate fast. In fast rotating stars most lines overlap and the simple picture of one line containing the velocity information disappears. If only applied to one or a few lines, it is *not* a method with a very high S/N as *very few photons are involved.*

3.3.2. *Doppler Imaging*

The broadening effect of the rotation and the oscillations on the spectrum can be derived using deconvolution techniques. By taking out the mean effect caused by rotation one is left with the combined effect of all modes. Given a time series of spectra one then generate a two dimensional Fourier diagram with time information in one direction and spatial information in the other. A beautiful example is presented for τ Peg at this meeting by Kennelly *et al.* (1996), where the final 2D diagram contains so many frequencies that it reminds one about a helioseismic $k - \omega$ diagram..

The power of the technique lies in the ability to use the whole spectrum. Blends of lines can be included without difficulty. A decent initial estimate of the spectrum is needed and can be obtained from a synthetic spectrum of a model atmosphere with effective temperature close to the temperature of the star being analysed.

Doppler imaging works well for high ℓ values, but it looses its sensitivity at low ℓ values. It is the only method that can detect high ℓ oscillations. So far, there is a gap between the low and high ℓ values, where none of the methods work well. This makes it difficult to relate frequencies for low degree photometric frequencies to high degree Doppler imaged oscillation frequencies.

An assumption being made in the Doppler imaging approach is that all lines react in the same way. Some difference can be accepted because the average behaviour is normally what one is looking for. But, in an example mentioned below, various metal lines do not react in the same way. For

δ Scuti stars the assumption is quite reliable, as the modes are well below the acoustical cutoff frequency.

3.3.3. Amplitude Ratio and Phase Difference Techniques

The use of phase shifts in multicolour photometry to identify modes has been suggested long ago. Even better diagnostics can be found by including more observables. The radial velocity decreases more toward the limb than the flux measured in various passbands. The same is true for the equivalent width of the Balmer lines. Each observable defines a spatial filter. Combining these filters with the spherical functions give different amplitudes for different ℓ (and m) values. When taking the ratios of the amplitudes for different observables the intrinsic amplitude of a mode cancels out as well as the phase. This ratio and phase difference then characterise the mode and can be used to label the mode. Some ratios are more sensitive than others and give a better discrimination between different ℓ values. Amplitude ratio and phase results for δ Scuti stars observed in the Strömgren system are given by Garrido et al. (1990). Calculations for β Cephei stars have been carried out by Cugier et al. (1994). They show the separation of modes with different ℓ's in a diagram of K/A_y vs. A_{u-y}/A_y, where K is the radial velocity amplitude and A_y and A_{u-y} the amplitudes of the magnitude and colour in the Strömgren system.

For a star with a spectrum of detected frequencies, one can derive the amplitude ratios for *given frequencies*. Then a good window function is not a necessity for obtaining precise ratios.

Some observables can be measured with great precision. Velocities can be obtained from many lines in several orders of the spectrum, and only a moderate resolution is adequate ($\Delta\lambda = 0.5\text{Å}$), as the rotational broadening is in general quite large. The strength of one or more lines (Balmer or metal) is another observable that can be measured in an efficient and accurate way.

Observations of this type are shown by Viskum et al. (1996) for the δ Scuti star FG Vir. The power spectrum of a time series in the Doppler velocity is presented as Fig. 3. All frequencies detected by Breger et al. (1995) are present. The data is only partially reduced and no information on mode identification is available yet.

Another exciting achievement, by Baldry et al. (1996) using Doppler studies, is the measurement of the velocity spectrum of the roAp star α Cir, where most of the peaks seen in photometric measurements (Kurtz et al., 1994) are present. The lowest amplitude detected so far is around 30 ms^{-1}. A very interesting observation, made during the reduction, was that the metal lines fell in two groups with opposite phase. The hypothesis is that the eigenfunction has a node in the atmosphere as the star is not heavily spotted. This has some interesting consequences. In the Sun the p modes are

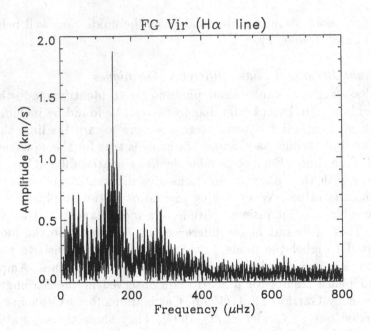

Figure 3. An amplitude spectrum of Doppler measurements of FG Vir. The white noise level above 400μHz is 64 ms^{-1}. Data from (Viskum *et al.*, 1996).

evanescent in the atmosphere and no nodes can be found. In the roAp star the reflecting boundary has to be situated very far out, if the interpretation of the data is correct. It is consistent with the fact that the modes are high overtones with a frequency very close to the acoustical cutoff frequency.

While discussing spectral techniques and the sensitivity of observables, it might be worthwhile to discuss the best technique to reach low noise levels. CCD photometry works fine for amplitudes down to 50-100ppm. A hard limit is reached, when scintillation noise is the dominating source of noise. In this situation it turns out, that one can do better in the same amount of time by using an observable, which is insensitive to scintillation. For bright stars, using velocity measurements, one might easily find that the S/N per time unit is smaller than by applying photometry. Even fainter modes can be detected than with CCD photometry.

A final word concerns the range of modes detectable by full disk measurements in slow as well as fast rotating stars. The velocity signal has a spatial filter function that still lets an $\ell = 3$ mode pass without too much attenuation. In equivalent width measurements indications of the $\ell = 3$ mode have been seen in the Sun (Bedding *et al.*, 1996). Observables probably exist that extend the range to even higher ℓ's. Doppler imaging, of course, goes even higher, but is limited to fast rotators.

4. The Future

Asteroseismology, based on observations of a rich set of modes, is now just around the corner. Data for solar type stars have been presented, but need confirmation.

Rich set of modes exist for a few δ Scuti stars and I expect soon to see additional data show up. The modelling of the stars and the inversion of these data should soon be possible giving a lot of new information about the interior of stars with masses in the range 1.5 to 2 $M_\odot$.

Considering the interest in setting up networks of instruments (DNS, STEPHI, SONG) to observe solar type stars as well as higher amplitude variables, an analysis of the best choice of instrument to use is of extreme importance. A lot of money and manpower is involved and a bit of thought on how to get most out of the investments is recommended.

The attempts of the detection of oscillations in solar type stars have led to lots of ideas about the most precise and efficient techniques. Observers of the classical variables should keep in mind that the same ideas apply when deciding about instrumentation for observing δ Scuti stars etc. White dwarfs are an exception due to the lack of spectral lines.

For bright stars, my personal opinion is that a simple spectrograph of intermediate dispersion, covering a wide spectral range in several orders, is the best bargain for a network instrument. The optimum solution might be to have both a multichannel photometer and a spectrograph, but if one has to choose one or the other, I believe a spectrograph is the prime instrument of the two. It is also the most costly, to build and to operate.

References

Aerts, C., De Pauw, M. and Waelkens, C. (1992) A&A 266, 294
Aerts, C. (1996) A&A 314, 115
Andersen, B.N. (1984) in Proc. of the 25th Liège Int. Astr. Coll., 220
Badiali, M., Catala, C., Favata, F., Fridlund, M., Frandsen, S., Gough, D.O., Hoyng, P., Pace, O., Roca-Cortés, T., Roxburgh, I.W., Sterken, C. and Volonté, S. (1996) STARS, Seismic Telescope for Astrophysical Research from Space, Report on the phase A Study, ESA report SCI (96) 4
Baldry, I.K., Viskum, M., Kjeldsen, H., Frandsen, S. and Bedding, T.R. (1996) This meeting, the poster proceedings
Balona, L.A. (1986) MNRAS 219, 111
Balona, L.A. (1986) MNRAS 220, 647
Balona, L.A. (1987) MNRAS 224, 41
Bedding, T.R., Kjeldsen, H., Reetz, J. and Barbuy, R. (1996) MNRAS 280, 1155
Bedford, D.K., Chaplin, W.J., Coates, D.W., Davies, A.R., Innis, J.L., Isaak, G.R. and Speake, C.C (1995) MNRAS 273, 367
Belmonte, J.A., Hernández, M.M., Michel, E., Álvarez, M., Jiang, S.Y. and the STEPHI network (1996) This meeting, the poster proceedings
Breger, M., Handler, G., Nather, R.E., Winget, D.E., Kleinman, S.J., Sullivan, D.J., Li, Z.-P., Solheim, J.E., Jiang, S.-Y., Liu, Z.-L., Wood, M.A., Watson, T.K., Dziem-

bowski,W.A., Serkowitch, E., Mendelsohn, H., Clemens, J.C., Krzesinski, J. and Pajdosz, G. (1995) A&A 297, 473

Brown, T.M., Noyes, R.W., Nisenson, P., Korzennik, S.G. and Horner, S. (1994) PASP 106, 1285

Brown, T.M., Kennelly, E.J., Noyes, R.W., Korzennik, S.G., Nisenson, P., Horner, S.D. and Catala, C. (1996) BAAS 188, 5902B

Butler, R.P., Marcy, G.W., Williams, E., McCarthy, C. and Dosanjh, P. (1996) PASP 108, 500

Cugier, H., Dziembowski, W.A. and Pamyatnykh, A.A. (1994) A&A 291, 143

Edmonds, P.D. and Cram, L.E. (1995) MNRAS 276, 1295

Edmonds, P.D. and Gilliland, R.L. (1996) ApJ 464, L157

Frandsen, S. (1984) in Proc. of the 25th Liège Int. Astr. Coll., 303

Garrido, R., Garcia-Lobo, E. and Rodríguez, E. (1990) A&A 234, 262

Gilliland, R.L., Brown, T.M., Kjeldsen, H., McCarthy, J.K., Peri, M.L., Belmonte, J.A., Vidal, I., Cram, L.E., Frandsen, S., Parthasathy, M., Petro, L., Schneider, H., Stetson, P.B. and Weiss, W.W. (1993) AJ 106, 2441

Harvey, J., Pilachowski, C., Barden, S., Giampapa, M., Keller, C. and Hill, F. (1996) BAAS 188, #59.03

Kennelly, E.J., Brown, T.M., Sigut, A., Noyes, R.W., Korzennik, S.G., Nisenson, P., Horner, S.D., Yang, S. and Walker, A. (1996) This meeting, the poster proceedings

Kjeldsen, H., Bedding, T., Viskum, M. and Frandsen, S. (1994) AJ 109, 1313

Kjeldsen, H. and Bedding, T.R. (1995) A&A 293, 87

Kjeldsen, H., Frandsen, S., Bedding, T., Dall, T. and Christensen-Dalsgaard, J. (1996) in SONGNews 1, http://www.noao.edu/song/

Kurtz, D.W., Sullivan, D.J., Martinez, P. and Tripe, P. (1994) MNRAS 270, 674

Mantegazza, L. and Poretti, E. (1996) A&A 312, 855

Mathias, P. and Aerts, C. (1996) A&A 312, 905

Noyes, R.W., Korzennik, S.G., Nisenson, P., Brown, T.M., Kennelly, T. and Horner, S. (1995) BAAS 187, #102.11

Noyes, R.W., Korzennik, S.G., Krockenberger, M., Nisenson, P., Brown, T.M., Kennelly, T. and Horner, S. (1996) BAAS 188, #59.06

Ronan, R.S., Harvey, J.W. and Duval, T.L. (1991) ApJ 369, 549

Viskum, M., Baldry, I.K., Kjeldsen, H., Frandsen, S. and Bedding, T.R. (1996) This meeting, the poster proceedings

ASTEROSEISMOLOGY FROM SPACE.

COROT and other projects.

A. BAGLIN
DESPA, Observatoire de Paris
92195 MEUDON CEDEX, FRANCE

AND

M. AUVERGNE
DASGAL, Observatoire de Paris
92195 MEUDON CEDEX, FRANCE

Abstract. The scientific objectives and the observational strategy of asteroseismology from space are presented. The projects proposed in different contexts are briefly reviewed, with a particular emphasis on the COROT experiment, now accepted for a launch in 2001 in the framework of the french "Petites Missions" program.

1. Introduction: scientific objectives and strategy

Since the discovery of solar oscillations and its brilliant success, the generalization of seismology to many stars has become a challenge, motivated by its powerful capacity of diagnostic of the physical state of the stellar matter.

Seismology of stars aims at determining the internal structure from the properties of the eigenmodes (frequencies, amplitudes, phases, lifetimes) , as described by Dziembowski (this colloquium). To do so, many modes are needed, for each individual object.

If one wants to understand stellar evolution and apply the theory, for instance, to the chemical evolution of the Universe, one cannot be satisfied to rely only on a single object. A new physical process tested on the Sun, as for example microscopic diffusion, has to be validated on a variety of objects of different ages, chemical compositions, state of rotation ....

But, the seismic information can be interpreted unambiguously only if the global parameters (mass, chemical composition, surface temperature) are known accurately.

J. Provost and F.-X. Schmider (eds.), Sounding Solar and Stellar Interiors, 345-356.

From the observational point of view, the aim is then to define instruments and observing conditions allowing us to detect and measure the properties of as many modes as possible on a variety of "well known"stars.

To do so, one has to look for extremely small periodic changes (a few ppm) of an observable quantity, as for instance irradiance or radial velocity.

Indeed, the success of helioseismology, both from the ground and from space missions, encourages to pursue and apply the tool to other stars. But, there are major differences between helio and asteroseismology, theoretical as well as observational ones.

The Sun is a peculiar object in many respects. Its structure is sufficiently simple so the asymptotic approximation for high frequencies is valid. This means in particular that the frequency of a mode is related in a straightforward way to its quantum number. In addition, its rotation rate is slow, and the multiplets are equidistant. These two properties allow us to identify easily the mode corresponding to a given frequency. This situation is restricted to the old low mass stars. For more massive and younger objects, even close to the main sequence, the existence of a convective core and of a large angular momentum, produce a much more complex spectrum of eigenmodes. For evolved stars, the shell structure leads to a very dense and messy spectrum.

The Sun is also the only star for which the global parameters (mass, radius, age) are directly and accurately measured. For other ones, age determinations rely on model fitting and so depend on the stellar evolution theory and physical assumptions. Radius is almost never measured and, in the best cases, masses are deduced from the orbital motion in double stars. Effective temperature and chemical composition come from detailed analysis if the target is bright enough.

Determinations of these parameters, needed to deduce a first approximation model, are then less precise than for the Sun. But, as seismology is a very sensitive tool to probe the internal structure, the frequency spectrum varies rapidly with the global properties of the star, as pointed already first by Gough (1987). Discrepancies between an observed spectrum and a predicted one can come either from an incorrect physical description, or from an incorrect estimate of the global properties.

This is why the first approaches in asteroseismology should choose their targets among the best known and the simplest stars.

From the observational point of view, the difficulties are also important. First, photons are lacking. The unavoidable photon noise, which decreases when the star brightens, limits the amplitude of the variations which can be detected, and favors the brightest targets.

Stars are seen without spatial resolution, which means that only low order modes can be seen. Fortunately, they are among the most interesting ones for stellar evolution, as they penetrate far inside the stellar cores and give information on the nuclear burning regions. But, the total number of observable modes remains small and limits the accuracy of the determination of the structure at least in the envelopes.

Finally, due to the Earth motion around the Sun, most stars are visible only for a few months from the ground, but sometimes for longer periods from space depending on the orbit.

2. The need to go to space

This subject has been largely documented already (see Harvey 1985). Frandsen (1996, this colloquium) has described what has been done from the ground already and how the situation could be improved.

Let us recall that, to perform an asteroseismology program and to reach the real diagnostic power of the seismologic tool, we need to measure the variations with time of a quantity affected by the oscillations, with a high accuracy on both frequencies and amplitudes. This translates into the following conditions:

1. Amplitudes $\geq$ 1ppm or 10 cm/s, relying on the solar values and theoretical estimates (i.e. Houdek 1995)
2. Signal to noise ratio in the amplitude spectrum ≥ 4
3. Frequency range [0.05, 20] mHz
4. Frequency resolution [0.1, 0.5] μHz
5. Visual magnitude down to $m_v \geq 10$

The performances in terms of detection threshold of the different techniques available on the ground, as discussed by Frandsen (1996), are still far from fulfilling all these requirements, up to now.

From the ground, the photometric accuracy is limited at high frequencies by the scintillation, which can be reduced only by increasing rapidly the size of the telescope and observing from high altitude. But, these improvements are limited, as the scintillation noise decreases only as $D^{-2/3}$, where D is the diameter of the telescope. The excellent results of Gilliland et al. (1993) indicate that for 2 to 4 meters telescopes in high altitude sites, the scintillation level is of the order of 6 ppm for a signal to noise ratio equal to 1, which means that to satisfy both conditions 1 and 2 extremely large telescopes are needed around the world.

In spectroscopy (either Doppler or Equivalent Width techniques), almost not affected by the transparency variations of the Earth atmosphere, the present results are promising. But, the detection of oscillations remains limited to very bright objects. Another difficulty arises when spectral lines are broadened by rotation. Though this broadening gives the possibility to observe high degree sectorial modes, the capacity to extend these methods to non-slowly rotating stars is not yet documented.

Let us show now how it is possible to satisfy the whole set of specifications with a photometer in space.

The tranquillity of the space environment, and in particular the absence of atmosphere, allows us to reach the unavoidable photon noise limit at high frequencies (see Buey et al. 1997), even on bright sources, though it implies strong constraints on all the instrumental sources of noise.

In this condition, the signal to noise ratio (SNR) of a pure sine wave of frequency f and relative amplitude a in a white noise of variance σ is given by Scargle (1982):

$$SNR = \frac{N_0^2 a^2 t}{4\sigma^2} \tag{1}$$

N_0 is the mean counting rate, $t = \inf(T, \tau)$ where T is the total observation time and τ the oscillation damping time.

The SNR value is determined by the instrumental stability through the σ value, all the other terms being imposed by the properties of the stars. As long as the noise is white, it is independent of the frequency.

Then, the detection threshold of the coherent variations is only determined by the number of photons collected, i.e. the diameter D of the telescope, for a given magnitude. For a standard photometer, using high quantum efficiency detectors, a detection threshold of 0.6 ppm is reached in 5 days when roughly

$$log(D/25) \geq 0.2(m_v - 6) \tag{2}$$

which shows that a reasonable entrance pupil allows us to observe a large set of stars.

Another very important condition to obtain information that can be used for diagnostic, is the accuracy on the frequency measurement. Let us rely on the estimate by Libbrecht (1992) in the case of a damped oscillation, showing that a SNR of the order of 15 (in the power spectrum) and a duration of observation of the order of more than 100 days are needed to reach the $1\mu Hz$ accuracy on the frequencies (Fig 1.).

And this condition is supplemented by a secondary requirement. To reduce the amplitude of the sidelobes of the observing window and to prevent from misidentification of frequencies, the duty cycle should remain larger than 90%.

Obviously, this very long duration of observation and high duty cycle needed to obtain the frequency resolution cannot be achieved with ground based networks.

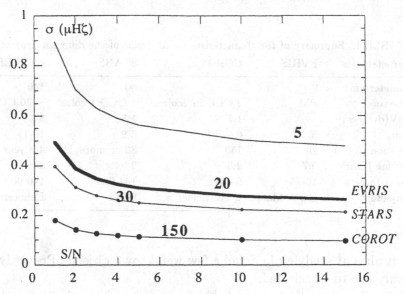

Figure 1. Accuracy on the frequency measurement from Libbrecht (1992), for the different projects.

3. The story of space projects.

The dream of asteroseismology in space has started in the early 80s with a proposal by the french group lead by F. Praderie and A. Mangeney (see Mangeney and Praderie 1984), followed by an Asteroseismology Explorer proposed by Hudson in the NASA context.

Different opportunities have been looked at since (see Noyes 1986), but none has gained a selection, except EVRIS, on board MARS96.

EVRIS, as a first attempt in the field, was designed as an exploratory mission, very small and cheap to be accommodated on an interplanetary spacecraft dedicated to the study of the MARS planet. It is described in Baglin et al.(1993), Buey et al. (1997) and Vuillemin et al. (1997). Its main scientific goal was to map the vicinity of the main sequence to determine the region where oscillations are sufficiently large. The main scientific return should concern the excitation and damping mechanisms, plus some insight in the internal structure, particularly for stars slightly more massive than the Sun (Michel et al. 1995)

Between the colloquium and the date of delivery of this manuscript, the MARS 96 mission has failed and EVRIS lies inside the ocean!

STARS, after PRISMA, has been submitted to ESA as a wide survey observing more than 100000 stars with a moderate frequency resolution, with particular attention to clusters (Badialdi et al. 1996).

COROT, presented in detail in the next section, has a different objec-

TABLE 1. Summary of the characteristics and status of the different projects.

characteristics	EVRIS	COROT	STARS	KEPLER
diameter(cm)	9	25	80	140
detector	PM	2 CCD in 2colors	4 CCD 1 color	~ 20 CCDs
FOV(degrees)		1.5	1.5	84
$< m_v >$	3.5	6	8-9	10-11
duration (days)	20	150	30 or more	4-8 year
lifetime (year)	0.7	2.5	2	4-8
nb. of targets	10-15	6+30	144 000	140 000
Trajectory	Earth-Mars	Quasi-polar	GTO	Heliocentric

tive. It aims at studying in detail a few well-known objects. Presently, it is the only one to be selected.

Let us mention that a new field of research is becoming interested in high precision relative photometry, i.e., the search for extrasolar planets through their transits across the stellar disk. The very similar instrumental constraints, as well as mission profiles (observing the same field for a long time) favors the idea of combining the two objectives. Several projects are being studied as for instance KEPLER submitted to the MIDEX NASA program, where the possibility of a seismological mode is under study.

4. COROT: stellar seismology, COnvection and ROTation

COROT was first proposed in 1994 in the framework of the french "Petites Missions" program (see Catala et al. 1995). Following EVRIS, it was dedicated to a detailed study of a few stars, specially chosen for their ability to provide precise tests on basic hydrodynamical processes, which are almost unknown in stellar interiors, focusing on convective regions (specially convective cores) and on the transfer and distribution of angular momentum.

After the EVRIS disaster, an updated version, including a program of detection of transit of extrasolar planets, has been selected by the CNES Science Program Committee on the beginning of December 96.

The present scientific program contains both the exploratory program, which should be realized by EVRIS, and the original COROT program.

4.1. SCIENTIFIC SPECIFICATIONS

They define the set of targets, the accuracy on the frequency measurement and the noise level. They are quite different for the two phases; constraints are softer for the exploratory program than for the main one.

4.1.1. *Exploratory program*

The targets are stars of spectral type B to G, of luminosity class V or IV. The objective is to detect their p-mode spectrum, and the splittings due to rotation. The corresponding scientific specifications are:

- Precision on measured frequencies better than $0.5\mu Hz$, which needs a duration of 20 days for one observation (Fig 1).
- Noise level less than 2 ppm over 5 days
- Duty cycle greater than 90 %

4.1.2. *Main program*

The best targets for this program are the F and early G main sequence stars, for several reasons.

The predicted amplitudes in late A and F stars close to the main sequence (Houdek et al. 1995) are larger than the solar ones, then easier to detect. These stars develop a convective core; its size, which is presently very badly known, provides information on the penetration of convective motions inside the stable layers; this process has important consequences on the evolution and in particular the ages. In addition, as these objects are still young, they rotate quite fast, and not yet rigidly; the rotational profile, derived from the splittings of the modes, will give tests of the processes of angular momentum transfer. Figure 2 shows how the rotation curve can be reconstructed from the measurement of the rotational splitting of the low order and low degree modes of a $1.8M_\odot$ star.

The late F and early G stars on the main sequence have extended outer convective zones. The concentration of Helium is unknown, as it cannot be derived from spectroscopy. Signature of the discontinuity at the bottom of the convective zone, and of the helium content are present in the behavior of successive frequency differences. Figure 3 shows how these frequency differences between two possible models can be measured, and interpreted to fix the most correct model which reproduces at best the observed spectrum. For instance, variations of 30% in the extension of the penetration region produce frequency differences of the order of $2\mu Hz$, easily detectable.

These two examples help to define the frequency resolution to be reached and the noise level. Then, the scientific specifications have been expressed as:

- Precision on measured frequencies better than $0.1\mu Hz$, obtained with a duration of 150 days for one observation.
- The noise level is fixed by imposing a SNR $\geq$ 4 and by the predicted amplitudes for the corresponding targets, i.e.
  * less than 0.6 ppm over 5 days for G stars
  * less than 2.5 ppm over 5 days for A and F stars.

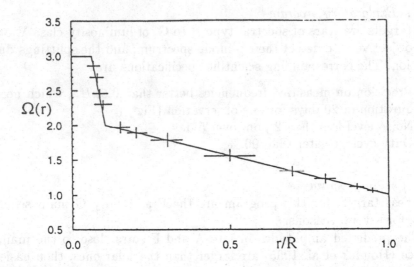

Figure 2. Determination of the rotational profile of a delta Scuti star, from Goupil et al. 1996. The full line represents the rotation curve introduced in the model to compute the splittings, whereas crosses indicate the results of the inversion of these data.

- Duty cycle greater than 90 % so that the side lobes of the window function remain negligible.
- Two colors photometry, to help the mode identification process.

4.1.3. *Secondary scientific objectives*

The observation of Jupiter in white solar reflected light is proposed as a secondary objective. The quantity which could be measured is the large separation ν_0.

As already mentioned, the mission profile of COROT allows us to propose a specific program of detection of Extra Solar Planets through the transits on the stellar disk. It will be realized by extending the field of view, to observe a neighboring region of the sky in a mode suited for this problem. This program is not described here.

4.2. DESCRIPTION OF THE EXPERIMENT.

4.2.1. *The spacecraft and the orbit.*

The "Petites Missions" Program provides the spacecraft called PROTEUS. It is an inertial platform, stabilized on three axes and designed for low orbit applications (maximum height at 1300 km). It includes a star tracker with a 20 degrees field of view, in charge of rough pointing. The mean power

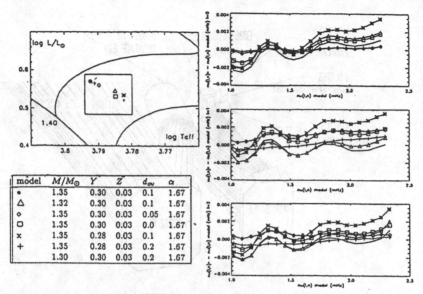

model	$M/M_\odot$	Y	Z	d_{ov}	α
•	1.35	0.30	0.03	0.1	1.67
△	1.32	0.30	0.03	0.1	1.67
◇	1.35	0.30	0.03	0.05	1.67
□	1.35	0.30	0.03	0.0	1.67
×	1.35	0.28	0.03	0.1	1.67
+	1.35	0.28	0.03	0.2	1.67
	1.30	0.30	0.03	0.2	1.67

Figure 3. Frequency differences of low degree p-modes between a reference model, and different models compatible with the position in the HR diagram of the star β *Virginis*, from Michel et al. 1995

supply delivered to the payload is 150 W. and the total mass of the payload cannot exceeded 300 kg.

The spacecraft can be launched by a large variety of rockets.

To fulfill the specification on the duration of the observing runs, an inertial polar orbit at 800 km has been chosen; the visible regions of the sky are located along the equator.

The telemetry data rate is on average 600 Mbit/day to download scientific and housekeeping data.

4.2.2. *The payload architecture.*

The set of targets of the main program are the closest and brightest F stars compatible with the constraints on the orbit (see next section). This means that their average visual magnitude lies around 6. To reach the specified noise level, a pupil diameter of 25 cm is needed (eq. 2).

The telescope consists in an off-axis scheme with the advantages of no occultation and compactness. It is composed of three mirrors as shown on figure 4. Mirror M2 is used to stabilize the image in focal plane. As the orbit plane is not always perpendicular to the Sun-Earth axis the spacecraft flies over Earth regions enlighted by the Sun. A large baffle reduces the scattered light to a level smaller than 0.01 photons per pixel.

Two focal planes, one receiving the blue light and the other the red one, are composed of four buttable CCD each with 2000×4000 pixels of size $13-14\mu m$. Half of each CCD is used for frame transfer to avoid a mechanical

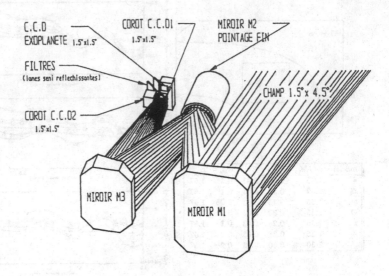

Figure 4. Scheme of the COROT telescope.

shutter. The CCD are thinned and backside illuminated. They operate in MPP mode and below − 50 C., to obtain a low radiations sensitivity. An aluminium shield of 12 mm. around the focal planes will cut all protons with energy less than 50 MeV and electrons with energy below 8 MeV. The readout noise will be less than $10e^-$ and dark current less than $2e^-$.

To avoid saturation on bright stars the focal plane is defocused giving a star image of 20 pixels diameter for the asteroseismology and 7 pixels for the exoplanets program.

4.2.3. *External perturbations.*
Most of the external perturbations are a consequence of the low polar orbit. They can be minimized and/or corrected. They are:

- Protons and electrons trapped in the SAA. They produce an aging of the CCD by increasing the dark current and reducing the charge transfer efficiency. The glitches in the image can be easily checked and corrected.
- Temperature variations due to the Sun occultations by the Earth. The quantum efficiency variation is typically 10^{-3} for one degree. Temperature variations will be known with a precision better than 0.01 K.
- Albedo variations during clouds and poles fly over. The Earth albedo can vary by a factor two around its mean value of 0.3. The scattered light induced variations will be checked on the CCDs.
- Coupling between the depointing and quantum efficiency variations from pixel to pixel. The angular stability of the image will be better

than 1 arc second.

4.3. MISSION PROFILE

In the first part of the mission, fields of view containing several stars of different types will be observed during approximately 20 days. Then, each field centered on a chosen target will be pointed during 150 days without interruptions. During this second phase, the exoplanet experiment will work permanently.

This long uninterrupted sequences can be achieved only for directions close to the equator and perpendicular to the orbit. But, it has been shown that a reasonable choice of adequate targets for the main as well as the exploratory programs satisfy these strong constraints. A launch in 2001 will permit also to reach Jupiter.

5. Conclusions

Several important results in stellar seismology can and will very probably be obtained from the ground, involving networks of moderate size telescopes. But, to achieve the overall scientific goals of this domain, photometry from space is certainly the best suited technique. With a quite small, simple and light photometer, one will have access to a wide variety of objects of different ages, chemical composition and internal structure.

Unfortunately, the difficulty to gain a selection among the different opportunities proposed by the space agencies, the failure of the MARS96 mission with EVRIS on board, and the unsuccessful STARS proposal in the "HORIZON 2000" context of ESA, have delayed the realization of such a program.

COROT is the next one to be launched in 2001, and let's hope that other more ambitious projects will appear.

There is certainly a strong need for an extended asteroseismology survey like STARS, even though no opportunity is presently foreseen.

Extremely high accurate photometry, as well as very long observing runs on the same field are also needed for programs of detection of transits of extrasolar planets, and combined missions are definitely an interesting challenge to look at.

References

Badialdi, M., Catala, C., Favata, F., Fridlund, M., Frandsen, S., Gough, D.O., Hoyng, P., Pace, O., Roca-Cortes, T., Roxburgh, I.W., Sterken, C. and Volonté, S. (1996) STARS, Seismic Telescope for Astrophysical Research from Space, Report on the phase A study, ESA report SCI(96), 4.

Baglin A., Weiss W., Bisnovatyi-Kogan G.S. (1993) in *Inside the Stars* , A. Baglin and W.W. Weiss eds., A.S.P. Conf. Ser. **40**, 758.

Buey J.T., Auvergne M., Vuillemin A., Epstein G. (1997) PASP **109**, 140.

Catala C., Auvergne M., Baglin A., Bonneau F., Magnan A., Vuillemin A., Goupil M.J., Michel E., Boumier P., Dzitko H., Gabriel A., Gautier D., Lemaire P., Mangeney A., Mosser B., Turk-Chièze S., Zahn J.P. (1995) in *4th SOHO Workshop: Helioseismology*, p. 549.

Dziembowski W. (1996) (this colloquium)

Frandsen S. (1996) (this colloquium)

Gough D.O. (1987) Nature **236**, 389.

Goupil, M.J., Dziembowski, W., Goode, P.R., Michel, E. (1996) *Astron. Astrophys.* **305**, 498.

Harvey J.W. (1988) in *Advances in Helio- and Asteroseismology*, J. Christensen-Dalsgaard and S. Franden eds., p. 497.

Houdek, G., Balmforth, M.J., Christensen-Dalsgaard, J. (1995) in *4th SOHO Workshop: Helioseismology*, p. 447.

Hudson, H.S., Brown, T.M., Christensen-Dalsgaard, J., Cox, A.N., Demarque, P., Harvey, J.W., Mc Graw, J.T., Noyes, R.W. (1986) A concept study for an Asteroseismology Explorer, Proposal submitted to NASA

Libbrecht K.G. (1992) *Astrophys. J.* **387**, 712.

Mangeney, A., Praderie, F. (1984) Proceedings of the workshop on Space Projects in Stellar Activity and Variability, Observatoire de Paris, p. 379.

Michel, E., Goupil, M.J.,Cassisi, S., Baglin, A., Auvergne, M., Buey, J.T. (1995) in *4th SOHO Workshop: Helioseismology*, p. 543.

Noyes, R.W., (1988) in *Advances in Helio- and Asteroseismology*, J. Christensen-Dalsgaard and S. Frandsen eds., p.257.

Scargle J.D. (1982) *Astrophys. J.* **263**, 835.

Vuillemin A., Tynok, A., Baglin, A., Weiss, W.W., Auvergne, M., Repin S., Bisnovatyi-Kogan, G. (1997) PASP, submitted

THREE YEARS OF ANTENA: WHAT WE HAVE DONE!

J. A. BELMONTE , M. M. HERNÁNDEZ,
F. PÉREZ HERNÁNDEZ, I. VIDAL AND T. ROCA CORTÉS
Instituto de Astrofísica de Canarias
38200 La Laguna, Tenerife, Spain
E. MICHEL, M. AUVERGNE , M. CHEVRETON,
M. J. GOUPIL, F. SOUFI AND A. BAGLIN
DASGAL, DAEC, DESPA, Observatoire de Paris-Meudon
92195 Meudon, France
S. FRANDSEN, M. VISKUM , H. KJELDSEN
AND J. CHRISTENSEN-DALSGAARD
Institut for Fysik og Astronomi, Aarhus Universitet
8000 Aarhus C, Denmark
F.X. SCHMIDER AND E. FOSSAT
Département d'Astrophysique, Université de Nice
06108 Nice, France
PH. DELACHE*, J. PROVOST , N. AUDARD
AND G. BERTHOMIEU
Observatoire de la Côte d'Azur
06304 Nice, France
M. PAPARÓ , G. KOVÁCS AND L. SZABADOS
Konkoly Observatory, Hungarian Academy of Sciences
1525 Budapest, Hungary

* Deceased October 1994. The rest of the authors wish to dedicate this paper to the memory of their colleague and one of the fathers of ANTENA, Philippe Delache.

Abstract. The European Union Network ANTENA started to work in October 1993. During these last three years, several collaborative projects have been undertaken. ANTENA has offered a very good opportunity for most of the European people doing asteroseismology to work together. The asteroseismological networks STEPHI and STACC have run within the framework of the project, obtaining fairly good results. New instrumentation has also been developed, such as the Four-Channel Stellar Photometer.

357

J. Provost and F.-X. Schmider (eds.), Sounding Solar and Stellar Interiors, 357-364.
© 1997 *IAU. Printed in the Netherlands.*

1. Introduction

The main objective of the groups included in the ANTENA network (A New TEchnology Network for Asteroseismology) has been to give a substantial impetus to the incipient discipline of asteroseismology from all points of view: the observational side, where networks such as STEPHI and STACC have been running with several campaigns; the theoretical side, obtaining improvements in the understanding of the interiors of pulsating stars; the technical side, developing instrumentation devoted to seismological purposes such as a new version of the Four-Channel Stellar Photometer; and the human side, giving the opportunity of exchanging information and manpower between the centres involved. In the following pages, we will offer a summary of the most remarkable results reached in the different fields developed within the framework of the ANTENA project.

2. Observational research: STEPHI and STACC networks

The observational network STEPHI (STEllar PHotometry International) (Michel *et al.*, 1995) has performed two campaigns on candidates belonging to the Praesepe cluster (BQ and BW Cnc, 1995; BS and BT Cnc, 1996) over the last three years. These new observations have the aim of completing the sample of Praesepe δ Scuti stars initiated with BN and BU Cnc in 1992 (Belmonte *et al.*, 1994). The Praesepe cluster, with 14 δ Scuti stars known so far, offers one of the best scenarios for performing asteroseismology, taking advantage of the common value of distance, age, and metallicity for all their components (Hernández *et al.*, 1996a). This allows a stronger restriction on the modelling of these stars. Theoretical results derived from these observations will be shown in the following section.

The synthetic spectra of the 8 Praesepe δ Scuti stars observed more extensively are shown in Fig. 1. Multiperiodic behaviour is observed in most of them, the pattern and range of excited frequencies changing with the evolutionary stage. On the other hand, drastic changes of amplitude in a short period of time, due to mutual interferences between very close peaks (~ 0.5 μHz), are also detected (Fig. 2 and Fig. 3).

The STACC network (Small Telescope Array with CCD Cameras) (Frandsen, 1992) has made important efforts in monitoring not well known open clusters in order to find the most suitable target for a campaign, considering hundred of stars at the same time. One of these surveys was made in the southern hemisphere where the cluster NGC 6134 was chosen for observation, providing promising results despite its faintness (Frandsen *et al.*, 1996). NGC 7245 (Fig. 4) was the second target discovered for a possible future campaign in the northern hemisphere (Frandsen *et al.*, 1995).

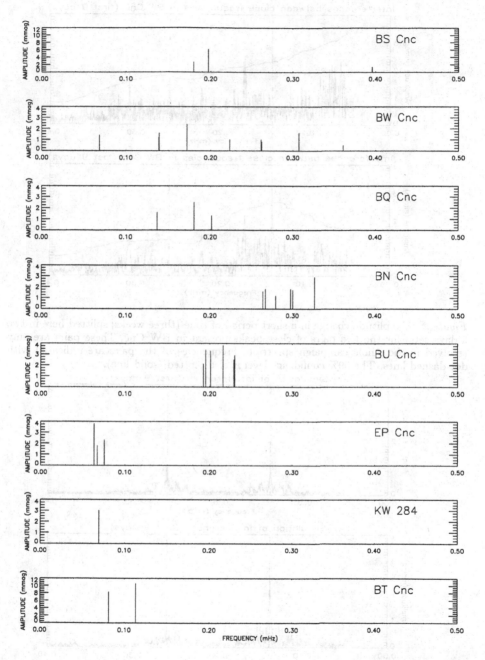

Figure 1. Amplitude spectra of the 8 δ Scuti stars more deeply observed in the Praesepe cluster. They are arranged in increasing luminosity from top to bottom. Detected modes have a confidence level larger than 90%.

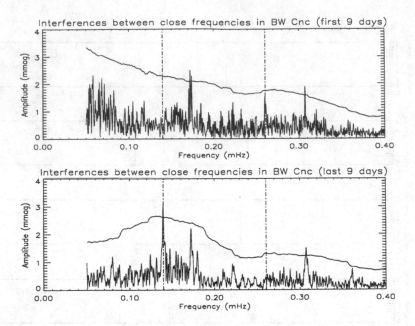

Figure 2. Amplitude change in a short period of time (three weeks, splitted here in two 9-days sets) for the two pairs of close peaks present in BW Cnc. These pairs are only resolved in the whole campaign spectrum. Frequencies of the pairs are indicated with dot-dashed lines. The 99% confidence level is also plotted (solid line).

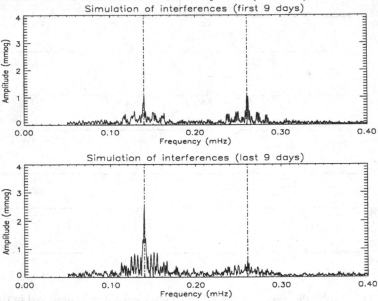

Figure 3. Simulation of the interference for both pairs of close peaks. Frequencies, amplitudes and phases of each peak are taken from the whole campaign spectrum. This shows that the reason of these changes are the mutual interference between the elements of each pair. When facing these situations, studies of secular amplitude variations must be used cautiously since they can be seriously affected by this sort of effect.

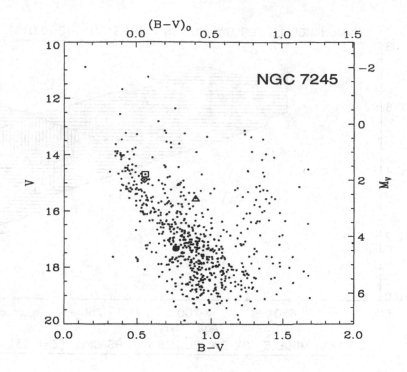

Figure 4.   HR diagram for NGC 7245. The four detected variables are indicated. The two stars represented by a square and a diamond are two δ Scuti stars situated at the hot border of the instability strip. Both stars seem to be multiperiodic with a main period of 0.1167 days and 0.0793 days, respectively. The star represented by a triangle is an eclipsing binary while the star shown as a black circle has a period of 0.2271 days and is probably a W UMa system.

3.   Theoretical results

The sample of 8 δ Scuti stars monitored extensively in Praesepe has been demonstrated to be extremely useful in terms of restricting the global parameters of the cluster. The ambiguity in the identification of the detected modes in every star separately is now partially removed. The fact of having the same global parameters (age, distance and metallicity) for every member of the cluster, and the little dependence of the ratios of the theoretical radial modes on these parameters, allows a congruent identification of pairs of radial modes in every star for those pairs that can be reproduced with the same parameters for all the stars (Hernández *et al.*, 1996b). In the case of Praesepe, a metallicity close to $Z=0.03$, an age of 640-660 Myr and a distance modulus of 6.4-6.5, match the candidates for radial pulsation frequencies of four cluster members (Fig. 5). In this analysis, an overshooting of $\alpha_{ov}=0.20$ has been considered.

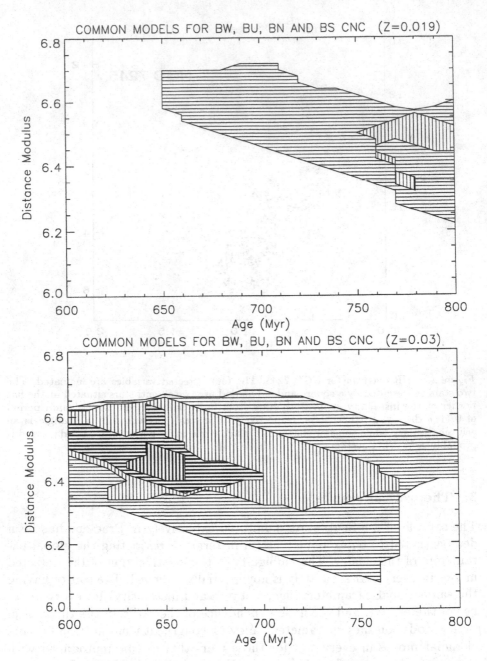

Figure 5. Valid models for 1 (horiz. thin), 2 (vert. thin), 3 (horiz. thick) and 4 stars (vert. thick) in Praesepe cluster with $Z=0.03$ for the only combination of potential pairs of radial modes with a common solution for all the stars. Common solutions with $Z=0.019$ for the same combination are also shown. Notice that, in this case, only two stars have common valid models. Blank spaces denote the models which cannot satisfy the observational and theoretical constraints on the cluster.

A similar case appears for the cluster NGC 6134 observed by STACC, where 6 δ Scuti stars were discovered. In this case, the lack (or even absence) of observational information about distance, metallicity or rotational velocity is responsible of the weak constraints which can be applied to the models. In Fig. 6, the frequency evolution versus the age for one of the variable stars and an attempt to identify their detected modes are shown. Despite of the uncertainties in the modelling, it can be seen that g-modes are necessary in order to explain the existence of some of the frequencies (Audard *et al.*, 1995, 1997).

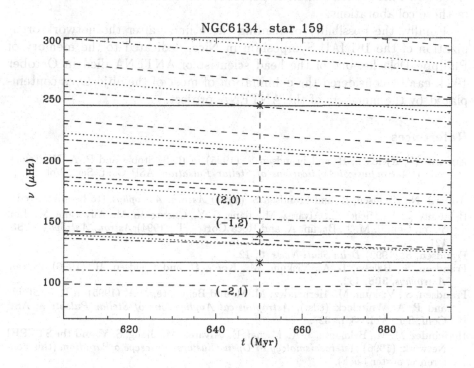

Figure 6. Variation with age of the theoretical frequencies for a model of 2.1 $M_\odot$ in NGC 6134. The dashed vertical line gives the direct comparison of the frequencies detected in the star 159 (solid lines) at an age of 650 Myr. Dotted lines are for radial modes $l=0$, while dashed and dot-dashed lines are for modes of $l=1$ and 2, respectively. The label (n, l) is indicated for three modes. Despite of the non-existence of a unique identification, the possibility of having g-modes oscillating is clear in this star.

4. Other achievements within ANTENA

From the technical point of view, the realization of a improved version of the Four-Channel Stellar Photometer designed to perform rapid photometry (suitable for any kind of moderate amplitude pulsators: white dwarfs, roAp stars, δ Scuti or λ Boo stars, etc.) in four channels simultaneously is the most representative achievement. This photometer has been successfully employed in the STEPHI 96 campaign. The Doppler Stellar Seismometer, designed to study solar-like oscillating stars with low amplitude ($1\sigma \sim 10$ cm s^{-1}) is also in progress.

The mobility of researchers between the different member institutions of ANTENA for collaborating, either in the theoretical or the observational field, has been initiated fruitfully during these three years. At the same time, three new PhD students and two postdoctoral researchers have participated in these collaborations.

Finally, the possibility of reading these lines, after the network organization of the 181 IAU Symposium in Nice, dedicated to the memory of Philippe Delache, one of the head scientist of ANTENA died in October 1994, can be considered the culmination of most of the objectives contemplated by the members of the ANTENA project.

References

Audard, N., Kjeldsen, H. and Frandsen S. (1995) in R. S. Stobie and P. A. Whitelock (eds.), *Astrophysical Applications of Stellar Pulsation*, ASP Conf. Ser., Vol. 83, p. 325.

Audard, N., Kjeldsen, H. and Frandsen S. (1997) *Astron. Astrophys.* (to be submitted).

Belmonte, J. A., Michel, E., Álvarez, M., Jiang, S. Y., Chevreton, M., Auvergne, M., Liu, Y. Y., Goupil, M. J., Baglin, A. and Roca Cortés, T. (1994) *Astron. Astrophys.* **283**, 121.

Frandsen, S. (1992) *Delta Scuti News.* **5**, 12.

Frandsen, S., Balona, L. A., Viskum, M., Koen, C. and Kjeldsen, H. (1996) *Astron. Astrophys.* **308**, 132.

Frandsen, S., Viskum, M., Hernández, M. M. and Belmonte, J. A. (1995) in R. S. Stobie and P. A. Whitelock (eds.), *Astrophysical Applications of Stellar Pulsation*, ASP Conf. Ser., Vol. 83, p. 327.

Hernández, M. M., Belmonte, J. A., Michel, E., Álvarez, M., Jiang, S. Y. and the STEPHI Network: (1996) *Asteroseismology in Open Clusters: Praesepe, a Paradigm* (this conference, poster book).

Hernández, M. M., Pérez Hernández, F., Michel, E., Belmonte, J. A., Goupil, M. J. and Lebreton, Y. (1996b) *Observational constraints on Praesepe cluster by Asteroseismology* (this conference, poster book).

Michel, E., Chevreton, M., Goupil, M. J., Belmonte, J. A., Jiang, S. Y., Álvarez, M., Suran M., Soufi F., Auvergne, M., Baglin, A., Liu, Y. Y., Vidal, I., Hernández, M. M. and Fu, J. N. (1995) in *Fourth SOHO Workshop: Helioseismology*, Pacific Grove, California, ESA SP-376, p. 533.

RESULTS ON SOME SELECTED
TYPES OF STARS

ASTEROSEISMOLOGY OF WHITE DWARF STARS

G. VAUCLAIR
Observatoire Midi-Pyrénées
14, Av. E. Belin 31400 Toulouse, France

Abstract. The theoretical potential of white dwarf asteroseismology is summarized. It is shown how one can derive fundamental parameters on the internal structure and evolution of these stars. The analysis of the non-radial g-modes permits in principle to determine the total mass, the rotation rate, the magnetic field strength. The mass of the outer layers, left on top of the carbon/oxygen core, can be determined as well as the structure of the transition zone between the core and the outer layers, giving an "a posteriori" unique information on the efficiency of the previous mass loss episodes. When measurable, the rate of change of the pulsation periods gives direct access to the evolutionary time scale and to the chemical composition of the core. These theoretical expectations are compared with the observations of variable white dwarfs in the three known instability strips for the planetary nebulae nuclei and PG1159 stars, for the DB and DA white dwarfs. Emphasis is put on results obtained from multi-sites photometric campaigns. Prospects on both theoretical developments and observations conclude the review.

1. Pulsating white dwarfs in the H-R diagram

At the end of their evolution, most of the stars become white dwarfs. Only massive stars, more massive than 6 to 8 $M_\odot$ do explode as supernovae. Less massive stars, after having exhausted their nuclear energy, cool down in radiating the thermal energy stored in their presumed carbon/oxygen core. They form in the H-R diagram a cooling sequence, going from the hot, luminous side, at an effective temperature exceeding 100 kK and a luminosity reaching $\log L/L_\odot = 4$, to the cool, faint end where they will eventually reach the invisibility limit when their internal temperature gets below the Debye temperature. In between these two extremes, they

367

J. Provost and F.-X. Schmider (eds.), Sounding Solar and Stellar Interiors, 367-380.

cross various instability strips, according to their effective temperature and chemical composition. These instability strips are precious windows which allow us to check many aspects of stellar evolution, dense matter physics, and galactic evolution.

Following the cooling sequence, the first instability strip to be found concerns the very hot stars of PG1159 type. Some of them are central stars of planetary nebulae still embedded in their nebulae while for others the nebulae is no longer visible. There is now a compelling evidence for an evolutionary link between the planetary nebulae nuclei and the white dwarfs. An interesting transition object is the hottest known pulsating PG1159 star, discovered during the ROSAT all sky survey, RXJ2117+3412 (Motch *et al.* 1993) whose very faint nebula was subsequently discovered (Appleton *et al.* 1993). But, another proportion of the hot white dwarfs comes from the evolution of the subdwarf stars, which evolve from the extended horizontal branch directly to the white dwarf sequence. What are the fractions of the white dwarfs coming from the Planetary nebulae or from the subdwarfs, and whether the final white dwarfs show any difference in mass and in structure is still not known. The asteroseismology of these objects may reveal some clues. Today one knows about 7 variable planetary nebulae nuclei (PNNV) and 4 variable PG1159 stars (also called DOV) and one transition object (RXJ2117+3412). The effective temperature range of the instability strip is quite large, from 150 kK for RXJ2117+3412, to 75 kK for PG0122+200 (Dreizler *et al.* 1995). Typical cooling time at this hot end of the cooling sequence is of the order of 10^6 years.

While both the PNNV and variable PG1159 stars show a very similar chemical atmospheric composition, made of a mixture of helium, carbon and oxygen, the chemical composition of the coolest white dwarfs differs significantly, with one helium rich family, the DB type white dwarfs, and one hydrogen rich family, the DA type white dwarfs. The origin of these two types is not entirely understood. But, because one PG1159 star was recently discovered to show hydrogen in its spectrum (HS2324+3944, Dreizler *et al.* 1996), it may be speculated that some of the PG1159 stars have still some hydrogen left in their atmosphere ("PG1159-Hybrid"), while the others have lost all their hydrogen in the previous phases of nuclear burning and mass loss. At the very beginning of the white dwarf sequence, diffusion processes become very efficient with typical time scales much shorter than the cooling time. PG1159 stars with hydrogen left would evolve into DA white dwarfs while those without hydrogen left would evolve into DBs. The following evolution is made more complex because of other intervening physical processes which may play a role in changing the atmospheric chemical composition: convective mixing, accretion, etc.

The DB white dwarfs are the next ones along the cooling sequence to cross an instability strip, at effective temperature around 23-25 kK. One knows 8 variable DB white dwarfs (DBV). At this effective temperature, the luminosity has decreased to about 10^{-1} $L_\odot$. The cooling time has increased to about 10^8 years.

The DA white dwarfs cross their instability strip at an effective temperature of about 11-12 kK. Their luminosity is of about 10^{-3} $L_\odot$ and their cooling time is of the order of 10^9 years. A total number of 26 variable DAs (DAV) is known (they are also called ZZ Cetis).

The instability mechanism triggering the pulsations in the three instability strips is the $\kappa - \gamma$ mechanism. In the PNNV and PG1159 stars, the instability is produced by the ionization of the heavy elements, mainly oxygen (Bradley and Dziembowski 1996, Saio 1996) following earlier suggestions by Starrfield and collaborators (Starrfield et al. 1983, 1984; Stanghellini et al. 1991). In the DBV the instability is related to the recombination of the helium (Winget et al. 1982b) while in the DAV it is related to the recombination of the hydrogen (Dolez and Vauclair 1981; Winget et al. 1982a).

Historically, the first variable white dwarf to be discovered was the DA HL Tau 76 (Landolt 1968), followed by the first PG1159 variable object PG1159-035 which gave its name to this new class of variable stars (McGraw et al. 1979). The last class of variable white dwarfs to be discovered, the DBV, is the only one which was theoretically predicted by pulsation theory previously to their observation (Winget et al. 1982b).

2. Non radial pulsations in white dwarfs

It is not my purpose to explain what non radial pulsations are to the participants of this symposium entirely devoted to helio- and asteroseismology. I just need to remind the very basic dispersion relation obtained after proceeding to all the simplifications one can think of and which takes the local form:

$$k^2 = \frac{1}{\sigma^2 v_s^2} \left(\sigma^2 - L_\ell^2\right) \left(\sigma^2 - N^2\right)$$

where L_ℓ is the Lamb frequency:

$$L_\ell^2 = \frac{\ell(\ell+1)v_s^2}{r^2}$$

N is the Brunt-Väisälä frequency:

$$N^2 = -g \left[\frac{d\ln\rho}{dr} + \frac{\rho g}{\Gamma_1 P}\right]$$

and k is the wave number of the mode, σ its frequency, and v_s the sound velocity.

The dispersion relation is useful to explain the differences between the non-radial modes in white dwarfs and in other types of stars. The non-radial modes can only propagate in those regions of the stars where the wave number k is real. This condition defines two types of waves and their related propagation regions. The modes whose frequencies are higher than both the Lamb and the Brunt-Väisälä frequencies are the p-modes, while those with lower frequencies are the g-modes. Because white dwarfs are high gravity, degenerate, stratified stars, their non-radial modes have some specific properties. The p-modes would have periods of the order of 1 s and less for higher harmonics: they are not observed in spite of serious and repeated efforts to detect them (Robinson 1984). The g-modes have periods in the range 100s<P<1500s. So, curiously, g-modes in white dwarfs have periods of the same order as p-modes in "normal", non degenerate stars.

In the degenerate core, the Brunt-Väisälä frequency drops to 0 with increasing depth. As a consequence, the g-modes can not propagate deep in the interior. Contrarily to "normal" stars where the g-modes give information on their deep interior, in white dwarfs they inform on their outer layers. The stratified chemical structure of the white dwarfs has interesting signature on the g-modes. It is predicted by stellar evolution that white dwarfs of average mass (about 0.6 $M_\odot$) have a C/O core. What surrounds this core depends on the previous evolutionary phases which are not yet satisfactorily understood. The outer layers are a mixture of He/C/O in PG1159 stars with some admixture of H in the PG1159-Hybrid stars, pure He in the DBs, while in DAs the He layers are surrounded by a pure H layer. Determining the mass of these remainings of previous evolution is one of the challenges of white dwarf asteroseismology. The chemical changes as a function of depth occur on short height scales, determined by the diffusion equilibrium. They produce discontinuities in the run of the Brunt-Väisälä frequency which in turn act as a selection mechanism between modes (trapping mechanism), as will be discussed below.

3. What can we learn from white dwarf asteroseismology?

If a large enough number of g-modes may be identified, one can derive some fundamental parameters for the star:

1 – the total stellar mass: the mass of the star may be derived from the period spacing. In the asymptotic regime, consecutive g-modes are regularly spaced in periods. For g-modes belonging to spherical harmonics of degree ℓ and order k, their periods are approximately given by:

$$P_{\ell,k} = \frac{\Pi_0}{\sqrt{\ell(\ell+1)}} (k + \varepsilon)$$

where

$$\Pi_0 = 2\Pi^2 \left(\int \frac{N}{r} dr \right)^{-1}$$

The integral has to be taken on the propagation zone. From this expression one deduces directly that because the Brunt-Väisälä frequency is proportional to the gravity, the period spacing Π_0 is inversely proportional to the gravity. As is well known, the mass and the radius are related for white dwarfs, a consequence of the degenerate equation of state. If a period spacing can be measured, it is a direct measurement of the mass. For the hot white dwarfs, there is a small dependence of Π_0 on the stellar luminosity, which reflects the variation of the propagation cavity as the star evolves along the cooling sequence: the degeneracy boundary moves towards the surface, pushing the propagation zone outwards; the integral of the Brunt-Väisälä frequency on the propagation region decreases as the star evolves; as a consequence Π_0 increases slightly for a given mass. For typical white dwarf masses, Π_0 varies between 38s for $0.5M_\odot$ and 25s for $0.7\ M_\odot$, with a typical value of 30s for $0.6\ M_\odot$, the white dwarf average mass (Kawaler, Bradley 1994). Of course, what is measured from the observation is the period spacing between modes $P_{\ell,k}$. Deducing Π_0 requires the identification of ℓ.

2 – the mass of the outer layers: the mass of the layer surrounding the degenerate core may be derived from the deviation of the period distribution from a regular distribution. In a chemically homogeneous star, successive g- modes are regularly spaced in periods, as discussed above. In a chemically stratified star, the period distribution is slightly changed because of the discontinuities in chemical composition which reflect into discontinuities of the Brunt-Väisälä frequency. In white dwarfs, such transition zones between regions of different chemical composition exist because of the stratification induced by the gravitational settling. In PG1159 stars, there is such a transition zone between the He rich surface layer and the C/O core. In DB white dwarfs the transition occurs between the pure He outer layer and the core (C/O). In DA white dwarfs there must be two such transition regions, one between the pure H outer layer and the underlying helium layer, and the second one between the He and the C/O core.

Those modes which have one node of their eigenfunction in, or close to, one of these transition zones keep a negligible amplitude below the transition zone; these modes have an appreciable amplitude only above the transition zone. They are trapped in the outer layers. For this reason this mechanism is called mode trapping. Other modes which do not behave like trapped modes and do have an amplitude below the transition zone are called confined modes. The trapped modes and their neighbours in terms of k values are more closely spaced in period than the average period spacing Π_0. The reverse is true for the confined modes. In a plot of the period differences between successive modes as a function of the period for modes of a given ℓ, the trapped modes are easily identified because they correspond to minima.

Such $\Delta\Pi$ versus Π diagrams contain interesting information on the structure of the outer layers. On one hand, they present regularly spaced minima corresponding to successive trapped modes. The differences of period between successive trapped modes, or trapping cycle, is related to the depth of the transition zone; the trapping cycle measures the mass of the outer layers. On the other hand, the amplitude of the variation of $\Delta\Pi$, around the average value Π_0 is related to the gradient of mean molecular weight at the transition zone. It depends on the difference in chemical composition on the two sides of the transition and on its width. Discussion and examples of mode trapping predicted in various models may be found in Brassard *et al.* (1992a, b) and in Bradley (1993, 1996).

3 – the stellar luminosity: once the total mass is known from the period spacing, the mass of the outer layers from mode trapping, with the effective temperature derived from spectrophotometry, the stellar model is entirely determined: one can then derive the luminosity. As an interesting by-product, the distance can be estimated with a much larger accuracy than through other methods. For instance, the distance of the planetary nebulae whose central star is a pulsating objet, may be measured. This may have important implications on the distance scale problem in improving the planetary nebulae distance scale.

4 – the rotation period: this can be deduced from the rotational splitting. It is known that rotation removes the degeneracy of the modes of identical degree ℓ and order k but different azimuthal index m. In the limit of slow rotation, which proves to be a good assumption in the case of the white dwarfs, the frequency of one mode of indices ℓ, k, m in the rotating case is related to the frequency of the mode in the non-rotating case by:

$$\sigma_{k,\ell,m} = \sigma_{k,\ell} + m\left(1 - C_{k,\ell}\right)\Omega + 0\left(\Omega^2\right)$$

where $\sigma_{k,\ell}$ is the frequency in the non-rotating case; $C_{k,\ell}$ comes from the Coriolis force term in the momentum equation and Ω is the rotation frequency. This is the classical development to the first order. In the asymptotic limit for g-modes, $C_{k,\ell}$ is simply related to the degree of the mode by:

$$C_{k,\ell} = \frac{1}{\ell(\ell+1)}$$

As m can take all values between $-\ell$ and $+\ell$, each mode of degree ℓ is split into $2\ell + 1$ components in the presence of rotation. The number of components allows the identification of ℓ, which in turn, allows the determination of the mass from the period spacing. The frequency separation between the components is a measure of the rotation period, once ℓ is unambiguously identified.

5 – the magnetic field: from the magnetic splitting. In a way similar to the rotation, the magnetic field removes the degeneracy of the modes. But, in contrast to the rotation, the magnetic field splits one mode into only $\ell + 1$ components. In addition, the frequency shift is proportional to m^2 (Jones et al. 1989).

6 – the evolutionary time scale: from the measurement of $\dot{\Pi}$.

As the white dwarfs evolve along the cooling sequence, their structure changes. The whole spectrum of their non-radial eigenvalues changes at the same rate. Roughly, the rate of change of the pulsation period, $\dot{\Pi}$, is related to the rates of change of the core temperature T_m and of the radius R, by the relation (Winget et al. 1983):

$$\frac{\dot{\Pi}}{\Pi} \simeq -\frac{1}{2}\frac{\dot{T}_m}{T_m} + \frac{\dot{R}}{R}$$

The first term corresponds to the rate of change in period induced by the cooling of the white dwarf and is a positive contribution, while the second term corresponds to the rate of change induced by the contraction and is a negative contribution. In the PG1159 stars, still in the phase of final contraction towards the cooling sequence, both the cooling and the contraction are acting on $\dot{\Pi}$. The trapped modes, whose eigenfunctions are concentrated in the outer layers, are more sensitive to the contraction of these layers: their $\dot{\Pi}$ are negative. On the contrary, for the confined modes whose eigenfunctions are more sensitive to the region underneath the transition zone, the $\dot{\Pi}$ is dominated by the cooling, due, for a large domain of luminosity, to the neutrino emission: their $\dot{\Pi}$ is positive. In DBV and DAV, the influence of the contraction on $\dot{\Pi}$ has become negligible, $\dot{\Pi}$ is dominated by the cooling and is positive for all modes.

A measurement of $\dot{\Pi}$ is of major interest as it is a direct measurement of the cooling time which, in turn, depends on the chemical composition of the degenerate core. It is a direct test of the validity of the predictions of the stellar evolution theory. Would the $\dot{\Pi}$ for the confined modes in PG1159 stars be accessible to measurement, it would be a test on the neutrino physics. Measuring $\dot{\Pi}$ of the pulsating white dwarfs in the three instability strips allows us to calibrate the cooling sequence age. This should give the age of the galactic disk in the solar neighbourhood (Winget et al. 1987).

4. Observations vs. theory

The full use of the asteroseismological tools developed in the framework of the pulsation theory requires that: 1) one can detect as many modes as possible, and 2) one can identify these modes. The first condition implies that one is able to reach a sufficiently high S/N ratio to detect small amplitude modes. To achieve that goal requires having access to large telescopes or going to space. Achieving the second condition requires getting long, uninterrupted, time-series. This is possible from the ground with the organization of multi-site observational campaigns, or again from space.

From the ground, some time of relatively large telescope (i.e. the CFHT 3.60m at Mauna Kea) has been devoted to the asteroseismology of white dwarfs and has demonstrated the impact that the use of such a large telescope may have in this field (Fontaine *et al.* 1991, Fontaine *et al.* 1995). However, it is not yet possible to use network of large telescopes well distributed in longitude to carry on multi- site campaigns. The Whole Earth Telescope (WET; Nather *et al.* 1990) is a compromise to fulfill at the best the two required conditions. WET is an international collaboration which organizes one or two multi-site campaigns a year. Telescopes of various size (from 60cm up to 3.6m) well located around the Earth observe as simultaneously as possible the same target.

Some examples of the results obtained in the field of the white dwarf asteroseismology will be discussed below. Because of the space available, this is of course a selection.

1. PNNV and PG1159 stars

Recent reviews on that part of the H-R diagram may be found in Werner et al (1996) and Vauclair (1996). The observations of the PNNV have been mainly performed through CCD photometry, because of the needed correction from the nebulae brightness (Bond and Meakes 1990; Bond and Ciardullo 1993; Bond 1995) with the exception of K1-16 whose dilute nebulae allowed the discovery of its central star pulsations through photo- multiplier (PM) photometry (Grauer and Bond 1984; Grauer *et al.* 1992). The same is true for the transition object RXJ2117+3412, whose nebulae is in the process of dispersing into the interstellar matter: its pulsations were discovered with classical PM photometry (Watson 1992, Vauclair et al. 1993). For cooler pulsating PG1159 stars, as for RXJ2117+3412, WET campaigns have been organized, which use PM photometers (Winget *et al.* 1991; Kawaler *et al.* 1995). The main results on the prototype star of this group, PG1159-035, are now summarized. The full results of the 1989 WET campaign on PG1159-035 may be found in Winget *et al.* (1991). A total of 264h of high-speed photometry was accumulated on this star with an excellent coverage. The resulting window function, together with a frequency resolution of 1μHz, allowed the detection of about 125 frequencies in the power spectrum. Most of them are members of multiplets: triplets and quintuplets, which are interpreted as modes of degree $\ell = 1$ and 2, respectively, split by rotation. When averaging the triplets and the quintuplets separately, it is found that the ratio of the frequency shifts between the components of the triplets and those of the quintuplets is 0.60, in excellent agreement with the expected theoretical value for the ratio of split modes $\ell = 1$ and $\ell = 2$ which is 0.61 in the asymptotic regime. The deduced rotation period is 1.38 days, a slow enough rotation rate justifying the approximation used to calculate the rotational splitting. A search for period spacing reveals two values: 21.5s and 12.5s. The first value is interpreted as the period spacing for modes $\ell = 1$, while the second value corresponds to the period spacing for modes $\ell = 2$. Their ratio, 1.72, is also in excellent agreement with the theoretical expected value of 1.73, in the asymptotic regime. The total mass derived from these period spacings is $M/M_\odot = .586 \pm .003$. Note the unusual accuracy on the mass determination of the asteroseismological method. The $\Delta\Pi$ vs. Π diagram shows the typical structure with minima, which is the signature of mode trapping induced by the chemical stratification. The fit with theoretical models leads to a fractional mass in the outer layers (above the C/O core) of $3 - 4 \times 10^{-3}$. From these values, one can derive the luminosity: $\log L/L_\odot = 2.29 \pm 0.05$, and a distance of 440 ± 40 pc.

The largest amplitude mode, with a period of 516s, has a measured time derivative of: $\dot{\Pi}/\Pi = (-4.8 \pm 0.1) \times 10^{-14}$ s^{-1}. This period corresponds to a trapped mode in the best fit model (Kawaler and Bradley 1994) with a period very close to the observed one. This is why $\dot{\Pi}$ is negative: this mode is sensitive to the still ongoing contraction of the outer layers. The characteristic contraction time measured by $\dot{\Pi}$ corresponds to 6.6×10^5 y. Unfortunately, the confined modes are expected to have much smaller amplitudes and the corresponding $\dot{\Pi}$ are not yet available.

2. DBVs

The prototype of this class of pulsating white dwarfs is GD358. It is also the best studied DBV. The results of the WET campaign on this object may be found in Winget at al. (1994). In this star, more than 180 modes were found, most of them identified as modes of degree $\ell = 1$. The analysis of the period spacing suggests a mass of 0.61 ± 0.03 $M_\odot$. From the deviation to the average period spacing (the $\Delta\Pi$ vs. Π diagram) the derived helium layer mass is determined as $M_{He}/M_* = 2.0 \pm 1.0 \times 10^{-6}$. From these values, a luminosity $\log L/L_\odot = -1.30$ was derived, which leads to a distance of 42 ± 3 pc, in agreement with the value derived from the parallax. The rotational splitting reveals a non uniformity which was interpreted as a differential rotation within the outer layers, with its outer part rotating about 1.8 times faster than its most interior part. This fact, together with the unexpected low value of the Helium layer mass, were intriguing enough to motivate a second WET campaign, conducted 2 years after the first one, still under analysis.

3- DAVs

In the DAVs, as in the DBVs, the location of the instability strip in the H-R diagram depends on the adopted description of the convection and on its efficiency. A contrario, a precise observational location of the instability strip may be used as a constraint on the convection theory. The ZZ Ceti instability strip is better defined observationally owing to the larger number of known DAVs and to the greater accuracy in the atmospheric parameters for pure hydrogen atmospheres. For these reasons, efforts are continuously made to improve both the statistics by searching for new DAVs, and the accuracy of model atmospheres.

In contrast with the two cases discussed above, most of the DAVs show only a small number of modes compared to their potential rich non radial g-modes spectrum. This prevents any determination of the total stellar mass from the period spacing since there is not enough periods available to derive such a period spacing. For instance, G117-B15A, one of the best studied ZZ Ceti stars, has only three independent modes in its power spectrum. All the other power present in the spectrum are due to linear combinations of these three frequencies. The pulsations of this particular star are followed for almost the last 20 years because the short period (215 s) of its dominant mode and the relatively clean power spectrum around the corresponding frequency, were thought to offer a good opportunity to measure $\dot{\Pi}$. However, the interpretation of the $\dot{\Pi}$ measurement in such stars (Kepler *et al.* 1991) may not be trivial as there may be some other effects than the simple thermal cooling to take into account. The physical process responsible for the efficient mode selection could also affect both the amplitude and the frequency of the observed modes. From high S/N ratio data, Fontaine *et al.* (1995), following Brassard *et al.* 1995, have shown how a linear perturbation in the surface temperature induced by the g-modes propagation lead to non linear perturbations in the emergent flux. They interpret that way the occurrence of small amplitude peaks in the power spectrum corresponding to linear combinations of the fundamental three independent modes. This interpretation leads to a hydrogen mass of $M_H/M_* = 10^{-6}$. But in other cases, the non linear effects could be more fundamentally present in the star, involving for instance interactions between the modes themselves. In such cases, it is not known what would be the frequencies and amplitudes of the finally observed modes.

The mode selection could result either from the earlier discussed mode trapping or from the interaction of the pulsations with the convection, or both. Mode trapping has been thoroughly explored by Brassard *et al.* (1992a, b) and Bradley (1996) in DA models. The "natural" assumption is that the observed largest amplitude modes are trapped modes. However, in some cases, in the best fit models, i.e. in the models where the calculated eigenvalues do correspond the best to the observed frequencies, the modes which are observed with the largest amplitude do not correspond to trapped modes. The other physical process able to act as a selection mechanism is the interaction with convection. Gautschy *et al.* (1996) have recently explored this effect for the ZZ Cetis and found that a large number of modes which would be otherwise unstable, are stable if one takes into account the interaction with convection. Clearly, the question of the mode selection mechanism needs more work.

Closer to the ZZ Ceti instability red edge, GD154 is again a white dwarf which has very efficiently selected its pulsation modes: only three independent modes are identified in the data obtained during a dedicated WET campaign, as well as during subsequent observations (Pfeiffer *et al.* 1996). The much longer period of the dominant mode in GD154 (1186 s) is suggesting a smaller hydrogen envelope mass, $M_H/M_* = 2 \times 10^{-10}$, than in G117-B15A.

Determining the age of the white dwarfs along their cooling sequence is a way of measuring the age of the galactic disk in the solar vicinity. The coolest white dwarfs observed are not old enough to have reached the Debye temperature in their core, a limit at which an abrupt drop in the white dwarf luminosity function is predicted by evolutionary models (Wood 1992). As the lack of faint, cool white dwarfs does occur at a higher luminosity than the one which would correspond to the Debye cooling, it is concluded that these coolest white dwarfs do indicate the age of the galactic disk (Winget *et al.* 1987). However, this age is made rather uncertain as a consequence of our incomplete understanding of the crystallization phase which takes place at the cool end of the white dwarf sequence. The latent heat produced during the crystallization slows down the global cooling in a proportion which depends on the core composition and on how it crystallizes in details. The luminosity at which the crystallization starts does depend on the white dwarf mass: a massive white dwarf starts crystallization at higher luminosity than a smaller mass white dwarf. As discussed above, if one is able to measure through the asteroseismological method, i.e. $\dot{\Pi}$, the age of white dwarfs in the process of crystallization, one could infer how the crystallization proceeds and how much it slows the cooling rate. Unfortunately, the large majority of the white dwarfs have a mass around 0.6 $M_\odot$ Such white dwarfs crystallize at much cooler effective temperature than the red edge of the relevant instability strip. However, one DAV was recently found to be much more massive than the average white dwarf: BPM 37093

(Kanaan *et al.* 1992) has a mass in the range 1.0-1.1 $M_\odot$ (Bergeron *et al.* 1995). Such a massive DA is predicted to start crystallization within the ZZ Ceti instability strip (Winget, private communication). It will be of course a first priority target of a next WET campaign.

5. Conclusions and prospects

The asteroseismology of white dwarfs has been shown to be a powerful tool in the field of stellar physics. It also provides unique constraints in other fields like the physics of the galaxy and the dense matter physics.

In stellar physics, knowing accurately the mass of the white dwarfs and the mass of their remaining outer layers provides constraints on various processes occurring in the previous evolutionary phases, like the mass loss efficiency. It also helps understanding the evolutionary links between various kinds of objects in the final stages of stellar evolution. The rotation rates deduced from the asteroseismology of white dwarfs confirm that the white dwarfs are slow rotators. Together with a similar information coming from the spectroscopy of bright white dwarfs, this result is interesting in itself as it points to the question of the fate of the angular momentum during stellar evolution.

The age of white dwarfs, hopefully deduced from the determination of the cooling time, via the measurement of $\dot{\Pi}$, leads to a direct exploration of the core composition. It is a unique check of the stellar evolution theory. The age-calibration of the white dwarf cooling sequence gives us the age of the galactic disk in the solar vicinity. This is of course of importance for our understanding of the galaxy formation and evolution. The analysis of the crystallization phase should give information on the way crystallization takes place in dense, degenerate matter, a still controversial subject in the field of dense matter physics.

The problem of locating the instability strips in the H-R diagram and of understanding the mode selection mechanism is intimately related to some long standing questions linked to thermal convection: its efficiency and its interaction with pulsation.

Now, that a large amount of data starts to be available on the pulsation properties of white dwarfs, it becomes evident that all kinds of variable white dwarfs do exhibit time dependent variations of their amplitudes. These variations are clearly related to the details of the excitation mechanism and/or to the non linear effects. To extract the information implicitly hidden in these variations will require long term observations and theoretical developments. It will also require the development of new tools for data analysis, better adapted to the new physics involved, like time-frequency analysis, etc (Dolez *et al.* 1996).

The asteroseismology of white dwarfs will continue to be a very active and growing field in astrophysics, taking full advantage of ground-based networks, and of the availability of larger telescopes in a near future. In the same time, the need for a space mission dedicated to the asteroseismology which has been expressed in many occasions during this symposium, is largely demonstrated by the limit of what can be achieved from the ground; it would largely contribute to the progress in our understanding of the pulsating stars.

References

Appleton, P., Kawaler, S., Eitter, J. (1993) *Astron. J.* **106**, 1973

Bergeron, P., Wesemael, F., Lamontagne, R., Fontaine, G., Saffer, R.A. and Allard, N.F. (1995) *Astrophys. J.* **449**, 258

Bond, H.E and Meakes, M.G. (1990) *Astron. J.*, **100**, 788

Bond, H.E. and Ciardullo, R. (1993) in *White dwarfs: Advances in Observation and Theory*, M. Barstow (Ed.), Kluwer Academic Publishers, p.491

Bond, H.E. (1995) in *3rd WET workshop*, E. Meistas and J.-E. Solheim (Eds.), Baltic Astronomy 4, p.527

Bradley, P.A. (1993) Ph.D. Thesis, Univ. Texas, Austin

Bradley, P.A. (1996) *Astrophys. J.*, **468**, 350

Bradley, P.A. and Dziembowski, W.A. (1996) *Astrophys. J.*, **462**, 376

Brassard, P., Fontaine, G., Wesemael, F. and Hansen, C.J. (1992a) *Astrophys. J. Supl.*, **80**, 369

Brassard, P., Fontaine, G., Wesemael, F. and Tassoul, M. (1992b) *Astrophys. J. Supl.*, **81**, 747

Brassard, P., Fontaine, G. and Wesemael, F. (1995) *Astrophys. J. Supl.*, **96**, 545

Dolez, N., Roques, S., Serre, B. and Vauclair, G. (1997) in *Sounding Solar and Stellar Interiors*, J. Provost and F.-X. Schmider (Eds.) in press

Dolez, N. and Vauclair, G. (1981) *Astron. Astrophys.*, **102**, 375

Dreizler, S., Werner, K. and Heber, U. (1995) in *White Dwarfs* D. Koester and K. Werner (Eds.), Lecture Notes in Physics **443**, Springer, Berlin, p.160

Dreizler, S., Werner, K., Heber, U. and Engels, D. (1996) *Astron. Astrophys.*, **309**, 820

Fontaine, G., Bergeron, P., Vauclair, G., Brassard, P., Wesemael, F., Kawaler, S., Grauer, A.D. and Winget, D.E. (1991) *Astrophys. J.*, **378**, L49

Fontaine, G., Brassard, P., Bergeron, P., Wesemael, F., Vauclair, G., Pfeiffer, B. and Dolez, N. (1995) Proceedings of the 4th CFHT User's

Grauer, A.D. and Bond, H.E. (1984) *Astrophys. J.*, **277**, 211

Grauer, A.D., Green, R.F. and Liebert, J. (1992) *Astrophys. J.*, **399**, 686

Gautschy, A., Ludwig, H.-G. and Freytag, B. (1996) *Astron. Astrophys.* , **311**, 493

Jones, P.W., Pesnell, W.D., Hansen, C.J. and Kawaler, S.D. (1989) *Astrophys. J.*, **336**, 403

Kanaan, A., Kepler, S.O. and Giovannini, O. (1992) *Astrophys. J.*, **390**, L89

Kawaler, S.D. and Bradley, P.A. (1994) *Astrophys. J.*, **427**, 415

Kawaler, S.D. *et al.* (1995) *Astrophys. J.*, **450**, 350

Kepler, S.O. *et al.* (1991) *Astrophys. J.*, **378**, L45

Landolt, A.U. (1968) *Astrophys. J.*, **153**, 151

McGraw, J.T., Starrfield, S.G., Liebert, J. and Green, R.F. (1979) in *White Dwarfs and Variable Degenerate Stars*, IAU Colloquium 53, H.M. Van Horn and V. Weidemann (Eds.), Univ. Rochester Press, p. 377

Motch, C., Werner, K. and Pakull, M. (1993) *Astron. Astrophys.*, **268**, 561

Nather, R.E., Winget, D.E., Clemens, J.C., Hansen, C.J. and Hine, B.P. (1990) *Astrophys. J.*, **361**, 309

380   G. VAUCLAIR

Pfeiffer, B., *et al.* (1996) *Astron. Astrophys.*, **314**, 182
Robinson, E. L. (1984) *Astron. J.*, **89**, 1732
Saio, H. (1996) in *Hydrogen-Deficient Stars*, C.S. Jeffery and U. Heber (Eds.), *ASP Conf. Ser.*, **96**, p.361
Stanghellini, L., Cox, A.N. and Starrfield, S.G. (1991) *Astrophys. J.*, **383**, 766
Starrfield, S.G., Cox, A.N., Hodson, S.W. and Pesnell, W.D. (1983) *Astrophys. J.*, **268**, L27
Starrfield, S.G., Cox, A.N., Kidman, R.B. and Pesnell (1984) *Astrophys. J.*, **281**, 800
Vauclair, G. (1996) in *Hydrogen-Deficient Stars*, C.S. Jeffery and U. Heber (Eds.), *ASP Conf. Ser.*, **96**, p.397
Vauclair, G., Belmonte, J.A., Pfeiffer, B., Chevreton, M., Dolez, N., Motch, C., Werner, K. and Pakull, M. (1993) *Astron. Astrophys.*, **267**, L35
Watson, T. (1992) IAU Circular 5603
Werner, K., Dreizler, S., Heber, U. and Rauch, T. (1996) in *Hydrogen-Deficient Stars*, C.S. Jeffery and U. Heber (Eds.), *ASP Conf. Ser.*, **96**, p. 267
Winget, D.E., Van Horn, H.M., Tassoul, M., Hansen, C.J., Fontaine, G. and Carroll, B.W. (1982a) *Astrophys. J.*, **252**, L65
Winget, D.E., Robinson, E.L., Nather, R.E. and Fontaine, G. (1982b) *Astrophys. J.*, **262**, L11
Winget, D.E., Hansen, C.J. and Van Horn, H.M. (1983) *Nature*, **303**, 781
Winget, D.E., Hansen, C.J., Liebert, J., Van Horn, H.M., Fontaine, G., Nather, R.E., Kepler, S.O. and Lamb, D.Q. (1987) *Astrophys. J.*, **315**, L77
Winget, D.E. *et al.* (1991) *Astrophys. J.*, **378**, 326
Winget, D.E. *et al.* (1994) *Astrophys. J.*, **430**, 839
Wood, M. (1992) *Astrophys. J.*, **386**, 539

OBSERVATIONAL ASTEROSEISMOLOGY OF δ SCUTI STARS

BREGER M. AND AUDARD N.

Institut of Astronomy, University of Vienna
Türkenschanzstraße 17, A-1180 Wien Austria

Abstract. Ground-based observational networks are now able to reliably detect more than 20 frequencies for individual δ Scuti stars. The ground-based Delta Scuti Network specializes in the intensive study of individual stars for several months with dedicated telescopes on different continents. With the new quantity and quality of data, the observational limit to extracting 50 or more frequencies in a single δ Scuti star is no longer determined only by the signal/noise ratio, but by the available frequency resolution.

Recent results on the δ Scuti stars FG Vir and 4CVn are presented. 4CVn exhibits some of the characteristics of δ Scuti stars: some close spacings are observed, pulsations occur in specific frequency regions with relatively sharp borders, and show large amplitude variations. This presents an important asteroseismological diagnostic, and can help us to improve our understanding of for example rotation and excitation mechanisms.

We also report a new Gamma Doradus variable with two excited g modes (HD 108100, F2V).

1. Introduction

The 14th DSN campaign was devoted to the δ Scuti star FG Vir. Observations were carried out in a combined photomultiplier campaign by DSN and WET (Whole Earth Telescope) networks. Furthermore, about one month of CCD data was collected. Two observational methods are used, depending on the observed frequency-range: for periods larger than about 30 minutes, the single-channel 3-star technique is adopted, while for periods smaller than 30 minutes, the high-speed photometry technique (one channel measurements on the target all the time) is adopted. The program PERIOD

J. Provost and F.-X. Schmider (eds.), Sounding Solar and Stellar Interiors, 381-386.

(Breger, 1990b) is used to detect the frequencies in the photometric data. These new observations confirmed all previously certain peaks, as well as many peaks, which were previously not found to be statistically significant. The data are currently being reduced, and the latest results reveal 23 frequencies. This is one of the largest numbers of detected frequencies for a single δ Scuti star. The presence of some gravity modes has already been reported by Breger *et al.* (1995) and Guzik and Bradley (1995). Some more g modes, with possible regular spacing, might have been detected in these new observations. These modes are particularly interesting for improving our knowledge of rotation (Goupil et al, 1996) and of processes such as convective-core overshooting (Dziembowski and Pamyatnykh 1991, Audard et al 1995).

The δ Scuti 4CVn was observed within the 15th DSN campaign, by Breger, Handler, Garrido, Audard, Beichbuchner, Zima, Paparo, Li Zhiping, Jiang Shi-yang, Liu Zong-li, Zhou Ai-ying, Pikall, Stankov, Guzik, Sperl, Kresinski, Ogloza, Pajdosz, Zola, Serkowitsch, Reegen, Rumpf, Solheim, Schmalwieser and Thomasson. 25 frequencies have been extracted in a preliminary analysis, among which we also find frequency combinations.

2. Specific frequency regions and close spacing: 4CVn

A summary of the results for 4CVn is represented Fig. 1. The pulsation occurs in a specific frequency region, between roughly 5 and 10 cycles/day, and the peaks at lower and higher period are frequency combinations. These features do not appear in FG Vir. The main frequencies have a dual nature, i.e. they are g modes in the deep stellar interior and p modes in the outer parts. Dziembowski (1997) states that they are g-like modes of high radial order in the deep regions.

Another property of some δ Scuti stars is the close spacing observed for some pulsations. In the case of FG Vir, two modes are separated by as little as 0.03 c/d (0.3 μHz). This spacing is too small to be due to rotational splitting, as found by for example Michel *et al.* (1996), Goupil *et al.* (1996), Pérez-Hernández *et al.* (1995). The pulsations can thus only correspond to modes of different degrees.

3. Amplitude variations

From 1966 to 1996, the amplitude of the pulsations of 4CVn varies drastically. Breger (1990a) states that this is more likely due to the Blazhko effect, rather than a beating phenomenom between two close frequencies.

We do not yet clearly know why the amplitudes vary, neither why only a small fraction of theoretically unstable modes are effectively observed. Some predictions are made for radial modes by e.g. Houdek *et al.* (1995).

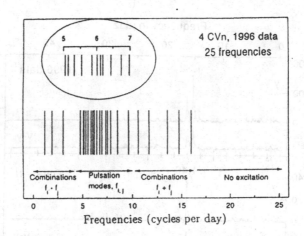

Frequencies (cycles per day)

Figure 1.   Observed pulsation spectrum of 4CVn (preliminary). The inset shows the peaks between 5 and 7 cycles/day.

The nonlinearity problem for radial and nonradial modes is addressed by Buchler *et al.* (1995), Goupil and Buchler (1994).

These varying amplitudes are particularly powerful for our understanding of the excitation mechanism, and of growth and decay rates.

4. γ Doradus stars

The variability of γ Doradus stars has been discovered only recently. These stars are early F stars, or are located just above the main sequence. They lie near the cool border of the instability strip of δ Scuti stars. They pulsate at very low frequencies, with a typical value near 1 c/d. The corresponding values of the pulsational constant, Q, are about ten times the value for the fundamental radial mode. Therefore, the pulsations can only be identified as high-order g modes. γ Doradus itself has been observed for example by Balona *et al.* 1996. Aerts and Krisciunas (1995) report results of 9 Aur. Breger and Beichbuchner (1996) give a general review.

Breger *et al.* (1996) have recently confirmed the variability of the γ Dor star HD 108100, within a coordinated campaign with the DSN and WET networks. Two pulsations of periods 17.1 and 18.1 hours have been detected (see Fig. 2).

At the present time, γ Doradus stars are a challenge to asteroseismology. The theoretical frequency spectrum of g modes is very dense, and an unambiguous mode identification is almost impossible. It is moreover made even more difficult by the effects of rotation. Therefore asteroseismology of γ Doradus stars is still waiting for accurate models with high-order g

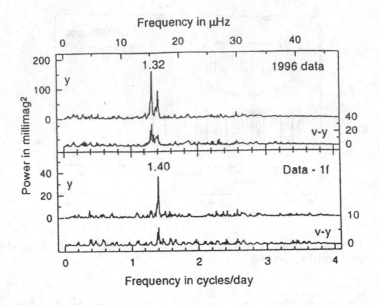

Figure 2.   Power spectrum of the new γ Doradus star HD 108100

modes.

5. Steps towards mode identification

5.1. FIRST MODE IDENTIFICATION

A first mode identification can be performed by determining the phase differences between y and v light curves. Garrido *et al.* (1990) have shown that these can be used reliably to discriminate between radial and nonradial modes of low radial order and low degree.

5.2. COMPARISON BETWEEN OBSERVATIONS AND THEORETICAL MODELS

Guzik and Bradley (1995) have computed models for FG Vir, and compared their results to those from Breger *et al.* (1995), obtained from Dziembowski's evolution and pulsation codes. The slight differences between Dziembowski's codes, and Guzik and Bradley's, result in nonradial frequencies which can differ by several μHz. As a result, Guzik and Bradley identified pulsations with the same radial orders and degrees (n, ℓ) as Breger *et al.* (1995), but with different azimuthal orders m. Moreover, they find that only models of 1.8 and 1.82 $M_{\odot}$ can match the 10 first detected frequencies of FG Vir found during the 1995 FG Vir campaign, and no model can match

all the 19 frequencies. Bradley and Guzik (1996) also report the effects of
the OPAL95 and OPAL92 opacities, and of different equations of state, on
the mode identification of the frequencies of FG Vir and the δ Scuti CD-24
7599.

This lack of uniqueness of models reveals the inaccuracies of parts of
the input physics and of the description of phenomena such as rotation and
convective core overshooting. Another source of uncertainty is our knowledge of the basic fundamental parameters of the stars, such as mass and
effective temperature. The uncertainties are often too large to constrain the
models, so that several models can match the same star in the HR diagram.

The observed frequencies of pulsation can be used to correct deficiencies
in the structure and the input physics of the evolution and oscillation codes
(opacities, equation of state, rotation, overshooting,..), if one or both of the
following two conditions are fulfilled: 1) modes are identified by amplitude
and/or phase studies of colours or spectroscopic data, 2) precise stellar
parameters are known for stars, where special techniques can be employed
to obtain these parameters (binaries, open cluster stars, very nearby stars).

6. Conclusion

We need more frequency resolution to resolve the peaks of the power spectra. Spectroscopic data of best quality are required to provide more constraints to the modelling. These can only be obtained with longer, multisite
campaigns.

Stellar models have to be improved, with a better description of processes such as rotation and convective-core overshooting.

With more than 20 frequencies for a single star, asteroseismology of δ
Scuti stars is a powerful tool for improving our knowledge of the interior
and evolution of stars.

References

Aerts C., Krisciunas K. (1996) MNRAS, **278**, 877
Audard N., Provost J., Christensen-Dalsgaard J. (1995) Astron. Astrophys., **297**, 427
Balona L.A., Hearnshaw J.B., Koen C., Collier A., Machi I., Mkhosi M., Steenberg C.
    (1994) MNRAS, **267**, 103
Bradley P.A., Guzik J.A. (1996) in Sounding solar and Stellar Interiors, poster book
Breger M., Beichbuchner F. (1996) Astron. Astrophys., **313**, 851
Breger M., Handler G., Garrido R., Audard N., Beichbuchner F., Zima W., Paparo M.,
    Li Zhi-ping, Jiang Shi-yang, Liu Zong-li, Zhou Ai-ying, Pikall H., Stankov A., Guzik
    J.A., Sperl M., Kresinski J., Ogloza W., Pajdosz G., Zola S., Serkowitsch E., Reegen
    P., Rumpf T., Schmalwieser A. (1996) Astron. Astrophys., in press
Breger M., Handler G., Nather R.E., Winget D.E., Kleinman S.J., Sullivan D.J., Li Zhi-
    ping, Solheim J.E., Jiang Shi-yang, Liu Zong-li, Wood M., Watson T.K., Dziembowski
    W.A., Serkowitsch E., Mendelson H., Clemens C., Krzesinski J., Pajdosz G. (1995)
    Astron. Astrophys., **297**, 473

Breger M. (1990a) *Astron. Astrophys.*, **240**, 308

Breger M. (1990b) *Comm. Asteroseismology* (Vienna) **20**, 1

Buchler J.R., Goupil M.J., Serre T. (1995) *Astron. Astrophys.*, **296**, 405

Dziembowski W.A. (1997) these proceedings

Dziembowski W.A., Pamyatnkh A.A. (1991) *Astron. Astrophys.*, **248**, L11

Garrido R., Garcia-Lobo E., Rodriguez E. (1990) *Astron. Astrophys.*, **234**, 262

Goupil M.J., Buchler J.R. (1994) *Astron. Astrophys.*, **291**, 481

Goupil M.J., Dziembowski W.A., Goode P.R., Michel E. (1996) *Astron. Astrophys.*, **305**, 487

Guzik J.A., Bradley P.A. (1995) *δ Scuti Newsletter*, issue 9

Houdek G., Rogl J., Balmforth N., Christensen-Dalsgaard J. (1995) in *GONG'94: Helio- and Astero-seismology from Earth and Space*, Ulrich R.K., Rhodes Jr E.J., Däppen W. (eds.), *PASPC* **76**, p. 641.

Michel E., Chevreton M., Goupil M.J., Belmonte J.A., Jiang S.Y., Alvarez M., Suran M., Soufi F., Auvergne M., Baglin A., Liu Y.Y., Vidal I., Hernandez M., Fu N. (1996) in *Fourth SOHO Workshop: helioseismology*, p.543.

Pérez Hernández F., Claret A., Belmonte J.A. (1995), *Astron. Astrophys.*, **295**, 113

PROBING THE INTERIORS OF ROAP STARS

Stellar Tubas That Sound Like Piccolos

JAYMIE M. MATTHEWS
University of British Columbia
Department of Physics & Astronomy
2219 Main Mall
Vancouver, B.C., V6T 1Z4, Canada

1. The overture: harmonies of the main sequence band

Chemically peculiar magnetic (Ap) stars in the Instability Strip shouldn't really pulsate (since the helium in their very stable atmospheres appears to have gravitationally settled below the He II ionisation zone). And if they did pulsate, they should have periods of a few hours like the δ Scuti variables, since the Ap stars have comparable mean densities and their atmospheric peculiarities wouldn't strongly affect the global eigenfunctions of low-degree p-modes.

As it turns out, some Ap stars *do* pulsate, but not in their fundamental or low-order resonances like δ Scuti stars. The rapidly oscillating Ap (roAp) stars (Kurtz 1990; Martinez & Kurtz 1995; Martinez 1996, this Symposium) vibrate in very high overtones of $n \sim 25 - 40$, like the solar oscillations seen in integrated light and velocity. Unlike the Sun, however, the roAp stars have coherent oscillations with amplitudes of millimagnitudes whose phases remain constant or drift only slowly over many years. In the symphony of stellar pulsators, the roAp stars are low-pitched tubas that are masquerading as high-pitched piccolos (while the Sun is a bassoon which sits idly on the floor, quietly resonating with the random vibrations of passing traffic).

Why are the high overtones so strongly driven in roAp stars? A very good question, which won't be answered here.[1] Regardless of the answer, we can still apply asymptotic pulsation theory (e.g., Tassoul 1990) and other techniques to these high-overtone p-modes to learn more about Ap stars.

Why should anyone want to learn more about Ap stars? Another good question, which will be answered (I hope) in this paper.

---

[1] Or anywhere else with certainty at this time, although Wojciech Dziembowski (private communication) is convinced hydrogen ionisation must be the key.

J. Provost and F.-X. Schmider (eds.), Sounding Solar and Stellar Interiors, 387–394.
© 1997 IAU. Printed in the Netherlands.

2. Using roAp stars as astrophysical laboratories

The strong magnetic fields and pronounced vertical and horizontal abundance gradients in Ap stars make them excellent testbeds for the theories of stellar magnetism (dynamos vs. fossil fields) and radiative diffusion. The oscillations of the roAp stars can serve as a diagnostic of these properties, as well as the global structures and ages of stars in this region of the H-R Diagram.

Since the high-overtone p-modes of roAp stars have most of their amplitude highly concentrated in the outer layers of the star, they should be very sensitive to atmospheric properties. However, there is also useful information about the interior, thanks to the second-order term δ_{02} in the frequency spacing which is highly sensitive to the sound-speed gradient in the core. A few examples of both types of diagnostics – featuring advances presented at this Symposium – are discussed briefly below.

3. In the atmosphere

The pulsation amplitude of an roAp star drops with increasing wavelength much more rapidly than expected for a pulsating blackbody (which is a reasonable approximation for classical pulsators like Cepheids and δ Scuti stars). This must be telling us something about what's going on in the atmosphere of an Ap star. *What* it tells us is still a matter of debate.

3.1. MULTICOLOUR PHOTOMETRY OF ROAP STARS: PROBING MODE DYNAMICS OR ATMOSPHERIC STRUCTURE?

Matthews *et al.* (1990, 1996) argued that this steep wavelength dependence could be explained by the weighting effect of limb darkening on the dominant $(\ell, m) = (1, 0)$ dipole mode of an roAp star, which always enhances the net amplitude integrated over the stellar disk. They used this to infer that the atmospheric $T - \tau$ gradient of the roAp star HR 3831 is much steeper than that for the Sun. This agrees with Shibahashi & Saio's (1985) mechanism to explain why some Ap stars oscillate in frequencies well above the acoustic cutoff for a grey atmosphere. It also agrees with the steeper atmospheric gradient employed by Muthsam & Stepień (1992) to model the line profiles of Ap spectra.

However, Medupe & Kurtz (1996, this Symposium; see also Kurtz & Medupe 1996) have argued analytically and through numerical simulations that limb darkening cannot enhance the amplitudes enough at short wavelengths (near 4500 Å) to account for the observations. Instead, they propose that the drop in photometric amplitude with wavelength is due to an actual drop in the local temperature amplitude ΔT of the mode with height in

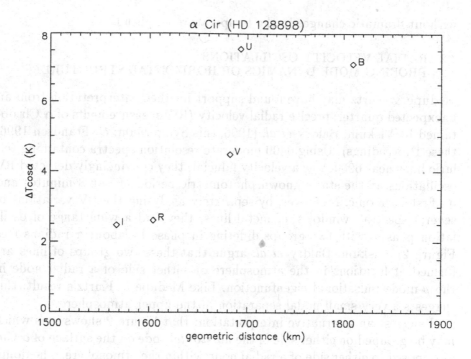

Figure 1.   Sampling the vertical wavelength of a high-order p-mode? The predicted temperature amplitude of the dominant dipole mode of α Cir as a function of atmospheric depth, based on its observed photometric amplitudes in five bandpasses. (From Medupe & Kurtz 1996, this Symposium.)

the atmosphere, where the continuum at longer wavelengths is produced at greater heights. Figure 1 shows the variation in $\Delta T cos\alpha$ (containing the unknown inclination α of the dipole mode) as a function of depth which can account for Medupe & Kurtz's multicolour photometry of the roAp star α Circini. Since the wavelength range of their $UBVRI$ filter set spans a range in geometrical depth in the Ap atmosphere of about 300 km, this implies that the radial node separation Δr_{node} in the upper layers is surprisingly small: of order 10^{-3} $R_{\star}$.

Medupe & Kurtz (1996, this Symposium) present a convincing case that limb darkening alone is insufficient to cause the observed amplitude decrease with wavelength in roAp stars. However, this may still prove to be a diagnostic of atmospheric structure. The steep μ gradients in abundance expected in the highly stratified upper atmosphere of an Ap star may cause mode trapping which would produce extremely small values of Δr_{node} implied by Figure 1. Medupe & Kurtz did not include compositional gradients in their models of $\Delta T cos\alpha$. Perhaps changes in flux with depth due to local changes in the metal opacity can weight the photometric amplitudes without dramatic changes in the temperature amplitude.

3.2. RADIAL VELOCITY OSCILLATIONS:
PROBING MODE DYNAMICS OR HORIZONTAL STRUCTURE?

Medupe & Kurtz may have found support for their interpretation from an unexpected quarter: precise radial velocity (RV) measurements of α Cir obtained by Viskum, Baldry et al. (1996, this Symposium; cf. Frandsen 1996, these Proceedings). Using 6400 moderate resolution spectra containing telluric lines near 6900 Å as a velocity fiducial, they convincingly detected RV oscillations at the star's known photometric period of $\sim$ 6.8 minutes and its first harmonic. Moreover, by separately analysing the RV variations of several spectral 'windows' of metal lines, they find a wide range of oscillation phases, with two groups differing in phase by about π radians (see Figure 2). Viskum, Baldry et al. argue that these two groups of lines are formed at locations in the atmosphere on either side of a radial node in the p-mode pulsational eigenfunction. Like Medupe & Kurtz's result, this suggests a very small nodal separation in the upper atmosphere.

I suggest an alternative interpretation: that Figure 2 shows ions which may be grouped on either side of the *horizontal* node on the surface of α Cir, as opposed to either side of a radial node within the atmosphere. The dominant dipole mode of the star means it is divided into two hemispheres which are pulsating in anti-phase, divided by a node along the pulsational equator (which is also presumed to be the magnetic equator in roAp stars). The distribution of elements on the surface of an Ap star can be highly inhomogeneous (e.g., Rice & Wehlau 1991) due to the effects of diffusion regulated by the magnetic field geometry. Different ions can be grouped around the magnetic poles, for instance. This situation is illustrated schematically in Figure 3.

Ions concentrated in opposite magnetic hemispheres would exhibit RV variations π radians out of phase with one another, as observed by Viskum, Baldry et al. In this scenario, the groups of lines in Figure 2 whose RV amplitudes are comparable to the noise would be located at the equatorial node (see Figure 3).

Frandsen (private communication) has pointed out that spectroscopy of α Cir does not seem to reveal strong starspots and Kurtz (private communication; cf. Kurtz et al. 1994) cautions that the equatorial node of α Cir may be located relatively close to the star's limb. Both suggest that these factors argue against the surface inhomogeneity scenario. However, the fact that the inclination and obliquity of α Cir may cause one hemisphere of the dipole to dominate the visible disk at all rotational phases would also reduce any rotational modulation due to polar spots or rings. Spectroscopically, lines due to ions raised high in the stellar atmosphere could have a

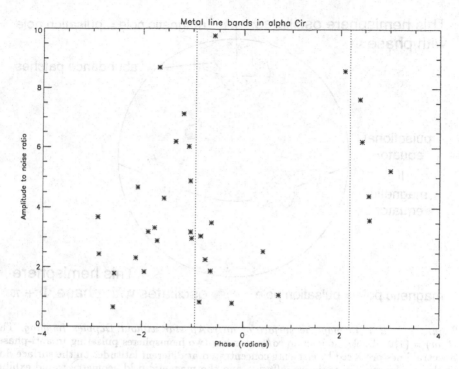

Figure 2. Déjà vu: Sampling the vertical wavelength of a high-order p-mode or mapping the surface of the dipole pattern? Radial velocity amplitudes (relative to the noise level) vs. oscillation phase for various windows of metal lines in the spectrum of α Cir. (Taken from Viskum, Baldry *et al.*, this Symposium.)

strong contribution from the limb, especially at wavelengths around 6500 Å where the limb darkening is relatively weak.

To settle the question, the ions corresponding to the different groups of 'anti-phase' lines must be identified and diffusion calculations performed to determine the depths and latitudes at which those ions would be most strongly concentrated.

3.3. MAPPING THE ATMOSPHERES OF AP STARS

Even if the Viskum, Baldry *et al.* (1996) results are not due to horizontal abundance inhomogeneities, this effect must be present in other roAp stars. Matthews & Scott (1995) and Scott & Matthews (1996, in preparation) have found evidence for line-profile variability with pulsational phase in the slowly-rotating roAp star γ Equulei. The projected rotational velocity *v sini* of this star is so low that its surface cannot be mapped by Doppler Imaging methods (e.g., Rice & Wehlau 1991). High-resolution spectroscopy sampling both the rapid pulsation cycle of the star and its rotational cycle

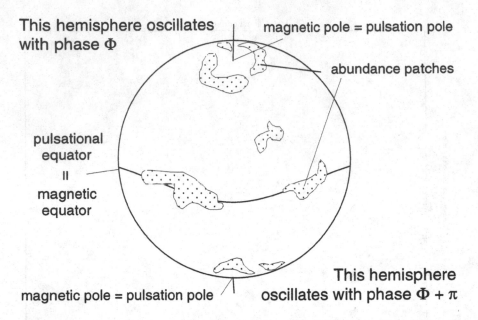

Figure 3.   How to map the surface of an roAp star without Doppler Imaging. The $(\ell, m) = (1, 0)$ dipole mode of an roAp star has two hemispheres pulsating in anti-phase. Spectral lines produced by elements concentrated at different latitudes on the surface due to the combination of radiative diffusion and the magnetic field geometry would exhibit RV variations of different amplitudes and phases, as well as distinctive profile variations.

could constrain the latitudes and longitudes of various elements on the stellar surface (again, see Figure 3). I have new high-quality CFHT coudé spectra of the roAp stars 10 Aql and HD 42659 which should be ideal for this purpose (Matthews 1997, in preparation).

4. In the interior

The principal mode spacing $\Delta\nu_0$ in the eigenfrequency spectrum of an roAp star can fix its radius $R_\star$ and luminosity $L_\star$ if the effective temperature T_{eff} is already known. This allows us to study the evolutionary traits of Ap stars. Soon we will be able to calibrate the results against luminosities of some roAp stars derived from Hipparcos parallaxes. The small separation δ_{02} in the eigenspectrum can also in principle yield the star's main sequence age (Christensen-Dalsgaard 1993) since δ_{02} is sensitive to the change in composition of the isothermal core brought about by H-burning. Audard & Provost (1993) have generated eigenspectra for $1 - 2M_\odot$ stars with convective cores which could be used for roAp stars (Matthews 1993).

Long-term frequency changes in some roAp stars (e.g., Kurtz et al. 1994) are reminiscent of changes seen in the Sun's eigenspectrum which correlate with the solar activity cycle. The latter have been interpreted as the effects of the solar dynamo. Could active dynamos be operating in Ap stars, despite their thin surface convection zones? Could a field generated in the star's convective core be strong enough to penetrate the surface and retain effective strengths of several kiloGauss? These are provocative questions whose impact may extend well beyond the regime of the Ap stars.

Rapid oscillations open another window on the internal magnetic properties of Ap stars through eigenfrequency perturbations. Dziembowski & Goode (1996) have modelled the effects of a curl-free dipole field on high-order p-modes and find perturbations $\Delta\nu_{mag} \sim 10 - 20$ μHz for modes and field intensities consistent with roAp stars. Unfortunately, these frequency shifts are comparable to the small separations δ_{02} (Audard & Provost 1993). Therefore, it is not a straightforward matter to derive the internal magnetic field strength and the main sequence age from the eigenfrequencies.

4.1. CAN WE UNTANGLE THE MAGNETIC FIELD LINES AND ISOCHRONES?

There may be a way out of the dilemma, if we can find an empirical link between the properties of the convective core of an Ap star and its magnetic field. Roxburgh (e.g., 1965) and Carlberg (1975) performed pioneering calculations on the influence of a magnetic field on the internal structure of a star. Das et $al.$ (1984) demonstrated that a poloidal field will reduce the mass fraction of the convective core in their simple polytropic models. They found reductions in core size of order a few $\times$ 0.1 pressure scale heights (H_p). If we can calibrate the effect of a field on the convective core mass and radius, then the spacings of the roAp eigenfrequencies $\nu_{\ell,n}$ could be solved simultaneously for the:

- a) effect of the convective core on $\nu_{\ell,n}$
- b) field strength consistent with the convective core from a)
- c) magnetic perturbations on $\nu_{\ell,n}$ consistent with b)

One way to simulate a reduction in core size due to a magnetic field is to turn off core overshooting in the stellar model. Audard et $al.$ (1995) have studied the seismic effects of core overshooting in models of $2M_\odot$ stars appropriate to the Ap mass range. The difference in eigenfrequencies between models with overshooting $\alpha \sim 0.2H_p$ (i.e., the non-magnetic case in my scenario) and no overshooting (i.e., a strong internal field) is of order 1 μHz. This is large enough compared to δ_{02} and $\Delta\nu_{mag}$ that the idea outlined here may indeed be feasible.

5. And the band plays on...

The papers on roAp stars presented at this Symposium reflect the rapid progress in the field, both observationally (e.g., spectacular detections of low-amplitude RV oscillations and precise multicolour photometry, each with time samplings of less than a minute) and theoretically (e.g., probes of the nodal structure of high-order p-modes, better treatments of the upper boundary of the acoustic cavity, and magnetic effects on the modes).

6. Acknowledgements

I am grateful to Thebe Medupe, Don Kurtz, Michael Viskum and Ivan Baldry for providing figures from their posters and for fruitful discussions. Thanks also to Nathalie Audard, Wojciech Dziembowski, Soren Frandsen, Alfred Gautschy, Peter Martinez and Hiromoto Shibahashi for insights.

References

Audard, N. & Provost, J. (1993) *ASPC* **40**, pp. 544–546
Audard, N., Provost, J. & Christensen-Dalsgaard, J. (1995) *A&A* **297**, pp. 427–440
Carlberg, R. (1975) *M.Sc. thesis*, University of British Columbia
Das, M.K., Kar, J. & Tandon, J.N. (1984) *ApJ* **281**, pp. 292–302
Christensen-Dalsgaard, J. (1993) *ASPC* **40**, pp. 483–496
Dziembowski, W. & Goode, P. (1996) *ApJ* **458**, pp. 338–346
Kurtz, D.W. (1990) *ARA&A* **28**, pp. 607–655
Kurtz, D.W. & Medupe, R. (1999), *Bull. Astron. Soc. India*, **24**
Kurtz, D.W., Sullivan, D.J., Martinez, P. & Tripe, P. (1994) *MNRAS* **270**, pp. 647-686
Martinez, P. & Kurtz, D.W. (1995) *ASPC* **83**, pp. 58–69
Matthews, J.M. (1993) *ASPC* **42**, pp. 303–316
Matthews, J.M. & Scott, S. (1995) *ASPC* **83**, pp. 347–348
Matthews, J.M., Wehlau, W.H., Rice, J. & Walker, G.A.H. (1996) *ApJ* **459**, pp. 278–287
Matthews, J.M., Wehlau, W.H. & Walker, G.A.H. (1990) *ApJL* **365**, pp. L81–L86
Muthsam, H. & Stepień, K. (1992) *Acta. Astron.* **42**, pp. 117-130
Rice, J.B. & Wehlau, W.H. (1991) *A&A* **246**, pp. 195-205
Roxburgh, I. (1965) in *Stellar and Solar Magnetic Fields*, ed. R. Lüst, North-Holland Publishing Company, Amsterdam, pp. 103–116
Shibahashi, H. & Saio, H. (1985) *PASJ* **37**, pp. 245-255
Tassoul, M. (1990) *ApJ* **358**, pp. 313–327

CONCLUSION

SOUNDING SOLAR AND STELLAR INTERIORS: CONCLUSIONS AND PROSPECTS

DOUGLAS GOUGH

Institute of Astronomy and
Department of Applied Mathematics and Theoretical Physics
University of Cambridge, UK

1. Introduction

I have been charged to summarize the mood of the symposium, although not to summarize the scientific details, for they are best learned by reading the proceedings. To the best of my ability, I shall comment on those details in a manner that Philippe Delache might spontaneously have done had it not been incumbent upon him to cover everything. Therefore I shall be somewhat personal. I shall make no attempt to offer a balanced view of the observations and calculations that have been reported, but instead concentrate on how they might be viewed. In so doing I bear in mind that Philippe was not one to tow the party line. He often ignored the democratic means by which unfortunately so-called scientific truth is too often established today, and instead showed more interest in the unusual. Although such behaviour can sometimes lead one into danger, as it did with regard to Philippe's proposed explanation of the infamous 160-minute oscillation of the sun (Arvonny, 1983), it is also the most common road to true discovery.

Philippe would have been pleased to observe that one of the most prominent features of the meeting is the large number of young people present. A substantial fraction of the invited scientific papers and contributed posters is the work of a new generation of scientists. It indicates that heliophysics, and particularly the seismology used to probe the internal structure and kinematics of the sun, and of other stars and planets too, is a growing vibrant discipline.

We have been treated with some excellent descriptions of how one makes inferences from helioseismic data (e.g. Basu, 1997; Sekii, 1997). Basically, one observes some variable v on the surface of the sun, be it a velocity

397

J. Provost and F.-X. Schmider (eds.), Sounding Solar and Stellar Interiors, 397-424.
© 1997 IAU. Printed in the Netherlands.

or intensity fluctuation, which one relates theoretically to the underlying structure X of the sun via an operator $\mathcal{O}$ such that $\mathcal{O}X = v$. The object of the investigation is to infer X from v. We know that the inverse $\mathcal{O}^{-1}$ of $\mathcal{O}$ does not exist, so it is necessary to replace $\mathcal{O}$ by some other operator $\overline{\mathcal{O}}$ which acts on some property $\overline{X}$ of X which can be obtained from v; inference then results from interpreting $\overline{X} = \overline{\mathcal{O}}^{-1}v$. The search for operators $\overline{\mathcal{O}}$ that lead to straightforwardly interpretable functions $\overline{X}$ is the art, rather than the science, of inversion. Typically $\overline{\mathcal{O}}$ is designed to render $\overline{X}$ a spatial average of X, which provides a somewhat blurred image of the actual structure X; however sometimes it is more expedient to attempt to answer specific questions about more global properties of X directly.

Since most of the progress reported at this symposium concerns the sun, I shall concentrate on helioseismic inference. The final goal of helioseismology is not merely to measure the solar interior. In common with other branches of science, measurement is merely one step towards our true scientific goal: to understand why the sun is as it is, and ultimately to understand the fundamental physics that controls the sun, the other stars, and also the planets. An integral part of the process of understanding the underlying physics is the construction of a theoretical model, for only then can we relate X to that physics. Necessarily, we start with the simplest model that embodies what we presume are the essential physical ingredients, and typically we add complexity only when forced to do so by the data. That model is spherically symmetrical and in hydrostatic equilibrium. It is also in thermal balance. Even though it changes with time, as a result of modifications (which one trusts are improvements) to externally calculated physical relations such as the equation of state or the dependence of nuclear reaction rates or opacity on thermodynamic state variables, the model is called the 'standard solar model'. Aspherical perturbations, including macroscopic motion, are treated as low-amplitude deviations from the structure of the standard model; their reaction on that structure is usually ignored. (This remark does not apply to thermal convection. However, the asphericity of convection is not taken explicitly into account in the models; instead, a formalism is adopted which provides a spherically symmetrical equation for the heat flux and, occasionally, the turbulent pressure.) The influence on the global structure of genuine temporal variations, such as the solar cycle, on timescales less than the nuclear evolution time is rarely considered. The reader is referred to Janine Provost's article in these proceedings for further information. Thus, the standard solar model provides a basis with which to compare and to appreciate reality. It is not to be believed: it is simply to be used.

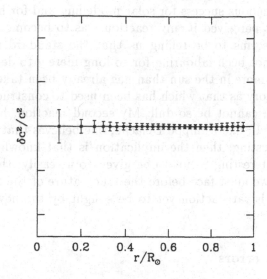

Figure 1. Generic (and somewhat inaccurate) representation of the relative difference $\delta\overline{w}/\overline{w}$ between localized averages $\overline{w}$ of any seismic variable w of the sun and the corresponding variable of the standard solar model. The filled circles represent the values of $\delta\overline{w}/\overline{w}$, and the error bars represent $\pm$ one standard deviation from them. The extent of the horizontal lines (not visible in this figure) through the filled circles measure the characteristic width of the weight function in the averages. The ordinate scale is a decreasing function of time, so I have not printed it as a protection against premature obsolescence. (Please note that no responsiblity for the accuracy of this figure should be laid on the shoulders of S. Basu or any of the others who have presented inversions at this symposium.)

2. The seismic structure of the sun

The seismic structure of the sun is that aspect of the stratification that is accessible to direct seismological investigation. Thus, it includes the variation with position of pressure p, density ρ and the first adiabatic exponent $\gamma_1 = (\partial lnp/\partial ln\rho)_{ad}$, and of any thermodynamical function of them.

It does not include the temperature, for example, because that is related to the seismic variables only through an equation of state which depends on the chemical composition. To infer temperature requires one to adopt additional, nonseismic assumptions, such as thermal balance coupled with the 'prior knowledge' of the mechanisms by which heat is transported. It is always useful to adopt the assumptions of the standard solar model in the first instance, but one must continually be mindful of the possibility that those assumptions may not apply to the actual sun.

To the best of my memory, the current status of the seismic investigations reported by Sarbani Basu in these proceedings is broadly summarized by Figure 1. Many scientists have hailed the immediate implication of that figure as a tremendous success for solar modelling and for helioseismology. But when I first perceived it, my reaction was to become profoundly depressed. For it seems to be telling us that the standard model is right. Surely we have not been labouring for so long merely to demonstrate that there is nothing more in the sun than has already been taken into account in so naive a theory as that which has been used to construct the standard models. The sun cannot be so dull. My second reaction, however, was to become cheerful. If, like Philippe Delache, one believes that indeed the sun actually is interesting, then the implication is that knowledge of what is perhaps most interesting is not to be given to us easily: there are further challenges that we must face before the true nature of the sun is revealed. Therefore there is satisfaction yet to be sought by the new generation of heliophysicists.

3. Systematic errors

It is a common feature of all the inversions that have been carried out to date that there is considerable uncertainty in the inferred structure of the energy generating core. The reason is that the results depend heavily on the frequencies of just the few low-degree p modes that penetrate into the innermost regions of the sun. Roughly speaking, the contribution of any region in the sun to the frequency of a mode is proportional to the time a propagating component of the mode spends in that region. Notwithstanding the stationarity of radial propagation in the vicinity of the lower caustic (turning point), the sound speed is so high in the core that core structure imparts only a minute signature on the frequency of a mode. Consequently there is a great danger that observational error, particularly any systematic component of it, will mask the true signature. This applies not only to multiplet frequencies, but to degeneracy splitting too.

Because of its extreme importance to core inference, I have always been interested in systematic errors. One of my earliest estimates of their influence was made with the encouragement of Philippe Delache during the 1983 conference on helioseismology in Catania (Belvedere and Paternò, 1984). Our concern there was with rotational splitting, and we were trying to reconcile the conclusions drawn from early disparate observations of low-degree modes. The error was presumed to result from the rotation of active regions on the sun, which biases the peaks in the power spectra of whole-disc Doppler data and of the projections of spatially resolved Doppler signals onto spherical harmonics. It was already known that solar activity induces a significant low-frequency signal in at least some Doppler data (Anderson and Maltby, 1983; Durrant and Schröter, 1983; Edmunds and Gough, 1983), but how should that bias the frequency splitting? The conclusion (which it was decided not to publish) was that the bias was small (actually miniscule) compared with contemporary differences between observers' reports. Therefore some other explanation of the observational discrepancies needed to be sought.

I returned to this problem of activity-induced bias nearly ten years later, this time in collaboration with Philip Stark, as a result of a statistical analysis which revealed that one of the sources of significant solar-cycle variation in the apparent rotational splitting measured by Libbrecht and Woodard (1990) was in the activity belt (Gough and Stark, 1993). The old calculation that had been carried out at Philippe's instigation was resurrected. Once again the bias was found to be small, and nothing other than a statement of the idea was published. The reason why the bias is small is that it is the product of two small quantities: one represents the relatively small contaminating influence that activity has on the frequencies, the other comes about because activity biases splitting by an amount proportional to the small difference between the rotation rate of the activity and the precession rate of the standing modes of oscillation.

In the meanwhile, interest in activity-induced bias in the inferred multiplet frequencies was stimulated by the emerging disparate inferences in core structure reported in 1988 at the helioseismology meeting in Tenerife (Rolfe, 1988). Philippe was there. The key point was that solar activity influences the even component of degeneracy splitting, which has a tendency to bias the centre of power, particularly in spectra of whole-disc measurements in which, for any degree l, substantial contributions from blended modes of all azimuthal orders m for which $l + m$ is even are present. Once again the outcome was found to be very small, and once again it was not published (this time, for reasons beyond the control of the author). Another source of bias is simply the global frequency shift due to solar-cycle changes in the spherically averaged structure of the outer layers of the sun. Although procedures to eliminate such effects on consistent data sets are incorporated into modern inversions, they cannot eliminate them entirely when two different data sets are combined, particularly if the data were analysed differently and, perhaps more importantly, if they were obtained in different epochs. Both of these sources of bias have been investigated recently (Gough, Kosovichev and Toutain, 1995) when combining the medium-degree data of Libbrecht, Woodard and Kaufman (1990) with low-degree data from IPHIR (Toutain and Fröhlich, 1992) and BiSON (Elsworth et al., 1991, 1994). Once again they were found to be insignificant.

The issue of multipole bias has been looked at again, quite recently, by Dziembowski and Goode (1997). Yet again, the influence on inversions was found to be small, and is (probably) currently insignificant, but now only marginally so. No doubt in the next generation of inversions it will have to

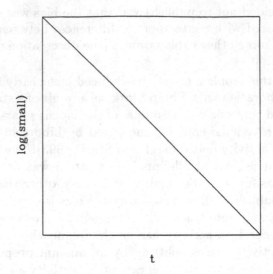

Figure 2. The logarithm of what is meant by 'small', plotted against time t.

be taken into account. Why should that be?

The reason is quite straightforward, and is illustrated in Figure 2. To some extent modern improvements in analysis techniques might be removing some of the bias, but more significant is the reduction of random errors. That changes what we mean by 'small', and once 'small' is smaller than the bias, the bias must be taken into account. Fortunately (for the inversions), almost all of the solar-cycle frequency change is produced by structural changes in the surface layers of the sun, and it can therefore quite easily be removed from the low-degree modes using the frequencies of modes of intermediate or high degree.

In what manner do the structural changes to the sun associated with the solar cycle modify the frequencies? Of course, there is the direct influence of what is presumably an augmented mean magnetic field at solar maximum in the outer layers of the convective envelope and in the atmosphere. That would increase the oscillation frequencies, by providing an additional restoring force. But there are indirect consequences of the changing magnetic activity, such as modification of the mean stratification of the upper convective boundary layer, and modifications to the convective fluctuations. The former cannot easily be reconciled with the frequency changes, and indeed appear to change them in the opposite direction to that observed (e.g. Gough and Thompson, 1988; Goldreich *et al.* 1992; Balmforth *et al.*, 1996). The effect of the latter is less clear. Brown (1984) discussed the effect just of advection by the velocity field, which decreases the frequencies, but the associated temperature fluctuations have the reverse effect. Indeed, it was another issue which interested Philippe Delache, who stimulated a rough assessment of the combined effect of all the convective fluctuations during the study programme at Santa Barbara (Gough and Toomre, 1991) using one of Nordlund's computer simulations. The outcome of our rough trial suggested, contrary to suspicions evoked by the usual convection scaling arguments, that there was a large degree of cancellation between the opposing influences, leaving the magnitude of the frequency shift substantially smaller than the earlier estimates, and its sign uncertain. It is encouraging that more serious work is now being undertaken in this important field, some of which has been reported at this meeting (e.g. by Rosenthal).

4. Gravity modes

It has long been known that much of our difficulty in measuring the structure of the core would vanish if we knew the frequencies of a few identified g modes. Gravity modes have a great advantage over p modes, namely:

Gravity modes are concentrated near the centre of the sun.

For that reason, unlike the case for p modes, the frequencies of g modes are very sensitive to the structure of the core. And for that reason, Philippe Delache expended a good deal of effort in their pursuit (e.g. Scherrer, these proceedings).

It is interesting that knowledge of only a few g-mode frequencies would enhance the inversions substantially. The reason is that in the delicate combinations of p-mode frequencies required to cancel out the influence of the structure of the solar envelope to infer even the crudest properties of the core, other information is lost in the noise. Therefore to relieve the p modes of this task leaves them free in other combinations to provide yet more subtle information. Indeed, the addition of only a single internal g mode to the data set would improve matters enormously. But, of course, a few would be better. We cannot hope for more.

As we have learned at this symposium from the latest reports on the current status of the networks and the helioseismic instruments on SOHO, g modes have not yet been found. The reason is that they suffer a grave disadvantage not shared by most of the p modes, namely:

Gravity modes are concentrated near the centre of the sun.

Consequently, their amplitudes at the photosphere, where we try to observe them, are on the whole small compared with the amplitudes of p modes of comparable degree having the same energy. According to Kumar (these proceedings), if interaction with the turbulence in the convection zone is the principal source of excitation, the expected energy in the mode varies only

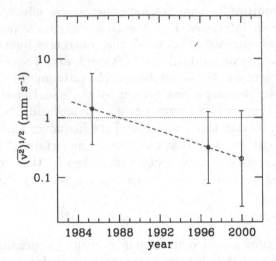

Figure 3. Estimates of the rms surface velocity amplitude of the gravest solar g modes, plotted against the time at which the results were announced. The filled circles represent theoretical estimates of the amplitudes of the modes (Gough, 1985; Kumar, 1997). The error bars were estimated from the uncertainties in the convective fluctuations estimated by Gough (1977) assuming that numerical parameters of the *calibrated* mixing-length formalism used to calculate the solar model are each uncertain by a factor 2 and that the uncertainties are independent. No contribution to the uncertainty from the approximations in the theory of excitation has been included. The open circle was obtained by linear regression, and its associated error bar was obtained assuming that the errors in the theoretical estimates (on the log plot) are independent and normally distributed. The horizontal dotted line is the ultimate threshold of detectability estimated by the observers. If that threshold is achieved by 1 January 2000, and if the assumptions upon which the extrapolation depends are valid (which is unlikely), then the probability of g-mode detection by the beginning of the year 2000 is 0.19.

weakly with frequency at low frequency. However, there is a rapid variation with frequency of the modal inertia, rendering the predicted power in the observations to be a rapidly increasing function of frequency.

It is, of course, of great interest to judge how likely it is that g modes will be detected, identified and have their frequencies measured. Therefore I compare the maximum expected amplitude with the anticipated threshold of measurement. For years George Isaak has assured us that an rms velocity in the solar atmosphere of 1 mm s^{-1} (and probably not much lower) is attainable, and Eric Fossat (e.g. 1985) concurs. Moreover, we have seen at this meeting that the latest observations are within a stone's throw of that value. How does that compare with theoretical expectation?

Until this meeting, to the best of my knowledge, there was only one estimate of the amplitudes of the gravest (i.e. lowest degree and order) modes (Gough, 1985). At this meeting Pawan Kumar has offered us a second, which provides us for the first time with a basis for extrapolation. The two estimates are plotted in Figure 3. It is difficult to know how the value will change as the theory is refined – I hope that the matter will be settled before there is time for much further refinement. However, linear extrapolation, although it is unreliable, suggests that the real amplitudes (to be calculable and, it is hoped, measurable in the future) are even lower. Fortunately, the observational upper bounds are decreasing somewhat faster. I do not represent them in Figure 3, however, but instead include the estimate of the ultimate threshold promulgated by the observers. Unfortunately, the intersection of the two lines does not tell us unambiguously whether a useful detection will be attained by the end of 1999, which for some of us is a criterion of considerable interest (Gough, 1995).

I should not end this discussion without mentioning a recent claim by Thomson *et al.* (1995) to have detected solar gravity waves propagating through the solar wind, not least because it is the result of an unusual investigation of a character that delighted Philippe Delache. There is a sequence of sharp peaks in the power spectrum of particle fluxes measured on the spacecraft Ulysses with frequencies corresponding to those of the solar oscillations – both p modes and g modes. A sophisticated statistical comparison of the frequencies of the fluctuations in the wind, after modifying them with a Doppler shift which it was hoped would account for the motion of the spacecraft relative to the spiralling wind, with the known solar p-mode frequencies led Thomson *et al.* to conclude that the two are causally connected. The argument is not wholly convincing. However, if one accepts it, then it should not be too unreasonable to accept that the peaks in the power spectrum in the g-mode frequency range are signatures of solar oscillations too. Not surprisingly, there has been considerable debate amongst scientists about this claim, much of it concerned with interpreting the frequencies as (extremely) high harmonics of the solar rotation, but there are yet few publishable conclusions. To convince most of us would require a substantially more extensive analysis of the data, particularly in the p-mode range where the presumed source of the waves in the wind has been measured directly.

5. Linewidths, excitation and damping

In his review of mode excitation, Pawan Kumar discussed the issue of the linewidths in the power spectra of acoustic oscillations. These constitute one of the important ingredients in the theoretical computation of the oscillation amplitudes. There have been several observational reports of linewidth measurements. Libbrecht and Woodard (1991) showed that, when scaled with the inertiae of the modes, they depend essentially on frequency alone.

This is as one would expect, since the strongest damping and excitation processes take place in the outer regions of the sun, mainly, we believe, in the upper superadiabatic boundary layer of the convection zone.

On the whole the linewidths increase with frequency ν, but there is an interesting dip in Libbrecht's data centred at about 3 mHz (Figure 4). Why should that be? It appears that it might be due to the influence of convection, since all calculations that ignore convection find the linewidths to be monatonically and smoothly increasing with ν. Christensen-Dalsgaard *et al.* (1989) compared the theoretical predications with Libbrecht's (1988) observations, and found tolerable agreement with a computation in which the modulation of the turbulent heat flux and Reynolds stress (turbulent pressure) by the oscillations had explicitly been taken into account, admittedly using only a very crude time-dependent local mixing-length theory. Subsequently, Balmforth (1992) obtained considerably better agreement using a nonlocal generalization.

Some solar physicists have believed – and perhaps still do believe – that the dip represents a transition between two different damping processes, one which predominates at low frequency and the other at high frequency. But the theory that predicted the dip had the same mechanism operating throughout. According to the theory, the properties of the modulation of the convective heat and momentum fluxes, which determine the energy exchange between the oscillation and the background state of the star, depend locally on the ratio σ of the characteristic turnover time of the convective eddies to the period of oscillation. Roughly speaking, the frequency range in which the damping rate is low compared with a smooth curve is that in which the spatial phase of maximal coupling of the eigenfunction with the convection occurs at a level in the star at which the value of σ is such as to cause minimal damping.

In his presentation, Pawan Kumar discussed none of the theoretical predictions, but instead offered an adjustment computed *post hoc* by Murray (1993), which is reproduced here in Figure 4. Perhaps that is because Murray's adjustment has attracted more attention in scientific conversations than the truly theoretical calculations, such as those carried out previously by Balmforth (1992). The reason, apparently, is that it is well known that the theory upon which Balmforth's calculations are based is highly uncertain. So what is the basis of Murray's computation? It is summarized in the caption to the figure (which could perhaps have been read only by those sitting in the front few rows). Murray considered damping from wave scattering by turbulence, and simply reduced the damping rate artificially in such a manner as to make the outcome appear to agree with the observations. I say this not to belittle the computation, but to make explicit what was actually done, in order that the result can properly be appreciated.

One can then recognize that Murray's result is important, for it confirms our view of how the line widths are controlled.

The manner in which Murray's adjustment was accomplished can be understood by translating frequency into eigenfunction phase. Any localized perturbation to the solar model, or its interaction with the oscillations, has an oscillatory signature with respect to frequency: as frequency varies, so does the phase of the eigenfunction at any specified location, and a perturbation at that location produces an oscillatory contribution to the (complex) frequency when that location lies in a propagating region of the mode (cf. Gough and Thompson, 1988; Vorontsov, 1988). For perturbations in the outer layers of the sun, the apparent 'frequency' of the oscillatory contribution is twice the acoustical depth of the location of the perturbation, namely $2(T - \tau)$, where $\tau(r)$ is the acoustical radius of the perturbation and T is the acoustical radius of the sun. It is evident from a comparison with observation of Murray's unadjusted theory that an oscillatory contribution with a 'period' of about 2.5 mHz is required. Therefore an appropriate reduction of the scattering at an acoustical depth of $(5 \text{ mHz})^{-1} = 200$ s does the trick. That location is at $\tau/T \simeq 0.94$, corresponding to $r/R \simeq 0.9965$ (cf. Gough, 1990), which is in the middle of the hydrogen ionization zone, about half a wavelength of a 3 mHz eigenfunction beneath its upper turning point. Murray accomplished the adjustment by multiplying the scattering by a positive power of $\gamma_1 - 1$, which is smallest in the hydrogen ionization zone.

It is important to realize that none of the calculations incorporates the whole story. Murray's calculation ignores nonadiabatic effects and the interaction of the oscillations with the perturbed Reynolds stress; Balmforth's calculation ignores scattering off spatial inhomogeneities. Since all three processes are weak compared with the dominant dynamics of the oscillations, to a first approximation one might simply add the contributions to the damping rate. Unless scattering is relatively unimportant at frequencies near 3 mHz, it seems likely that agreement with observation will not be achieved without at least some adjustment of the scattering process discussed by Murray, because Murray's *ab initio* theoretical estimate of the contribution to the linewidth by scattering alone exceeds the observed values.

It is important to realize also that there is considerable uncertainty in the magnitude and form of the turbulent convective fluctuations, even after the formalism for calculating the heat and momentum transport has been calibrated to yield a complete theoretical model of the sun having the observed luminosity and radius (Gough, 1977). Therefore there is no doubt adequate leeway to reconcile theory with observation. It is my opinion that in the fullness of time the observations will be used in the reverse sense: to calibrate turbulence theories. Rosenthal's contribution in these proceedings is a step towards that goal.

Another indication of the properties of the turbulent fluctuations, and how they interact with the oscillations, is provided by the statistical distribution of mean power. This was discussed by Yvonne Elsworth in her review. The long BiSON data set has provided a valuable confirmation that on the whole the energy in the modes satisfies the Boltzmann distribution (Elsworth et al., 1995; Chaplin et al., 1995), which is as one would expect if the modes are excited stochastically by the turbulence in the convection zone. But also, it reveals that the frequency of occurrence of highly energetic modes is greater than Boltzmann. There are two posters at this conference addressing the issue. Relative fluctuations about the expected frequency of the common low-energy occurrences are small, and the data adhere well to the most likely distribution, irrespective of the details of the excitation process. But in the high-energy tail there remains some legacy of the excitation, and because that process is superexponential, the distribution that is realized lies above the Boltzmann line. Once again in the fullness of time, further study of this distribution might lead us to a better understanding of the convective motion.

Another way to calibrate convection theories is by studying the peaks in the power spectrum at frequencies above the critical cutoff frequency of the atmosphere. This is perhaps a more direct method, for, unlike the trapped modes, these waves suffer almost no reflection in the surface layers of the sun, and instead propagate directly away. Because the refracting properties of the sun are such that waves emanating from any local region are divergent almost everywhere, there is no substantial coherent interference to mask the excitation process. The peaks in the power spectrum arise simply as a result of observational filtering. As discussed by Pawan Kumar at this symposium, the positions of the peaks provide an estimate of the depth at which the excitation occurs. His most recent calibrations yield a value of about 140 km beneath the photosphere, which now agree with earlier local studies by Goode et al. (1992).

Another very interesting calibration can be carried out by matching the variation with frequency of the amplitudes of the supercritical peaks. It determines the spectral index of the turbulent energy, which Kumar (1994) has shown agrees with the Kolmogorov law. This is one of the few cases in nature in which evidence, albeit indirect, for the existence of inertial-range turbulent cascade has been found.

Returning to the linewidths of the trapped p modes, it is interesting to observe that there is not always obviously a local minimum near 3 mHz. Figure 4 shows three examples, one from Libbrecht and Woodard (1991), one from BiSON and the other from VIRGO. Libbrecht and Woodard's data

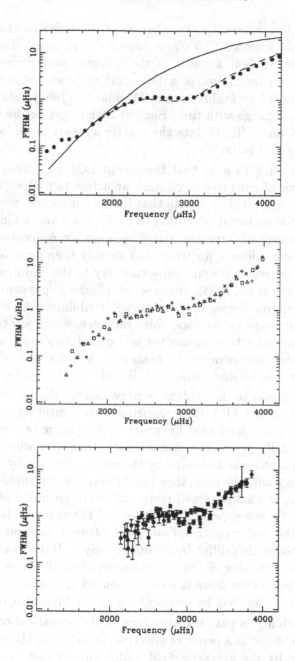

Figure 4. Full width at half maximum of p-mode power, plotted against cyclic frequency ν: top panel from Libbrecht and Woodard (1991), middle panel from BiSON (Elsworth, these proceedings), bottom panel from VIRGO (LOI and SPM) (Fröhlich, these proceedings). The filled circles in the top panel represent data from low-l modes (l probably between 5 and 60) in 1989 scaled by the modal inertiae to $l = 0$. The continuous curve is Murray's (1993) estimate of the contribution to the damping rate from scattering. The dashed curve was obtained from the continuous curve by Murray by artificially decreasing the assumed turbulent eddy correlation length in the hydrogen ionization zone by about 30 per cent.

exhibit a minimum near 3 mHz, whereas in the BiSON data there may be only a point of inflexion. No discussion of the apparent disparity has yet been put forward. Is it a result of the different procedures for analysing data, or is it an indication of a temporal variation related, perhaps, to magnetic activity? In Libbrecht and Woodard's (1991) data one can see a slight hint of a change with time. Such change might be the answer to why in the BiSON and VIRGO data the scatter appears to be locally maximal in the vicinity of 3 mHz.

It is interesting to note that the recent LOI data from VIRGO show slight evidence of enhanced excitation at active latitudes (Fröhlich *et al.*, these proceedings). This suggests that in the convection zone there is perhaps an extra component of motion to excite the waves which is associated with the presence of intensive active-region-scale concentrations of magnetic field, some evidence for which has already been provided by Haber *et al.* (1988). That result is apparently contrary to the findings of the BiSON group (Elsworth *et al.*, 1993), however, who find a 35 per cent augmentation of the power in low-degree modes near solar minimum. One is reminded of the similar disparity in the total solar radiance, which is decreased locally by the presence of active regions yet is a global maximum at the epoch in the solar cycle corresponding to maximum activity (e.g. Fröhlich, 1994). That too is yet to be understood (cf. Balmforth *et al.*, 1996).

A final thought on linewidths is provoked by the interesting autocorrelation analysis of GOLF data reported at this meeting by Gérard Grec. It is commonly assumed that linewidth is a measure of mode lifetime, by which one usually means the characteristic time over which a mode oscillates before its phase is destroyed by the random forcing by the turbulence. If that were actually the case, then the autocorrelation should decay on this timescale too, at least for small temporal displacements τ of the signal. At large values of τ one expects the envelope of the autocorrelation to tend to a constant, the major contribution coming from components of the signal excited at epochs that differ by approximately τ. But Gérard Grec reports that the observed value of that constant significantly exceeds expectation, which is evidence that there is a component of the signal with longer-term coherence. How can that be reconciled with the linewidth data?

One possibility is that the structure of the acoustical cavity is varying with time, perhaps as a result of effective distortion of the upper boundary of the cavity by the magnetic field, either directly, or by sound-speed or convective velocity variations that are modulated by the magnetic field. To be sure, we know that magnetic activity changes oscillation frequencies (Woodard *et al.*, 1991; Bachmann and Brown, 1993). This would cause a deviation of phase from that of a purely sinusoidal oscillator, thereby contributing to the spectral line width, but not necessarily destroying totally the autocorrelation. The result reported by Gérard is therefore very exciting, for, if this suggested explanation is correct, it would indicated that genuine damping is actually less than had previously been suspected, opening up the possibility, with suitable analysis (cf. Chang, Gough and Sekii, 1995), of measuring oscillation frequencies more accurately, and thereby improving our helioseismological inferences about the structure and kinematics of the solar interior.

6. Angular velocity

The reviews by Sasha Kosovichev, Takashi Sekii and Michael Thompson have shown quite clearly that our early inferences of the variation of angular velocity Ω in the sun are more-or-less correct. Broadly speaking, it appears that the latitudinal variation of Ω observed at the surface of the sun is maintained essentially to the base of the convection zone, and then there is a sharp transition to uniform rotation in the radiative interior. What happens deep in the core is not yet settled.

As Spiegel and Zahn (1992) have discussed, Ekman circulation currents are set up beneath the base of the convection zone in the rotational shear layer, which they call the tachocline. These currents exchange material with the convection zone. I shall return to this point later. But I remark now that evidently, if the tachocline extends deeply enough for lithium and beryllium to be destroyed by nuclear reactions, or if it has done so in the past, the tachocline must have an important impact on the issue of the subcosmic photospheric solar abundances of these light elements.

Why the convection zone rotates as it does is a matter that is not understood, but which is presumably intimately associated with the anisotropy of the turbulent motion imposed by Coriolis forces, and the meridional flow it induces. The characteristic timescale for adjustment to external influences is expected to be of the order of only a year or so, which implies that the convection zone determines its own rotational structure. The adjustment of the radiative interior takes much longer. Why should it be rotating uniformly, at least in the outer layers?

The suggestion by Spiegel and Zahn (1992) was that instability of the shear in the stably stratified boundary layer induces small-scale two-dimensional turbulence in horizontal surfaces: the turbulence acts on the large-scale flow like a viscosity (in horizontal surfaces), and thereby leads to rigid rotation at the base of the tachocline. The equilibrium rotation of the interior is therefore also rigid, provided Eddington-Sweet circulation is ineffectual. This view is not unnatural, but it should be remarked that, as Michael McIntyre (personal communication) has emphasized, two-dimensional turbulence in the Earth's atmosphere tends to render potential vorticity rather than angular velocity uniform, at least away from the polar regions. One cannot be sure, therefore, that uniform rotation should be the natural state of the radiative interior under these conditions, unless the interior is pervaded by a magnetic field.

It is also necessary to explain how the equilibrium state of the rotation of the radiative interior is attained. In his review of the dynamics of stellar rotation at this symposium, Jean-Paul Zahn put forward the suggestion that the uniformity of rotation is established by angular-momentum transport by gravity waves generated in the lower boundary layer of the convection zone. This idea is very interesting to me, partly because, to the best of my knowledge, angular-momentum transport by gravity waves in the sun was first discussed at the IAU Colloquium organized in 1976 by Philippe Delache (and Roger Bonnet). However, the conclusion then was that the process was probably not important for the global distribution of angular momentum (Gough, 1977). The reason was that waves that resonate with the largest convective eddies at the base of the convection zone dissipate before they penetrate very deeply into the radiative interior, and waves that resonate with smaller eddies arising from a turbulent cascade and which are able to propagate more deeply have insufficient amplitude to transport an interesting amount of angular momentum. Furthermore, the higher-frequency grave low-degree modes, which might be generated near the top of the convection zone, hardly dissipate at all. Only if the core of the sun were convective would there be dissipation enough to communicate a substantial stress.

Several studies have been carried out since, with broadly similar results, but the result of Jean-Paul Zahn is different. It appears that Jean-Paul's estimate of normal wave dissipation is substantially lower than mine, but it is not clear to either of us why that is so. I, for one, will certainly be interested in any further illucidation of the process.

7. Magnetic fields

Measurement of the characteristics of solar oscillations is reaching a level of precision sufficient to enable us to detect the influence of magnetic fields. That magnetically active regions absorb or scatter acoustic waves in the sun has been known for a decade (e.g. Braun et al., 1987), and there is some evidence also for emission by major flares (Haber et al., 1988). Magnetic activity on the surface of the sun also augments p-mode frequencies (Woodard et al., 1991; Bachmann and Brown 1993), reducing the acoustic volume of the cavity either by effectively lowering the level of the upper turning point, thereby diminishing the physical size of the cavity, or by directly augmenting the wave propagation speed. The relation between the oscillation frequency perturbations and the intensity of magnetic activity appears to be essentially independent of the timescale of variation of the activity, which is consistent with the idea that the frequency perturbation is the direct effect of the presence of the magnetic field, B, and is influenced only weakly by the large-scale thermal adjustments associated with the solar-cycle radiance variation. This is consistent with the conclusions of Gough and Thompson (1988b), Goldreich et al. (1991) and Balmforth et al. (1996), who found that potential magnetically induced modifications to the thermal stratification of the upper boundary layer of the convection zone of a magnitude compatible with the observed radiance variation is insufficient, and of the wrong sign, to account for the oscillation frequency shifts.

There have also been hints of deviations from the standard theoretical models of the hydrostatic stratification of the sun in the vicinity of the base of the convection zone, which Dziembowski and Goode (1989) have interpreted as being evidence for a toroidal B. And at this symposium, Haber et al., using local spectral ring analysis, and Ryutova and Scherrer, using time-distance analysis, provide evidence for anisotropic wave propagation which they interpret as being a possible signature of B. These results are very encouraging for those with an interest in the dynamics of magnetically active regions and the maintenance of the solar cycle. But one must Beware! Anisotropic wave propagation can result also from horizontal thermal inhomogeneity. Such inhomogeniety in the outer layers of the sun can distort the shapes of spectral rings and impart a directional dependence on time-distance analysis. To separate it from B requires careful piecing together of the results of analyses over a network of local areas. Likewise, the signature of the putative toroidal B inferred by Dziembowski and Goode might actually have been produced by a latitudinal variation in the structure of the tachocline, and not by a magnetic field at all.

It is only where the magnetic stresses are comparable (within no more than just a few powers of ten) with the gas pressure that there is some hope of a direct seismological detection. Otherwise one must resort to indirect methods. In my opinion it is therefore unlikely that a direct measurement of the large-scale field that pervades the radiative interior of the sun will be made, at least in my lifetime. My reason is simply that my estimate of that field (Gough, 1990b), which presumably has decayed ohmically from an initial poloidal field of characteristic intensity B_0 of about 1T, provides a stress of at most 10^{-6} of the gas pressure in the radiative core and envelope. But of course my estimate could be widely wrong. It is therefore incumbent upon us to ask the question (using Jamie Matthews' licence to talk in terms of H rather than B):

What is the value of H_0?

That is a question of interest to a community much wider than the participants of this symposium. It is evident that any value even remotely close to my estimate is sufficient to transport angular momentum from the core to the convection zone in less than the age of the sun – actually 10^{-9} T is sufficient – and hence maintain the radiative envelope in a state of almost uniform rotation provided that there is no perturbation on a significantly shorter timescale. One wonders whether it is no more than an interesting curiosity that a field intensity of order 1T would cause a response of the sun to a torsional perturbation on a timescale comparable with the characteristic period of the solar cycle.

8. Telechronohelioseismology

One of the highlights of this symposium was the report by Sasha Kosovichev on his work with Tom Duvall on measuring the flows in the upper layers of the convection zone. The earliest results of time-distance techniques essentially reproduced what we had learned already from analysing the presumed eigenfrequencies of free oscillation modes. But here we have witnessed a great leap forward in the diagnosis of lateral inhomogeneity. This, no doubt, is but a first jump beyond what we have achieved by our older methods, and I am quite sure that there is a great deal more kilometrage yet to be gained from the technique. Indeed, the technique is likely to provide the most powerful tool available for studying the dominant energy-containing eddies in the upper reaches of the convection zone.

In principle, telechronoseismology can be used for diagnosing any structural or kinematical aspect of the sun. Whether it is more prudent to use this technique or, for example, a normal-mode technique will depend partly on the relative difficulty in carrying out the observations and the data analysis, and partly on the extent to which the results are contaminated by noise. My guess is that, at least at first, the technique comes into its own in diagnosing near-surface inhomogeneities. It could be particularly powerful when those inhomogeneities are not immediately beneath part of the visible hemisphere. For example, one can construct an acoustical lens, by cross-correlating signals with suitable time lags from diametrically opposite portions of circular annuli on the visible surface of the sun and averaging around the annuli, to focus on an antipodal region of the solar surface. In this way one should expect to detect active regions forming on the far side of the sun, and thereby have a truly global image of at least the sun's outer layers. Coupled with observations discussed by Judit Pap and Anne Vigouroux at this symposium of changes in radiance, apparent solar diameter and other surface properties, it should help us gain a deeper understanding of the overall mechanism of solar variability. An obvious immediate potential use to others might be as an early warning for space storms. Of perhaps less immediately obvious importance to us might be, especially if g modes are not detected, its use for monitoring the emergence of major perturbations to the surface of the sun's acoustic cavity. It may then be possible to correct for the frequency modulation of the p modes of low and intermediate degree, in order that we might improve the frequency estimates and thereby measure more accurately the structure deep in the solar interior.

9. A few remarks on extrasolar observations

I shall not dwell long on asteroseismology or planetary seismology, because both disciplines are in their infancy and the current status of them has been well discussed by the reviewers. Both disciplines are healthy, which is evident from the optimism of the speakers, even though resources may not be as copious as one would like. Indeed, from Søren Frandsen we learned that in two years from now the observational situation will have improved to the extent that asteroseismology will have become a real subject.

The sun has been an excellent playground on which to have learned the basic techniques of seismological inversion. We are thus well equipped to venture into the wider arena, where oscillation amplitudes are sometimes greater or where deviations of the basic structure of the star from the relatively simple spherically symmetrical state are large. Wojtek Dziembowski announced that the nonlinearity experienced in large-amplitude oscillators is a disaster. On this point I cannot agree, firstly because our younger colleagues need challenging obstacles against which to pit their wits, and secondly because nonlinear dynamical behaviour is always richer with information than is its linear counterpart. Of course, in order to extract that information, whether it be from linear or nonlinear oscillations, we must necessarily be able to identify the modes of oscillation that are being observed, as Nathalie Audard, Wojtek Dziembowski and Jaymie Matthews, amongst others, have emphasized. Large-amplitude deviations from spherical symmetry also produce interestingly richer behaviour. An example is rotational splitting of g modes in stars whose angular velocity is comparable with the frequency of oscillation. Another example is the existence of only prograde modes in some rotating stars, which Søren Frandsen reported. I have always suspected that that might be a result of critical-layer absorption of the retrograde g modes, but I've never worked hard enough on the problem to demonstrate whether or not the idea is plausible. A third example is the rapidly oscillating Ap stars, which appear to have large magnetic spots in which the stratification is presumably quite different from that in the quiet regions. As Jaymie Matthews explained, the magnetic field

may have a major influence on the structure of the core too, as well as on
the dynamics of the oscillations. The situation is apparently quite rich in
possibilities, and interesting inferences are bound to emerge once the various influences on the pulsation frequencies have been unravelled. Finally,
amongst the stars, are the white dwarfs. As Gérard Vauclair has explained,
a period-mass-luminosity relationship enables one to determine their distances, analogously to the Cepheids. Measurements of rotational splitting
of PG 1159 and pulsating DB dwarfs are reaching a level of detail to permit
differential rotation in those stars, when it is present, to be discernible. All
these and many more new advances will supply asteroseismologists with
plenty to think about in the years ahead.

Finally, I come to Benoît Mosser's discussion of the seismology of giant
planets. I draw attention to this subject partly because there is some interesting physics to be learned from its study, and partly because the seismic
observations were instigated by Philippe Delache (Schmider *et al.*, 1991).
The greatest interest at present is in Jupiter, because, unlike Saturn, there
is good positive observational evidence that it is pulsating. There seems to
be less known about the structure of Jupiter than, for example, the structure of the sun. To be more precise, I suppose I should really say that we are
less confident in the theoretical models of Jupiter, and that the physics of
those models has been less thoroughly studied. Observationally, we known
the mass and the radius, of course, as we do of the sun, together with some
moments of inertia deduced from multipole moments of the gravitational
potential. Theory suggests there has been chemical differentiation leading
to a rocky core, possibly surrounded by a layer of ice; and acoustically mid-
way between the centre and the surface there is probably a plasma phase
transition between metallic and molecular hydrogen, which may or may
not occur in a discontinuity. Moreover, substantial gravitational settling of
helium appears to have taken place, judging from the low abundance of helium observed in the surface. How extensive that settling is is uncertain. So
also is the extent of the settling of heavier elements, so the very existence
of the rock and ice cores, with distinct boundaries, is a matter of debate.

If the cores do exist with sharp boundaries, and also if the boundary
determined by the plasma phase transition is abrupt, then there must be
oscillatory signatures imparted on the oscillation eigenfrequencies, whose
frequencies (with respect to frequency) and amplitudes depend respectively
on the acoustical radii of those boundaries and on the magnitudes of the
discontinuties in the equation of state across them (Provost *et al.*, 1993;
Gough and Sekii, 1995). However, unravelling those signatures will require
extremely accurate frequencies. The reason is twofold. First the plasma
phase transition does not perturb the frequencies by very much. And second, the core boundaries are acoustically so close to the centre of the planet

that very little of the oscillatory perturbations, perhaps less than a period, can fit into the frequency range, which is presumable bounded above by the critical cutoff frequency of the atmosphere. Consequently it will be very difficult to unravel the two periodicities, if indeed they are even present. However, the situation is mitigated by the fact that the magnitudes of the discontinuities in the core are quite large, leading to greater complexity in the oscillatory signatures, which carry more information per datum than sinusoids do. Of course if the transitions are not sharp, the signatures will be smoothed out, and inference will be correspondingly more difficult.

Further observations are planned for the future, of both Jupiter and Saturn, and it is to be hoped that soon we shall have a seismic spectrum to study. I might point out that David Thomson (personal communication), whose claim that I mentioned earlier in connexion with g modes to have discovered waves in the solar wind produced by solar oscillations (Thomson *et al.*, 1995), says he now has evidence also of a Jovian signal. If that evidence is convincing, perhaps we shall have data earlier than we anticipated.

10. The golden path to happiness

The consolidation of the data from the well established ground-based networks BiSON and IRIS, and the rapidly accumulating data both from the ground-based GONG, TON and the other new networks and from the seismic instruments on SOHO, all of which have been reviewed in Session I of this symposium, must surely lead to an escalation of publications of a host of inferences. At first we shall commonly see inversions of the kind we have witnessed in the past, but no doubt improved by virtue of the increase in both quality and number of data and of new refinements in the techniques to analyse them. From those inversions will be raised new questions. And then it will often be the case that subsidiary methods of data analysis will need to be developed in order to answer them.

It is easy to ask questions if one adopts the following maxim:

Always overinterpret the data.

Indeed, it is a scientist's responsibility to do so, provided that it is not done irresponsibly. To discuss hints of phenomena at the threshold of detectability whets the appetite for improving the data and their analysis so that the threshold is pushed back and the phenomena are revealed (or not, as the case may be). So I conclude by indulging in the activity, as Philippe Delache was wont to do. One must be aware, however, that the indulgence is in speculation, not in deduction. Therefore its purpose is simply to provide discussion, not to make claims.

The motivation is to raise one's spirits in the face of the depression that some of us have suffered at the sight of Figure 1. A more optimistic view

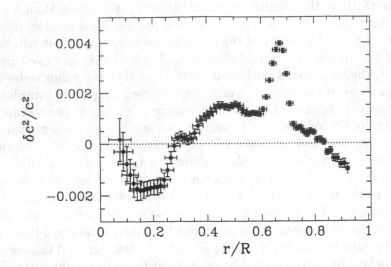

Figure 5. Relative differences $\delta c^2/c^2$ between the squares of the sound speed in the sun and in a theoretical reference model, using data from the SOI/MDI instrument on SOHO (from Kosovichev *et al.*, 1997). The meaning of the symbols is the same as in Figure 1. The reference model is the same as that used for the GONG structure inversions reported in these proceedings. It is a standard model in which some account is take of the effects of gravitational settling of helium and heavy elements.

is provided by Figure 5, which is a recent sound-speed inversion of MDI data (Kosovichev *et al.*, 1997). The structural inversions of GONG data are superficially similar (Anderson *et al.*, these proceedings). To be sure, the deviation of the sound speed c from that in the reference theoretical model is typically less than a part in a thousand; that value is exceeded in only about 5 per cent of the radius range, and then by no more than a factor of two or so. Therefore, one might argue, the theoretical solar model is very nearly correct, which is the message that Figure 1 was intended to convey.

Is that conclusion actually correct? The important point to notice about Figure 5, which differs qualitatively from the false view offered by Figure 1, is that the inversions actually indicate a very significant deviation from the reference model. Indeed, in places that deviation exceeds 20 standard deviations of the formal random errors. So clearly there is a sense in which either the standard reference model or the inferences from the data are very bad indeed. The errors must be explained. How do we go about it?

The boring way to try to account for the disparity between the sun and the reference model is to argue that minor adjustments need to be made to the theory, by tweaking the equation of state, the opacity and the nuclear reaction rates. With such small deviations to reproduce, one might not unreasonably think that that route might lead to the desired end. And

I am quite sure that in due course the modelling industry will carry out the necessary calculations. But it is much more interesting, and no doubt more realistic, to take a more global view of the situation.

Before proceeding it might be worth determining what the desired end is. After all, one needs to be able to tell whether or not one has arrived at one's destination. There are some, possibly lucky, people who would be content once Figure 5 really does look like Figure 1. They want merely to find a theoretical model that can account for the data. I recall that two decades ago this was a common attitude amongst those in search of resolving the solar neutrino problem: some of the experts in the field took the view that if a phenomenon was not 'needed' for reducing the neutrino flux its role should not be entertained when trying to 'explain' the workings of the sun. And it seems that that view has not yet completely disappeared. However, I presume the universal view at this symposium is different. The objective is to understand the sun, and, if two models that differ in structure both reproduce the observations that are available, new distinguishing observations must be sought. And, of course, one must continue to try to refute the surviving unique theory, once we find it.

Let us begin our scrutiny of Figure 5 with the greatest anomaly, situated immediately beneath the convection zone. I have already mentioned that the Ekman meridional circulation in the tachocline, discussed by Spiegel and Zahn (1992), exchanges material with the convection zone, and, if penetration into the radiative interior is sufficient, reduces the photospheric abundances of lithium and beryllium by nuclear transmutation. It must also reverse some of the gravitational settling of helium from the convection zone, which in the theoretical models is impeded only by diffusion. Replenishment of the relatively hydrogen-rich material of the convection zone would reduce the mean molecular mass in the tachocline, thereby increasing the sound speed. Indeed, measuring the sound-speed anomaly in Figure 5 is at present probably a more accurate way of measuring the extent of the tachocline than is fitting to the rotational splitting data a functional form for the angular-velocity transition. I must point out, however, that the observations indicate that the hydrogen has not been completely replenished, which is consistent with the helium settling rate in the tachocline being comparable with advection. The resulting composition gradient exerts a retarding negative buoyancy force on the Ekman flow, which has not yet been taken into account in the dynamical studies. Nevertheless, present indications are that substantial lithium depletion may result, although more refined dynamical and seismological studies will be required before we can be sure.

It should also be pointed out that Stokes drift associated with the gravity waves generated near the base of the convection zone can also transport material. I discussed a simple estimate of the effect some years ago (Gough, 1988), and found it probably to be unimportant on a global scale. But in the light of the more complicated coherence in the transport discussed by Knobloch and Merryfield (1992), and the discussion at this symposium by Jean-Paul Zahn, the matter needs to be reexamined.

An important feature of the anomaly immediately beneath the base of the convection zone is its thinness. This suggests that the cause is not thermal, such as one might find with a simple modification (e.g. one whose sign does not change) to opacity, for opacity changes tend to produce a broad response (e.g. Christensen-Dalsgaard, 1996). Of course, one can always contrive an artificial opacity perturbation of a form that would reproduce the anomaly, and such a perturbation should not be ignored. However, because model builders have on the whole followed a path determined by physics, even though they have been led largely by the observations, one expects that the final route to reality is unlikely to involve contrivance. A localized change to the $p - \rho - T$ relation determined by the equation of state can produce a more localized sound-speed response, and my suggestion is that that has been brought about via a change in the helium abundance. I should point out, however, that if the sun were to have lost sufficient mass during its main-sequence evolution, there could have been a similar outcome: the material immediately beneath the convection zone would previously have been at greater depths where helium-abundance augmentation by gravitational settling was slower (and lithium and beryllium destruction more rapid). Indeed, an appropriate amount of mass loss reduces the disparity between the theoretical model and the sun substanially (Gough *et al.*, 1996), though probably not entirely. Therefore there are at least two candidate mechanisms for the proposed abundance anomaly, and both must be investigated. One must bear in mind, however, that the candidates do not have the same footing. There is little evidence for the sun having lost a substantial amount of mass during its main-sequence evolution phase, whereas the dynamical evidence for the existence of a tachocline circulation is overwhelming. What is uncertain on theoretical grounds, however, is how deeply that circulation should penetrate. It is encouraging to observe that the apparent thickness of the tachocline – roughly 5 per cent of the solar radius if indeed essentially the whole of the anomaly is due to it – is similar to the value suggested by Spiegel and Zahn (1992) prior to its measurement.

I conclude by drawing attention to the second most prominent anomaly, namely that in the energy-generated core. Hence I really show my prejudice by finding that anomaly to be more interesting, and hence more prominent in my mind, than the discrepancy of similar magnitude in the radiative envelope beneath the tachocline: $0.3 \le r/R_\odot \le 0.65$. But that interest was shared with Philippe Delache, as was the dynamical result I am about to remind you of, which is one of the reasons why Philippe worked so hard looking for g modes. So my ending is not unfitting.

According to Figure 5, the sound speed in the central region of the solar core appears to be greater than that in the reference model, and that in the surrounding shell is lower. Of course, this anomaly too would no doubt be accommodated by an appropriately designed modification to the opacity and the nuclear reaction rates, and perhaps such modifications do play a role. But as in the tachocline, there are dynamical considerations which might lead one to suspect that the matter is not so straightforward.

The cores of solar models are linearly unstable to grave gravity modes (e.g. Christensen-Dalsgaard, Dilke and Gough, 1974). Although the property has been known for more than two decades, it is largely ignored by the solar modellers. If the sun has suffered such instability too, then the nonlinear development of the instability must have influenced the structure of the core in a manner that is not incorporated in the standard solar models. The important issue with regard to Figure 5 is whether or not the magnitude of the effect is great enough to have a significant bearing on the mean stratification. There is, of course, also the more general issue of understanding the dynamical development of the instability. There have been a variety of studies, with a variety of suggested outcomes, including transition to direct convection and nonlinear limiting of the oscillations at an amplitude great enough to play a significant role in the nuclear evolution of the core, and also nonlinear limiting, via a triad interaction with a pair of stable g modes, at a structurally insignificant amplitude. However, all of the calculations are too idealized for one to carry over the results to the real sun. The issue is still open.

The issue is also currently topical, as a result of a recent provocative investigation by Cumming and Haxton (1996). The work was motivated by the desire to address the apparently discordant solar neutrino measurements. One can broadly rationalize the results of the ^{37}Cl detection, the Kamiokande experiment and the SAGE/GALLEX ^{71}Ga experiments by asserting that the pp flux is apparently equal to the theoretical prediction, that the ^8B flux is about 40 per cent of that of current solar models, and that the ^7Be flux is essentially nonexistent. Studies based on a broad range of spherically symmetrical solar models, in all of which the reaction chains are presumed to be very nearly in balance, reveal no adjustment that can reproduce this result. Therefore, it has been concluded, neutrino transitions must take place irrespective of astrophysical considerations. Indeed, the conclusion now supports a small industry studying neutrino mass ratios and mixing angles, the possible influence of helicity flipping and other exotic phenomena. That is very healthy, for one must never become too complacent with the standard models, be they of electroweak particle interactions or of the sun. However, the strengths and weaknesses of the arguments against an astrophysical solution to the problem must not be forgotten. In particular, if one admits that the balance between the chains in the nuclear reactions can be broken, the present case for the necessity of neutrino transitions collapses. Neutrinos might still be massless.

Dynamical instability is precisely the mechanism that unbalances the nuclear reactions, by imparting a time dependence that is more rapid than nuclear equilibration, or by advecting intermediate products of the reactions away from their sites of creation. Indeed, the original discussion of the instability was motivated by this very fact. Now Cumming and Haxton have produced a new twist. They point out that if the instability gives way to direct convection with a not implausible asymmetry in the geometry of the flow, then the results of the different neutrino measurements might be reconciled. The toy model they discuss, if taken literally, is obviously in conflict with the helioseismic inferences, and no doubt there will soon be a paper to point that out.* However, there may be variants of the model that represent the actual conditions in the solar core more faithfully than the standard models do, even if they don't actually resolve entirely all the issues concerned with the neutrino observations. Indeed, the reversal in the gradient of $\delta c^2/c^2$ near $r/R = 0.3$, where the ^{3}He abundance is greatest, is indirect evidence for an advecting shell. Care must be exercised when using seismic comparisons with spherically symmetrical models when making deductions about reality. In this context it is interesting to note that the initial nonlinear development of the direct ^{3}He-driven convective mode, if it were to be unstable, does not influence the sound speed, so it may be necessary to look for the motion more carefully than one might have suspected. Indeed, actuality may be significantly different from the standard models of both the sun and electroweak interactions, even though a change in only one of them may be required to erase the disagreement between theory and neutrino-flux observations.

I am sure that it will not require enormous ingenuity to find some minimal modification to the standard solar model to eliminate the disparity plotted in Figure 5. Our real challenge will be to distinguish between the

---

*After the presentation, Jørgen Christensen-Dalsgaard told me that such a paper is already in preparation (Bahcall et al., 1997). Notwithstanding the enormously statistically significant disparity between the sun and the current standard solar models (illustrated in Figure 5), it is argued that the remarkable agreement between standard 'predictions' and helioseismological observations essentially rules out solar models with temperature or mean molecular mass profiles that differ significantly from the standard profiles. It is also pointed out that standard models 'predict' the measured properties of the sun more accurately than is required for applications involving solar neutrinos. No case against spherical asymmetry and associated nuclear imbalance is made.

many ways of doing so. That was the challenge that Philippe Delache always had in mind.

I am very grateful to T. Sekii for his assistance with producing the figures, and to G.R. Isaak for providing the middle panel of Figure 4. I am also grateful to J.W. Leibacher for pointing out some typographical errors and some awkward phrases in the original manuscript.

References

Anderson B.N. and Maltby P. (1983) *Nature*, **302**, 808.

Anderson, E. *et al.*, these proceedings.

Arvonny (1983) *Le Monde*, **12039**, 1

Bachmann, K.T. and Brown, T.M. (1993) *Astrophys. J.*, **411**, L45.

Bahcall, J.N., Pinsonneault, M.H., Basu, S. and Christensen-Dalsgaard, J. (1997) *Phys. Rev. Lett.*, **78**, 131.

Balmforth, N.J. (1992) *Mon. Not. R. astron. Soc.*, **255**, 639.

Balmforth, N.J., Gough, D.O. and Merryfield, W.J. (1996) *Mon. Not R. astron. Soc.*, **278**, 437.

Basu, S. (1997) these proceedings.

Belvedere, G. and Paternò L. (1984) in *Oscillations as a probe of the sun's interior, Mem. Soc. Astron. Italiana*, **55**, N° 1-2.

Bonnet, R.M. and Delache, Ph. (1977) in *The energy balance and hydrodynamics of the solar solar chromosphere and corona* (ed. R.M. Bonnet and Ph. Delache, G. De Bussac, Clermont-Ferrand) *Proc. IAU Colloq.*, **36**.

Braun, D.C., Duvall, Jr, T.L. and La Bonte, B.J. (1987) *Ap J. Lett.*, **319** (1), L27.

Brown, T.M. (1984) *Science*, **226**, 687.

Chang, H-Y., Gough, D.O. and Sekii, T. (1995) in *Fourth SOHO Workshop: Helioseismology* (ed. J.T. Hoeksema, V. Domingo, B. Fleck and B. Battrick, ESA SP-376(2), Noordwijk) p. 175.

Chaplin, W., Elsworth, Y., Howe, R., Isaak, G.R., McLeod, C.P. and Miller, B.A. (1995) in *Fourth SOHO Workshop: Helioseismology* (ed. J.T. Hoeksema, V. Domingo, B. Fleck and B. Battrick, ESA SP-376(1), Noordwijk), p. 335.

Christensen-Dalsgaard, J. (1996) in *The structure of the sun* (ed. T. Roca Cortés and F. Sánchez, CUP, Cambridge) p. 47.

Christensen-Dalsgaard, J., Dilke, F.W.W. and Gough, D.O. (1994) *Mon. Not. R. astron. Soc.*, **169**, 429.

Christensen-Dalsgaard, J., Gough D.O., and Libbrecht, K.G. (1989) *Astrophys. J. Lett.*, **341**, L103.

Cumming, A. and Haxton, W.C. (1996) *Phys. Rev. Lett.*, **77**, 4286.

Durrant, C.J. and Schröter, E.H. (1983) *Nature*, **301**, 589.

Dziembowski, W.A. and Goode, P.R. (1989) *Astrophys. J.*, **347**, 540.

Dziembowski, W.A. and Goode, P.R. (1997) *Astron. Astrophys*, submitted.

Edmunds M.G. and Gough, D.O (1983) *Nature*, **302**, 810.

Elsworth, Y., Howe, R., Isaak, G.R., McLeod, C.P. and New, R. (1991) *Mon. Not. R. astron. Soc.* **251**, 7p.

Elsworth Y., Howe, R., Isaak, G.R., McLeod, C.P. Miller, B.A., New, R., Speake, C.C. and Wheeler, S.J. (1993) *Mon. Not. R. astron. Soc.*, **265**, 888.

Elsworth, Y., Howe, R., Isaak, G.R., McLeod, C.P., Miller, B.A., New. R., Speake, C.C. and Wheeler, S.J. (1994) *Astrophys. J.*, **434**, 801.

Elsworth, Y., Howe, R., Isaak, G.R., McLeod, C.P., Miller, B.A. and Wheeler, S.J. (1995) in *GONG'94: Helio and asteroseismology* (ed. R.K. Ulrich, E.J. Rhodes Jr and W. Däppen) *Astron. Soc. Pac. Conf. Ser.*, **76**, 318.

Fossat, E. (1985) in *Future missions in solar, heliospheric and space plasma physics,* (ed. E. Rolfe and B. Battrick, ESA SP-235, Noordwijk) p. 209.

Fröhlich, C. (1994) in *The sun as a variable star* (ed. J.M. Pap, C Fröhlich , H.S. Hudson and S.K. Solanki, CUP, Cambridge), p. 28.

Fröhlich, C. *et al.* (1997) these proceedings.

Goldreich, P., Murray, N., Willette, G. and Kumar, P. (1991) *Astrophys. J.*, **370**, 752.

Goode, P.R., Gough, D.O. and Kosovichev, A.G. (1992) *Astrophys. J.*, **387**, 707.

Gough D.O. (1977) in *The energy balance and hydrodynamics of the solar solar chromosphere and corona* (ed. R.M. Bonnet and Ph. Delache, G. De Bussac, Clermont-Ferrand) *Proc. IAU Colloq.*, **36**, 3.

Gough, D.O. (1985) in *Future missions in solar, heliospheric and space plasma physics,* (ed. E. Rolfe and B. Battrick, ESA SP-235, Noordwijk) p. 183.

Gough, D.O. (1988) in *Solar-terrestrial relationships and the Earth environment in the last millennia* (ed. G. Castagnoli-Cini, Soc. It. Fisica, Bologna) p. 90.

Gough D.O. (1990a) in *Progress of seismology of the sun and stars,* (ed. Y. Osaki and H. Shibahashi, Springer, Berlin), p. 283.

Gough, D.O. (1990b) in *Phil. Trans. R. Soc. Lond.* A, **330**, 627.

Gough, D.O. (1995) in *Fourth SOHO Workshop: Helioseismology* (ed. J.T. Hoeksema, V. Domingo, B. Fleck and B. Battrick, ESA SP-376 (1) Noordwijk) p. 181.

Gough, D.O. and Sekii, T. (1995) in *GONG'94: Helio- and asteroseismology from the Earth and Space* (ed. R.K. Ulrich, E.J. Rhodes Jr and W. Däppen), *Astron. Soc. Pac. Conf. Ser.,* **76**, 374.

Gough, D.O. and Stark, P.B. (1993) *Astrophys. J.,* **415**, 376.

Gough, D.O. and Thompson, M.J. (1988a) in *Advances in helio- and asteroseismology* (ed. J. Christensen-Dalsgaard and S. Frandsen), *Proc. IAU Symp.*, **123**, 175.

Gough, D.O. and Thompson, M.J. (1988b) in *Advances in helio- and asteroseismology* (ed. J. Christensen-Dalsgaard and S. Frandsen, Reidel, Dordrecht) *Proc. IAU Symp.*, **123**, 155.

Gough, D.O., Kosovichev, A.G. and Toutain, T. (1995) *Sol. Phys.,* **157**, 1.

Gough, D.O., *et al.,* (1996) *Science*, **272**, 1296.

Haber, D.A., Toomre, J., Hill, F. and Gough, D.O. (1988) in *Seismology of the sun and sun-like stars* (ed. E.J. Rolfe, ESA SP-286, Noordwijk) p. 301.

Knobloch, E. and Merryfield, W.J. (1992) *Astrophys. J.*, **401**, 196.

Kosovichev, A.G. and the MDI team (1997) *Solar Phys.*, in press.

Kumar, P. (1997) these proceedings.

Libbrecht, K.G. (1988) *Astrophys. J.*, **334**, 510.

Libbrecht, K.G. and Woodard, M.F. (1990) *Nature*, **345**, 779.

Libbrecht, K.G. and Woodard, M.F. (1991) *Science*, **253**, 152.

Libbrecht, K.G., Woodard, M.F. and Kaufman, J.M. (1990) *Astrophys. J. Suppl.* **74**, 1129.

Murray, N. (1993) in *GONG 1992: Seismic investigation of the sun and stars,* (ed. T.M. Brown), *Astron. Soc. Pac. Conf. Ser.* **42**, 3.

Provost, J., Mosser, B. and Berthomieu, G. (1993) *Astron. Astrophys.,* **274**, 595.

Rolfe, E.J. (ed.) (1988) in *Seismology of the sun and sun-like stars* (ESA SP-286, Noordwijk).

Rosenthal, C.S. (1997) these proceedings, (vol. 2).

Scherrer, P.H. (1997) these proceedings.

Schmider, F-X., Mosser B. and Fossat, E. (1991) *Astron. Astrophys.,* **248**, 281.

Sekii, T. (1997) these proceedings.

Spiegel, E.A. and Zahn, J-P. (1992) *Astron. Astrophys.,* **265**, 106.

Thomson, D.J, MacLennan, C.G. and Lanzerotti, L.J. (1995) *Nature*, **376**, 139.

Toutain, T. and Fröhlich, C. (1992) *Astron. Astrophys.,* **257**, 287.

Vorontsov, S.V. (1988) in *Advances in helio- and asteroseismology* (ed. J. Christensen-Dalsgaard and S. Frandsen, Reidel, Dordrecht) *Proc. IAU Symp.*, **123**, 151.

Woodard, M.F., Kuhn, J.R., Murray, N. and Libbrecht, K.G. (1991) *Astrophys. J.*, **373**, L81.

LIST OF POSTERS

*published in the Poster Volume**

1. Helioseismology: from ground to space, a worldwide cooperation

1.1 An unbiased average rotational splitting from VIRGO/SPM ?
   T. Appourchaux, D.O. Gough, T. Sekii and T. Toutain
1.2 Detection of solar p-modes in the guiding signals of the luminosity oscillations imager
   T. Appourchaux and T. Toutain
1.3 Time/frequency analysis of solar p modes from GOLF data
   F. Baudin, A. Gabriel, R.A. Garcia, T. Foglizzo, V.G. Gavryusev, E.A. Gavryuseva, D.O. Gough, R. Ulrich and the GOLF team
1.4 GOLF data and homomorphic deconvolution
   F. Baudin, C. Régulo, A. Gabriel, T. Roca Cortes and the GOLF team
1.5 Power spectra products - A new technique for isolating persistent oscillations applied to GOLF spectra
   L. Bertello, R.K. Ulrich, R.A. Garcia, S. Turck-Chièze and the GOLF team
1.6 Lorentz fitting of whole-disk multiplets
   H.-Y. Chang, D.O. Gough, T. Sekii
1.7 The current status of the Birmingham Solar-Oscillations Network (Bi-SON)
   W.J. Chaplin, Y. Elsworth, G.R. Isaak, C.P. McLeod, B.A. Miller, R. New and H.B. van der Raay
1.8 The solar cycle dependence of low-degree p-mode eigenfrequencies from recent BiSON data
   W.J. Chaplin, Y. Elsworth, G.R. Isaak, C.P. McLeod, B.A. Miller and R. New
1.9 Limitations of the IRIS network performance
   S. Ehgamberdiev, S. Khalikov, O. Ladenkov, A. Serebryanski, Y. Tillaev, E. Fossat, G. Grec, B. Gelly, F.-X. Schmider, P. Pallé, C. Régulo, M. Lazrek and T. Hoeksema

---

*The volume is published by the *Observatoire de la Côte d'Azur* and the *Université de Nice* and can be ordered by sending an e-mail to *poster-IAU181@irisalfa.unice.fr*

1.25 A search for g-modes in earth-based velocity spectra
  L. Martín, P.L. Pallé, F. Pérez Hernández and H.B. van der Raay
1.26 Comparisons on fitting ring diagrams
  J. Patrón, I. Gonzalez Hernandez, D.-Y. Chou and the TON team
1.27 Velocity and/or intensity in the GOLF one wing signal ?
  C. Régulo, T. Roca Cortés, P. Boumier, R. García,
  J.M. Robillot, S.Turck-Chièze, R. Ulrich and the GOLF team
1.28 Aligning the GONG network
  C.G. Toner and J. Harvey

2. Internal structure and rotation: Seismic inversions

2.1 Structure inversion hare-and-hounds
  H.M. Antia, S. Basu, J. Christensen-Dalsgaard, J.R. Elliott,
  D.O. Gough, J.A. Guzik, and A.G. Kosovichev
2.2 Solar internal sound speed as inferred from combined BiSON and LOWL oscillation frequencies
  S. Basu, W. J. Chaplin, J. Christensen-Dalsgaard, Y. Elsworth,
  G.R. Isaak, R. New, J. Schou, M.J. Thompson and S. Tomczyk
2.3 Equation of state effects in helioseismic inversions
  S. Basu and J. Christensen-Dalsgaard
2.4 Seismic calibration of solar envelope models with using the "differential-response" technique
  V.A. Baturin and S.V. Vorontsov
2.5 Sensitivity of p and g-modes on specified physical processes of solar modeling
  S. Brun, I. Lopes, P. Morel, and S. Turck-Chièze
2.6 3D hydrodynamic simulation of the solar radiation-convection transition region
  K.L. Chan et Y.C. Kim
2.7 Rotation of the solar core
  W.J. Chaplin, J. Christensen-Dalsgaard, Y. Elsworth, R. Howe,
  G.R. Isaak, R. New, J. Schou, M.J. Thompson, S. Tomczyk
2.8 The solar core: new low-l p-mode fine spacing results from BiSON
  W.J. Chaplin, Y. Elsworth, G.R. Isaak, C.P. McLeod, B.A. Miller
  and R. New
2.9 The solar rotation rate from inversion of the first GONG datasets
  T. Corbard, G. Berthomieu and J.Provost
2.10 The solar core rotation from LOWL and IRIS or BiSON data
  T. Corbard, G. Berthomieu, J. Provost and E. Fossat
2.11 The current situation of the solar equation of state
  W. Däppen and A. Perez

3. Other inputs from helioseismology to solar physics

3.19 Steady parts of rotation and magnetic field in the solar interior
   K.M. Hiremath and M.H. Gokhale

3.20 Effects of solar activity on amplitude of acoustic oscillations
   A.G. Kosovichev and V.V. Zharkova

3.21 Simultaneous measurement of radial oscillations and optical thickness oscillations in the chromosphere
   M. Missana

3.22 On the origin of solar photospheric turbulence: Granular velocity shear
   A. Nesis, R. Hammer and H. Schleicher

3.23 Study of line asymmetries of solar oscillations above and below the acoustic cut-off frequency
   R. Nigam, A.G. Kosovichev and P.H. Scherrer

3.24 Does the solar internal rotation rate can be inferred from full-disk magnetograms?
   D.I. Ponyavin

3.25 Long period oscillations in solar diameter
   P.C.R. Poppe, N.V. Leister, M. Emilio and V.A.F. Martin

3.26 New method for diagnostics of solar magnetic fields and flows
   M. Ryutova and P. Scherrer

3.27 Implication of long-term frequency variation of low and intermediate degree modes
   H. Shibahashi

3.28 Solar irradiance variations during the activity minimum in 1996
   M. Steinegger, J.A. Bonet, M. Vazquez and A. Jiménez

3.29 Flow around sunspots from frequency shift measurements
   M.-T. Sun, D.-Y. Chou, C.-H. Lin and the TON team

3.30 Solar velocity noise measurements from recent BiSON helioseismological data
   C.J. Underhill and G.R. Isaak

3.31 Excitation of torsional waves and an hypothesis on the origin of the solar cycle
   Y.V. Vandakurov

3.32 Solar irradiance variations and active regions observed by VIRGO experiment on SOHO
   C. Wehrli, T. Appourchaux D. Crommelynck, W. Finsterle, C. Fröhlich and J. Pap

4. Asteroseismology: Theory and methods

4.1 Asteroseismological calibration of stellar clusters and convective-core overshooting
   N. Audard and I.W. Roxburgh

4.18 Numerical problem to calculate stellar nonradial oscillations : *FILOU*
code
F. Tran Minh and L. Léon

5. Results on some selected type stars

5.1 Search for periodicity in the variations of the H α line of the Be star
o And
Briot D., J. Chauville, J.P. Sareyan, G. Guerrero, L. Huang,
X.Z. Guo, Y.L. Guo, J.X. Hao, V. Desnoux, F. Morand

5.2 Interpretation of the observed oscillations of τ Pegasi
T.M. Brown, J. Christensen-Dalsgaard, E.J. Kenelly and
M.J. Thompson

5.3 Observational constraints on Praesepe cluster by asteroseismology
M.M. Hernandez, F. Pérez-Hernández, E. Michel, J. A. Belmonte,
M.J. Goupil and Y. Lebreton

5.4 Rapid photospheric variability in the Be star 48 Per
A.M. Hubert, M. Floquet and the MUSICOS 1989 team

5.5 The instability strip of fundamental mode RR Lyrae stars
J. Jurcsik

5.6 The "universal" frequency seen in δ Scuti pulsating stars
V.A. Kotov and S.V. Kotov

5.7 Mode typing of the dominant mode in the δ Scuti star X Cae by line
profile moments
L. Mantegazza and E. Poretti

5.8 The p-mode spectra of the roAp stars
P. Martinez

5.9 Determination of $\Delta T \cos \alpha$ as a function of atmospheric depth in roAp
stars
R. Medupe

5.10 Whole Earth Telescope observations of G185-32
P. Moskalik, S. Zola, G. Pajdosz, J. Krzeiński, D. O'Donoghue,
M. Katz, G. Vauclair, N. Dolez, M. Chevreton, M.A. Barstow,
A. Kanaan, S.O. Kepler, O. Giovannini, J.L. Provencal, S.D. Kawaler,
J.C. Clemens, R.E. Nather, D.E. Winget, T.K. Watson, K. Yanagida,
J.S. Dixson, C.J. Hansen, P.A. Bradley, M.A. Wood, D.J. Sullivan,
S.J. Kleinman, E. Meištas, J.E. Solheim, A. Bruvold and
E.M. Leibowitz

5.11 Synchronous change in the frequencies of the double mode RR Lyrae
star, V26 in M15
M. Paparó, S.M. Saad, B. Szeidl and M.S. Abu elazm

AUTHOR INDEX

438